1 MONTH OF
FREE
READING

at

www.ForgottenBooks.com

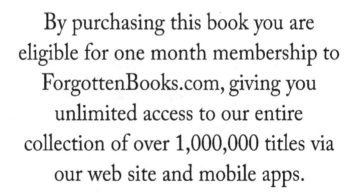

By purchasing this book you are eligible for one month membership to ForgottenBooks.com, giving you unlimited access to our entire collection of over 1,000,000 titles via our web site and mobile apps.

To claim your free month visit:

www.forgottenbooks.com/free1201969

ISBN 978-0-331-50333-3
PIBN 11201969

PROCEEDINGS

OF THE

YORKSHIRE

Geological and Polytechnic Society.

NEW SERIES, VOL. XIV.

1900—1902.

WITH SIXTY-NINE PLATES.

EDITED BY

WILLIAM LOWER CARTER, M.A., F.G.S.,

AND

WILLIAM CASH, F.G.S.

Leeds:

CHORLEY & PICKERSGILL, THE ELECTRIC PRESS.

1902.

TABLE OF CONTENTS.

LIST OF PLATES.

[Vol. XIV.]   [Part I.

PROCEEDINGS

OF THE

YORKSHIRE

GEOLOGICAL AND POLYTECHNIC SOCIETY.

Edited by W. LOWER CARTER, M.A., F.G.S.,

and WILLIAM CASH, F.G.S.

1900.

THE UNDERGROUND WATERS OF NORTH-WEST YORKSHIRE.

PART I.   THE SOURCES OF THE RIVER AIRE.

I.   INTRODUCTION.

BY J. H. HOWARTH, F.G.S.

Since Camden in 1590 wrote that "The Aire has its sources in the roots of Pennyghent," the sources of that river and the underground waters about Malham have been subjects of much interest and some mystery. What is the problem to be solved? It may be briefly stated thus:

From the valley of the Ribble by Stainforth, Neals Ing, Capon Hall, and Malham Tarn runs a long strip of Upper Silurian rocks which die out under the Carboniferous Limestone just eastwards of Upper Gordale where the beck leaves the moor and takes to the gorge. This strip is brought up by the North Craven Fault, and its southern edge is sharply marked

B

by the line of fault and may be easily traced the whole distance. On the north this strip is bounded by the overlying Carboniferous Limestone, which has been very unevenly denuded and presents a long sinuous front to the Silurian, here rising into sharp escarpments as at Great Close and behind Malham Tarn House, and there receding in long gentle slopes as on Chapel Fell, Knowe Fell, and about West Side House. In a line across by Neals Ing and Cattrigg the Silurian is nearly two miles wide, half a mile west of Capon Hall it narrows to zero, widens to about a mile at Malham Tarn, and narrows again to where it disappears eastwards near Gordale Beck.

The area under consideration is included in a line drawn from near Capon Hall on the west, Knowe Fell on the southerly side of Fountains Fell on the north, round by Middle House and East Great Close (under Hard Flask) to High Stoney Bank on the east, and Kirkby Malham in the Aire Valley on the south. This area forms the upper watershed of the River Aire.

What happens is this. All the rainfall on the limestone on the north side of the Silurian, and several springs which rise up the slopes of Knowe Fell towards Gentleman's Gate and Fountains Fell, sink into the limestone and are brought to the surface again at the edge of the Silurian rocks which are tilted at a high angle. These waters flow across the Silurian rocks, either in streams or through Malham Tarn, only to sink again on reaching the limestone on the south side.

To this rule Gordale Beck has been regarded hitherto as the only exception, but it now transpires that Gordale stream is only partially an exception, and is itself undergoing absorption into the limestone. These waters, sinking south of the North Craven Fault, reappear below the great escarpments about Malham and Gordale formed by the Mid Craven Fault, with all the rainfall and springs on the limestone area lying between the two lines of fault. The limestone on the south side of the Silurian absorbs all surface waters just as it does on the north side.

A twenty-five acre field called Hensit (or Hensetts) by
n Hall lies just on the watershed between the Aire and
Ribble, and supplies water to both rivers.   On the Aire side
his field are two never-failing springs, and in a small croft
d "The Coppy," close by, is a third.*   These springs appear
on the northern edge of the Silurian, and are doubtless fed
i the limestone above.   They form the stream which runs
n through the Whyes (or Withes) and the Streets, and
:h, on reaching the North Craven Fault at the Smelt Mill,
ppears in the limestone.   In the meadows about Capon Hall
strong springs, carried off by drainage, which are similarly
ight up by the Silurian rocks from the north side limestone.
the Whyes adjoining Black Hill is a stream which in a
:t distance sinks in a pot-hole close to the main fault.

The streams and springs on Grisedales and Outside sink in
holes in the limestone along the subsidiary fault which brings
Yoredale Shales of Clattering Sykes down against the lime-
ie of the Ha on the east of Outside.   These waters, although
ig in the watershed of the river Ribble, find their way into
Aire by the same route as the Smelt Mill stream.

Malham Tarn is fed by springs from the limestone on the
thern edge of the Silurian area.   A small stream rises in the
dows of High Trenhouse Farm, under Knowe Fell, and flows
Tarn Moss into the lake.   The principal springs, however,
at Water Houses, where there are three or four of some
ortance ; and here, doubtless, most of the water sinking on
we Fell reappears.   On this line "Bursts" have been known
:cur on Chapel Fell just above Water Houses, when these
ngs were unable to discharge an abnormal supply.   There is
oring also on the lake side (N.W.) between the old boat
ie and the moss, and several on the east side from under
it Close and about Ha Mire which find their way into the
i.   These all appear near the junction of the limestone with

* Mr. James Howarth, who lived at Capon Hall for many years, tells
hat in dry seasons he has known this spring ebb and flow.

the Silurian, but there is one at Spiggott Hill at the South West corner of the Tarn out of the Silurian, and another in the lower Streets.   Malham Tarn rests on the upturned and denuded edges of the Silurian rocks, and is dammed in as to its sides with drift, and deepened by an artificial embankment at its outlet.   The lake discharges at its south-east corner, and the outflow soon crosses the North Craven Fault on its line from the Smelt Mill eastwards, and sinks in the limestone at the well-known sinks.

Further east under the escarpments about Middle House, the east side of Great Close, and on High Stoney Bank, rise several streams which together form Gordale Beck.   These are on the limestone, but, owing perhaps mainly to drift, they keep to the surface.   They flow over the narrow strip of Silurian near where it is dying out; and, now in one stream, they cross the North Craven Fault and proceed down Gordale.   By the present investigation we now learn that the stream always undergoes considerable absorption into the limestone in mid-Gordale, and during the summer of 1899 was entirely absorbed.

There are thus three principal sinks, viz.: At the Smelt Mill, Malham Tarn Sinks, and in Gordale; while curiously enough there are three principal outlets, viz.: At Malham Cove, Aire Head, and in Gordale bottom on the east of the main stream between the falls and Gordale House.

To trace each sink to its outlet, and conversely each outlet to its source, is the problem.

How long it is since the waters of Malham Tarn ceased to flow regularly down Comb Scar, and over the top of Malham Cove, but took instead to disappearing in the limestone at the several "sinks," is not known.

The earliest information available is of a time when the Cove itself was only over-run by the stream on rare occasions during exceptional floods.   More frequently the water overflowed Comb Scar, but sank at its foot; whereas now it fails even in a flood to get as far down the valley as Comb Scar.

The early records are highly interesting as showing, not only the changes that have occurred in the surface flow, but also that the facts connected with the underground streams were fairly well known at the close of the last century.

Thomas Hurtley, who lived at Malham, published in 1786 a pamphlet entitled "Natural Curiosities of the environs of Malham-in-Craven," in which he describes the Cove ; and, referring to its foot, says "from whence issues a strong current of water having traversed upwards of a mile from the Tarn in its subterranean caverns." He also speaks of cascades from the top after "rainy and tempestuous weather when the water-sink at the southern extremity of the Tarn is unable to receive the overflow of the Lake."

The Rev. Thomas Dunham Whitaker, LL.D., F.S.A., who began his History of Craven about the close of the last century and published it in 1805, says that "in rainy seasons the overflowings of the lake spread themselves over the shelving surface of the rocks below, and, precipitating from the centre of the Cove, form a tremendous cataract of nearly 300 feet." He says that the inhabitants of Malham "plead" that the waters of the Tarn appear at Aire Head, and further that "it is well known that a collection of springs rising in the Black Hills, the Hensetts, and Withes is swallowed up in a field called the Street, and from the turbid quality of the water, very unlike that of the Tarn, there is little doubt that, after a subterraneous course of more than two miles, this is the stream which here emerges again."

Mr. William Howson, in his "Guide to the District of Craven," published in 1850, says that "twice within the last forty years the swollen waters of the Tarn have made their way over the Cove." Referring to *Comb Scar he continues : "In a flood the Tarn water not unfrequently rushes over here and forms a second Gordale, but it is commonly prevented from

* Mr. Howson spells it "Coomb."

reaching the Cove by sinking at the foot of this pass through the shattered and fissured stratum with singular noise and rapidity."

From these records it is evident that as the joints in the limestone at the "sinks" were slowly widened by the solvent action of the water they gradually absorbed the stream, until it was only in floods that the water reached the Cove. As the process continued the stream dwindled, and retreated from stage to stage. From what Mr. Howson says of the flood-water sinking at the foot of Comb Scar there was probably a pause in the retreat there of some duration. Now, however, flood-water no longer reaches Comb Scar, and the old track is what mid-Gordale (where exactly the same process of stream absorption is now going on) will one day become—a dry valley.

With respect to the underground streams it will be seen that Hurtley considered the Cove outflow was supplied from the Tarn, while Dr. Whitaker ascribes it to the "Smelt Mill" sinks. The Aire Head Springs it appears were known by the villagers to be connected with the Tarn even then, and one wonders whether they had arrived at that conclusion by experiment or mere guesswork. It was more or less natural to assume the Cove to be the outlet for both the old surface stream from the Tarn and the Streets water. Where else could they come? But the same process of reasoning would not apply to the Aire Head Springs. Anyhow they appear to have known at the close of last century all that was known until eighty years later in 1879.

Mr. Morrison says there is a tradition, but not a clear tradition, that Lord Ribblesdale put chaff in at Malham Tarn Water Sinks and that it came out at Aire Head.

Farmers and others have made attempts from time to time for many years back to ascertain what the facts really were by inserting chaff and similar materials at the sinks, but without result. The media employed were never seen again at any of the presumed outlets, and so an air of mystery surrounded the question and has helped to keep the interest alive.

It is worthy of record that some forty to fifty years ago or more, when ore was washed in the mines about Pike Daw, the stream issuing from the Cove was discoloured.   The writer has often been told this by villagers who had seen it.

In 1879 Mr. Walter Morrison, the late Mr. Thomas Tate, of Leeds, and other members of the Yorkshire Geological and Polytechnic Society,* made certain experiments with the result that they concluded that Malham Cove and the springs at Aire Head were both connected with the Malham Tarn water-sinks. They further concluded that the Smelt Mill water-sinks in the Streets on Malham Moor were not connected with Malham Cove, and the outlet was not discovered.

The nature of these experiments may be more fully described. Chaff was tried at the Smelt Mill, and both chaff and bran at the Tarn water-sinks, but, although nets were set in the outlet streams, none was found to emerge. This is not to be wondered at, as a preliminary trial of bran had shown that it became waterlogged in the stream within forty yards. An attempt was made also at the Smelt Mill to stain the stream by introducing an aniline dye (magenta), but in a preliminary trial up-stream all colour traces disappeared in a distance of seven or eight yards.   These several insertions, therefore, produced only negative results.

The principal experiments were in flushing the streams by means of the sluice at Tarn Foot.

The stream flushed at 4 p.m. united all the sinks at 4.25, and ten minutes later it began to over-flow the sinks and to follow the old stream bed towards Comb Scar. A few minutes after 5 p.m. this overflow was 150 yards long.

"At 5.25, one hour and twenty-five minutes after leaving the Tarn, the water began to creep over the half submerged

* Proceedings of the Yorkshire Geological and Polytechnic Society, N.S., Vol VII., p. 177 (1879).

pebbles at Aire Head." Twenty-five minutes later (5.50) the stream, 4 feet wide, had risen at its junction with the Aire 4¼ inches; later in the evening it rose to nearly a foot.

Up to 5.45, one hour and forty-five minutes after leaving the Tarn, there was no change at Malham Cove, but on returning after dinner (time not stated) the Cove Stream, 31½ feet wide, had risen two inches.

At 10 p.m. the Tarn sluices were closed again, "all but an inch or so to keep the fish down stream alive."

Next day, at 1 p.m., the sluices were opened again. Owing to the diminished volume of the stream between the Tarn and the sinks the water was 25 minutes in reaching the sinks, instead of 18 as on the previous day. The Aire Head Spring was affected at 2.32, being 1 hour 32 minutes against 1 hour 25 minutes the previous day, so that, allowing for the 7 minutes lost before the water-sinks were reached, the results were the same on both days as between the sinks and Aire Head. By 5 p.m. the water at Aire Head had risen 13 inches.

The Cove on this day was affected at 3.10, that is 38 minutes after Aire Head. "At 3.15 it had risen ¼ inch; at 3.20 ½ inch; at 3.35 1 inch; at 4.5 2 inches; at 4.45 2½ inches; and by 5.30 the stream had risen 2¾ inches, its outlet being 31½ feet wide."

On the third day "the water was permitted to flow freely from the Tarn during the day, and the high-water level was maintained both at Malham Cove and Aire Head. At 5 o'clock the Tarn sluices were closed again as before."

On the fourth day "the streams at the Cove and Aire Head gradually subsided, and by noon the water at the Cove stood ¾ inch below its normal level. At 5.45 p.m. the water in the Tarn had not quite risen again to the level of the sill of the overflow, so that this sinking of the outflow at the Cove and Aire Head confirms the experiment of the rise on the first and second days."

The investigators considered the connection of the Smelt Mill with the Cove "sufficiently refuted by their analyses," since their table showed the hardness of the water at the Cove to be 10·8 as compared with 13·1 at the Smelt Mill, and they regarded a loss of 2·3 in hardness in passing through a mile and three quarters of limestone strata impossible, besides which "the volume of the Smelt Mill Sike was not a twentieth that of the Cove Stream." They thought "Pike Daw still less likely to be the source of the Cove waters," but to probably drain "to a lower point in the river."

They agreed with a suggestion of Professor Boyd Dawkins[*] that a water-cave of greater or less extent probably existed behind Malham Cove.

Before these experiments Mr. Leather, C.E., had tried flushing the Tarn stream and reported his results to Mr. Tate. He considered his experiments "proved conclusively that the water flowing out at the foot of the Cove is not the water which sinks into the ground and disappears some distance above the Cove. The water comes out at Aire Head."

At the British Association Meeting in 1890, at Leeds, Professor Sylvanus Thompson, F.R.S.,[†] reported having introduced one and a quarter pounds of uranin into one of the Malham Tarn Sinks, but without any result within three hours at Aire Head, or anything distinctive at Malham Cove, and he concluded either there was a considerable body of water at some intermediate spot between the water-sink and the Cove, or that the Aire Head spring communicated with some other water-sink than that marked on the Ordnance Survey maps.

In the handbook prepared for the same meeting, Professor L. C. Miall, F.R.S., says that when the Tarn waters are suddenly

---

[*] "Cave Hunting" (1874).

[†] British Association Report, 1890.

discharged the floods affect both Aire Head and Malham Cove; but the latter, although a mile and a quarter nearer, half an hour after the former. He says that the passage to Aire Head may be along a vertical fissure, and that to the Cove for part of its course along a wide, shallow, and almost horizontal fissure. If so, the increased friction in the passage to the Cove may explain the retardation of the water. Professor Miall further says that measurement of the issuing streams shows that the Cove discharges more than half of the water which issues from the Tarn.

During the summer of 1899 the Yorkshire Geological and Polytechnic Society took the matter up again at the instance of Mr. George Bray, of Leeds, who generously undertook to provide material for the experiments. This came about in a somewhat remarkable manner.

At Easter, in 1899, these several experiments were the subject of discussion at the Cove by a party on a walking tour including the writer and Mr. F. Swann, B.Sc., of Ilkley. Mr. Swann suggested that the employment of some harmless medium such as common salt, which could be readily detected by simple chemical tests, and would be more likely to meet with success. By a happy and curious coincidence this party, three days later, met four strangers at Clapham who were discussing the underground waters of Gaping Gill Hole and Ingleborough. The conversation became general upon the subject of the underground waters of North West Yorkshire, and one of the strangers remarked that if any reputable and capable Yorkshire Society would take the matter up he would be glad to provide material for the experiments.

One of the walking party, happening to be a member of the Council of the Yorkshire Geological and Polytechnic Society, said that if the gentleman were in earnest he would be glad to have his name and address. These were promptly forthcoming, and out of this chance meeting came the present investigations.

The Council of the Yorkshire Geological and Polytechnic Society accepted this generous offer on the part of Mr. Bray (who has taken a personal and active interest in the inquiry), and appointed a large and representative Committee to conduct the investigations.

Further points of interest and inquiry may be stated, such as : Why does the Tarn water flush Malham Cove sometimes and not always? Why does the Tarn water reach Aire Head before Malham Cove, which is a mile and a quarter nearer? Why were the chemicals introduced so long in transit? Are there underground caverns between the sinks and the outlets?

II. ENGINEERING REPORT.

BY C. W. FENNELL, F.G.S., AND J. A. BEAN, C.E.

In reference to the gauging of the flow of water at the streams near Malham Tarn, on the 22nd June, 1899, we have carefully considered the information then obtained, and have to report as follows :—

On the 17th June we visited Malham, and after very considerable difficulty, owing to the rocky bed of the streams, we were able to select three positions for the gauges marked 1, 2, and 3 on the annexed sketch plan. (Plate XII.).

No. 1 gauge on the outlet of the Malham Tarn sluice.

No. 2 on Malham Beck, above Malham village.

No. 3 on the stream from Aire Head.

We were thus enabled to gauge the volume of water going down the Malham Tarn Sinks, and the increase in volume, if any, at the Cove and at Aire Head. We also gauged the water at the Smelt Mill Sink, the flow being constant at the rate of 19,800 gallons per day. The gaugings at the outlets, viz., the Cove and Aire Head, were taken every 15 minutes from 7 a.m. until 5.45 p.m., and the observations are carefully plotted on the diagram accompanying this report (Plate I.). The results of the three gauges may be readily seen on reference to this diagram.

The time chosen was particularly favourable
out experiments on underground water, as there had
cally no rainfall for the previous three weeks.
had fallen in the early hours of the 22nd, and prob
Aire Head at 10.15 a.m.; it caused, however, a
increase in the yield. This increase was not show
Malham Beck. Reference to the diagram (Plate
that the effect of a steady flow of water at t
nearly 700,000 gallons a day, started through the s
Tarn at 10.15 a.m. and allowed to continue unt
began to exhibit itself at Aire Head at 12.15,
steady rise until 2.45 when the first supply was
by the much greater volume of water, which w
from the Tarn at 1 p.m. During the whole day, as
the diagram, the gauging at Malham Beck near
remained constant. The great flush of water let off
Tarn for 45 minutes affected the Aire Head stream
hours. From this we learn that the stream bed and o
insufficient to deal with this large volume of water, an
was backed up. No doubt if the flush had continued for
period, or had been greater in volume, the strata wo
got charged with water to such a height as to cause an
into the Malham Beck through some fissures or over som
ground sill, causing it to rise as noticed when Mr. Morr
Mr. Tate experimented in 1879.

We are unable to give any information as to t
drained by the tributaries of the Aire above Kirkby
without further water-gaugings, which should be taken
period of many weeks.

The following, however, may be arrived at from the
experiments, viz. :—That the area drained by the stream at
Head is probably distinct from the area drained by the a
from Malham Cove or Malham Beck, because the flow at M
Beck was constant the whole time that the experiments
carried on; but this is not proven, as the gaugings were

The time chosen was particularly favourable for carrying out experiments on underground water, as there had been practically no rainfall for the previous three weeks. A little rain had fallen in the early hours of the 22nd, and probably affected Aire Head at 10.15 a.m.; it caused, however, a very slight increase in the yield. This increase was not shown at all in Malham Beck. Reference to the diagram (Plate I.) shows that the effect of a steady flow of water at the rate of nearly 700,000 gallons a day, started through the sluice at the Tarn at 10.15 a.m. and allowed to continue until 1 p.m., began to exhibit itself at Aire Head at 12.15, showing a steady rise until 2.45 when the first supply was overtaken by the much greater volume of water, which was let off from the Tarn at 1 p.m. During the whole day, as shown on the diagram, the gauging at Malham Beck near the Cove remained constant. The great flush of water let off from the Tarn for 45 minutes affected the Aire Head stream for three hours. From this we learn that the stream bed and outlet were insufficient to deal with this large volume of water, and that it was backed up. No doubt if the flush had continued for a longer period, or had been greater in volume, the strata would have got charged with water to such a height as to cause an overflow into the Malham Beck through some fissures or over some underground sill, causing it to rise as noticed when Mr. Morrison and Mr. Tate experimented in 1879.

We are unable to give any information as to the area drained by the tributaries of the Aire above Kirkby Malham without further water-gaugings, which should be taken over a period of many weeks.

The following, however, may be arrived at from the above experiments, viz. :—That the area drained by the stream at Aire Head is probably distinct from the area drained by the stream from Malham Cove or Malham Beck, because the flow at Malham Beck was constant the whole time that the experiments were carried on ; but this is not proven, as the gaugings were not

<--Measure Binding

Adjust Cradle Gap

SCALED RULER FOR MOVING CRADLE --> SCALE = CM

continued for a sufficiently long period to show whether the flow from Malham Tarn had been drained by various fissures, joints, and faults towards Malham Beck as well as Aire Head. Also that the area through which the Malham Tarn water flows is capable of storing, probably in cracks and fissures, a large volume of water. The total area drained by Malham Beck and Aire Head at the junction with Goredale Beck is approximately 6,000 acres.

In our opinion very little definite information can be gathered from experiments carried on during a few hours, therefore should the Committee desire to continue them in order to ascertain the movement of underground water below the sink holes at Malham Tarn (which it will be noticed are closely connected with the North Craven Fault), we consider that it will be necessary to adopt a more elaborate plan and have gauges fixed not only in the positions above mentioned but on Gordale Beck and other smaller streams in the district, and to continue the gaugings over a long period of time. It would be necessary also to watch the streams along the line of the North and South Craven Faults to see whether the experiments affect them in any way.

III. REPORT OF THE CHEMICAL SUB-COMMITTEE.

BY F. W. BRANSON, F.I.C., AND W. ACKROYD, F.I.C.

Sub-Committee meetings were held at Malham on the 26th and 27th of May; on the evening of the 26th, at Lister's Arms. There were present: Messrs. Ackroyd, Bingley, Bray, Branson, and Swann. The prospecting work for the following day was arranged, and it was resolved, on going over the ground, to take samples of water for the determination of the chlorine figures to be used as data in future salt experiments, and to make a preliminary trial of an alcoholic solution of fluorescein as an agent for tracing the flow of underground waters.

continued for a sufficiently long period to show whether the flow from Malham Tarn had been drained by various fissures, joints, and faults towards Malham Beck as well as Aire Head. Also that the area through which the Malham Tarn water flows is capable of storing, probably in cracks and fissures, a large volume of water. The total area drained by Malham Beck and Aire Head at the junction with Goredale Beck is approximately 6,000 acres.

In our opinion very little definite information can be gathered from experiments carried on during a few hours, therefore should the Committee desire to continue them in order to ascertain the movement of underground water below the sink holes at Malham Tarn (which it will be noticed are closely connected with the North Craven Fault), we consider that it will be necessary to adopt a more elaborate plan and have gauges fixed not only in the positions above mentioned but on Gordale Beck and other smaller streams in the district, and to continue the gaugings over a long period of time. It would be necessary also to watch the streams along the line of the North and South Craven Faults to see whether the experiments affect them in any way.

III.   REPORT OF THE CHEMICAL SUB-COMMITTEE.

BY F. W. BRANSON, F.I.C., AND W. ACKROYD, F.I.C.

Sub-Committee meetings were held at Malham on the 26th and 27th of May; on the evening of the 26th, at Lister's Arms. There were present: Messrs. Ackroyd, Bingley, Bray, Branson, and Swann. The prospecting work for the following day was arranged, and it was resolved, on going over the ground, to take samples of water for the determination of the chlorine figures to be used as data in future salt experiments, and to make a preliminary trial of an alcoholic solution of fluorescein as an agent for tracing the flow of underground waters.

On the 27th, the Sub-Committee visited the Water Sinks below the Tarn, and the Smelt Mill Water Sink.

The samples collected gave the following figures on subsequent analysis:—

				Chlorine. Parts per 100,000.
Smelt Mill Stream	...	...	...	1·1
Malham Tarn, outlet	...	...	...	1·0
Malham Cove	...	...	...	0·95
Aire Head ...	...	...	...	1·0
Average		...	...	1·01

One part of chlorine represents 1·647 parts of common salt.

As a trial experiment a solution of 4 oz. of fluorescein was put in the Water Sink below the Tarn outlet at 3.15 p.m. and observers were stationed at the Cove and at Aire Head for several hours after. No evidence of the presence of fluorescein was obtained at either of these outlets up to the evening of the 27th of May.

At a meeting of the Chemical Tests Sub-Committee, held in Leeds on June 9th, present Messrs. F. W. Branson, G. Bray, and B. A. Burrell, it was decided to use salt at the Smelt Mill Water Sink, ammonium sulphate at the Tarn Water Sinks, and fluorescein dissolved in 10 % aqueous potassium carbonate at Tranlands Beck, these chemicals being all discernible in minute proportions in the presence of each other, and being also comparatively innocuous to fish or cattle.

To the second Committee Meeting, held at Malham, a general invitation to members of the Society was issued to proceed with the work under the guidance of Messrs. Wm. Ackroyd, F.I.C., and F. W. Branson, F.I.C. The members met on June 21st at the Buck Hotel, and after dinner a meeting was held, at which the next day's work was planned, Mr. Walter Morrison, M.P., occupying the chair. There were present :—Messrs. W. Ackroyd, G. Bingley, F. W. Branson, G. Bray, B. A. Burrell, F.I.C., S. W. Cuttriss, C. W. Fennell, Rev. J. Hawell, R. M. Kerr, W. Stewart,

'. Swann, B.Sc., and G. White.   Messrs. Carter and Kendall were
lso present during the investigations of the following day.

The general plan of work resolved on was as follows:—

(1.) To sample Malham water the same evening so that the
llorine figures obtained could be compared with those of May
7th, and on the morrow :

(2.) To continue the gaugings of water already commenced
y Messrs. Fennell, Bean, Cuttriss, and Stewart.

(3.) To put salt into the Smelt Mill Water Sink at 5 a.m.

(4.) To put ammonium sulphate into the stream below the
arn outlet at about 10 a.m.

(5.) To examine the outlets at the Cove and Aire Head
ir the chemicals introduced in the morning, and

(6.) To carry on the necessary quantitative and qualitative
nalyses required at the temporary laboratory in the Buck Hotel.

These various duties were undertaken and carried out by the
iembers as follows:—Mr. Bray superintended the putting in of
he salt at the Smelt Mill Water Sink, Mr. Kerr saw the stream
elow the Tarn outlet charged with ammonium sulphate.   At
0. a.m. the sluice at the Tarn was partially drawn so as to
llow of a flow of water at the rate of 69,500 gallons per
4 hours.   This rate of flow was maintained until 1 p.m., when
much larger volume (not gauged) was sent down; the sluice
/as closed about 2 p.m.

Messrs. Bingley and Hawell took their posts by the Cove
hroughout the day, dividing their attention between testing for
xcess of chlorine, presence of ammonium sulphate, and register-
ag the height of the water at the gauge : similar duties were
erformed by Messrs. Branson and Cuttriss at Aire Head, and
rom both these places samples were despatched by messengers
very half hour to the Buck Hotel, where they were tested
uantitatively by Messrs. Ackroyd, Burrell, and Swann.

The chlorine figure of the sample taken the previous evening
1 Malham village was found to be 1·1, only a small variation
-om the results obtained on May 27th.

The samples received at the Buck Hotel throughout the 22nd of June remained normal. This appeared very remarkable and gave rise to several hypotheses which were subsequently discarded as the work progressed. The water at Aire Head increased in volume, as was anticipated from Messrs. Morrison and Tate's experiments in 1879,* about two hours after the sluice at the Tarn had been raised, but no rise of water occurred at the Cove on this occasion, nor did the chemicals appear at either outlet. The collection of samples was therefore continued until August 2nd, and the necessary analyses were performed by Messrs. Ackroyd, Branson, Burrell, and Swan.

SALT.

In the search for excess of salt in the samples sent from Malham, June 22nd to August 2nd, the following quantitative estimations of chlorine were made :—

Malham Cove	64
Aire Head	45
Gordale Beck	42
Scale Gill Spring and Mill	40
Hanlith Bridge	7
	198

To present the bearing of these analyses at a glance, the chlorine estimations are plotted as a curve in the accompanying diagram (Plate II.), and in constructing the curve the average is used where there are many observations on one date.

On June 22nd one ton of salt was put into the Smelt Mill Water Sink, one ton more was put in on the 23rd and also a third ton on the 24th; up to July 2nd little or no alteration had been observed in the Cove, but on the 4th the chlorine figure rose to 6·05 per 100,000, and after that it gradually fell to normal on the 13th and 14th. The connection between the

---

* Proc. Yorks. Geol. and Polytec. Soc., 1879.

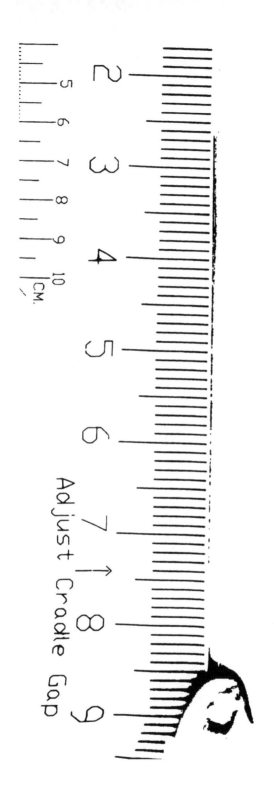

Adjust Cradle Gap

melt Mill Water Sink and the Malham Cove outlet is there-
re abundantly demonstrated.

The suddenness of the rise in combined chlorine suggests
mething of the nature of a flushing out of underground
annels, and it was therefore decided to compare these results
ith the rainfall.

Mr. Morrison has kindly supplied the record of the Malham
infall over the period covered by the experiments, these results
e plotted in the diagram (Plate II.) from which it will be seen
at heavy rainfall shortly preceded the maximum chlorine
servations. No sample of water appears to have been taken on
ie 3rd of July, so that we are not in a position to say that
ie chlorine figure on this day was not higher even than on the
h. The conditions preceding the rainfall were briefly these :—

On the 22nd of the month 19,800 gallons of water per
iy were sinking at the Smelt Mill Water Sink (Plate III.),
carefully gauged by Mr. Fennell. After this date the water
nking would be a diminishing quantity as there was no rainfall
) to the 27th, then there is a temporary rise in the rainfall
irve and a further and larger rise on the 30th of June and
ie 1st of July. The waters of Aire Head, Scale Gill Spring,
id Gordale Beck remained normal throughout the entire series
observations, showing that the Smelt Mill Water Sink under
ie conditions investigated is not connected with any of the
st three outlets.

AMMONIUM SULPHATE.

The search for the ammonium sulphate put into the Water
nk below the Tarn outlet (Plate V.) was qualitative, being the
nple addition of Nessler's reagent. Very decided or marked
sults were looked for, and such results were obtained with the
iter from Aire Head from July 4th to 11th. Distinct traces
the presence of the ammonium compound were also present
the Cove water on July 4th, and for a week after.

c

FLUORESCEIN.

On the 22nd June an aqueous potassium carbonate solution of fluorescein was put into the Tranlands Beck, S.S.W. of Malham, and it showed itself early on the following morning half a mile away at Scale Gill Spring. It was also evident in the samples of water collected at Hanlith Bridge on this date, the river Aire being coloured to a point three miles away. It was, therefore, with considerable confidence that fluorescein was put into the Smelt Mill Water Sink on the 24th June. Having failed to make its appearance at any of the outlets, a further solution of 1 lb. of fluorescein was emptied into this Water Sink on the 27th, and 2 lbs. more on the 28th, but not until the evening of July 4th (the date of the chlorine maximum) did the fluorescein show itself at the Cove. It had become less marked on the 8th, and was somewhat uncertain in samples obtained on the 10th. Mr. Swann, who saw the waters of the Cove on the 6th, describes the fluorescence as "most profound and intense." The samples received by the analysts also gave results in the search for fluorescein which accorded with Mr. Swann's observations, that is : the presence of the compound at Malham Cove on the dates mentioned, and its absence at Aire Head. It may be added that the Sub-Committee find that one part of fluorescein in 40 million parts of water may be readily detected.

On August 26th a new set of experiments was commenced on the same lines as before, viz., the simultaneous use of ammonium sulphate, salt, and fluorescein. A solution of 1½ lb. fluorescein was emptied into the bottom of Grey Gill Cave and washed down by a copious supply of water. This operation was a work of some difficulty and danger : a water barrel had to be placed in the middle of a slippery hill side, whence the water was piped over the rock into the cave, and this barrel was supplied by piping from the hill top. Seven hundredweight of ammonium sulphate was emptied into Upper Gordale Beck by the foot-bridge, and 18 cwt. of salt was subsequently emptied into the "burst" on the side of Cawden. From the 26th of August to

Photographed by Godfrey Bingley, Headingley, Leeds.   SMELT MILL, WATER SINK.

Proc. Yorks. Geol. and Polytec. Soc., Vol. XIV., Plate III.

end of October 168 samples of water were taken at
ous points in the search for these compounds, involving
quantitative estimations of chlorine and 336 qualitative
rvations for the fluorescein and ammonium sulphate. No
ence of the fluorescein has been observed. The ammonia
ın to be excessive in amount at New Laithe Bridge on the
of September. The chlorine figures of the samples from
e's Barn, Gill Flats, had risen from the normal unit to 11·1
the 23rd of September, and on subsequent dates were as
ws:—September 24th, 12·5; September 25th, 12·8; September
ı, 7·6; September 27th, 2·9, and on the 28th the chlorine
fallen to normal (Plate IV.). It is noteworthy here that
rising chlorine was shortly preceded by heavy rain-
as was the previous experience in the Malham Cove
rvations. These results appear to be confirmatory of the
ımittee's hypothesis as to the course of the underground
ers in this particular area of the Malham district. There
ə present at Malham on August 26th:—Messrs. Ackroyd,
gley, Branson, Howarth, and Kendall. The analyses were
ıequently made by Messrs. Ackroyd, Branson, and Burrell.

LIST OF CHEMICALS USED, AND RESULTS.

399.			
: 22 ...	13 cwt. ammonium sulphate.	Malham Tarn Beck, below sluice, 10 a.m.	Showed at Aire Head and Malham Cove, July 4th to 11th.
: 22, 24	3 tons of salt.	Smelt Mill Water Sink.	Showed at Malham Cove, July 4th to 11th.
: 24, 28	3 lbs. of fluorescein.		
: 22 ...	1 lb. of fluorescein.	Tranlands Beck.	Showed at Scalegill Mill, June 23rd.
26 ...	7 cwt. ammonium sulphate.	Upper Gordale Beck, by foot bridge.	Showed at Springs below Gordale Scar, Sept. 7th.
26 ...	1½ lbs. fluorescein.	Grey Gill Cave.	No results up to October 31st.
18 ...	18 cwt. salt.	Cawden "Burst."	First noticed at Mire's Barn, Gill Flats, Sept. 23rd to Sept. 27th.

SUMMARY OF ANALYSES MADE DURING THE INVESTIGATION.

		Quantitative.	Qualitative.
At the Buck Hotel, June 22nd...		68	62
Up to Aug. 2nd ... Malham Cove and Beck ...		64	109
Gordale Beck		42	84
Aire Head		45	71
Hanlith Bridge...		7	14
Scale Gill Spring and Mill ...		40	80
Aug. 6th to Oct. 31st. Gordale Beck Springs ...			
New Laithe Bridge			
Cow Gill, Scale Gill Mill ...		83	336
Aire Head, Malham Beck ...			
Mire's Barn			
		349	756
Total Analyses		1105	

The conclusions to be drawn may be summarised as follows:—

(1.) The unexpected delay in the appearance of the chemicals at the outlets has been due to a quiescent waterflow, and their appearance has usually succeeded a comparatively heavy rainfall.

(2.) Under conditions of summer flow such as prevailed on June 22nd :—

(a) The water descending at the Smelt Mill Sink emerges at Malham Cove and not at Aire Head, Gordale Beck, or Scale Gill Spring.

(b) The Sinks below Malham Tarn are connected with Aire Head and not with Gordale Beck; and under certain conditions, as detailed in the report, some of the ammonium sulphate put in at the Tarn Sinks emerges at the Cove.

(3.) The Tranlands Beck Water Sink is connected with Scale Gill Spring and not with Aire Head.

(4.) The Gordale Beck during the very exceptional drought of the present year (1899) disappeared in the stream bed about a quarter of a mile above the Scar Waterfall and reappeared at the springs between Gordale Waterfall and Laithe Bridge.

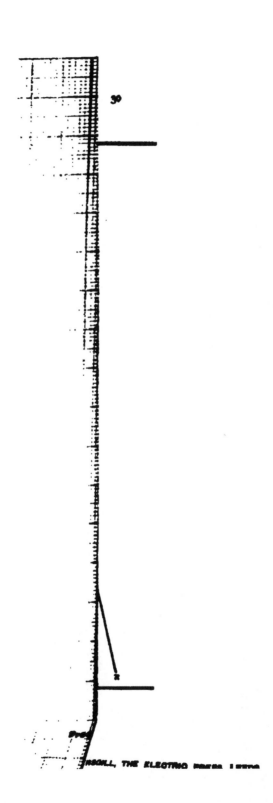

30°

x

(5.) A solution of salt poured in Cawden "Water Burst" appeared at Mire's Barn Spring.

(6.) The fluorescein solution put into Grey Gill Cave on August 26th had not been observed up to October 31st.

REGISTER OF RAINFALL AT MALHAM TARN.

BY THOS. COULTHARD.

From July 14th to October 31st, 1899.

Date.	July. in.	Aug. in.	Sept. in.	Oct. in. (·54)
1		—	·52	·05
2		—	·17	·62
3		—	·01	·05
4		—	—	—
5		·53	—	
6		·03	-	
7		—		
8		—	07	
9		—	—	
10		—	·04	·01
11		—	—	·59
12			·02	·05
13		—	·04	—
14	—	·06		—
15	—		·86	—
16	—	·0	·03	—
17	—		·80	—
18	·41	·07	·54	—
19	—	·03	·64	
20	·30	—	·09	—
21	—	—	1·05	—
22	·14	—	·24	—
23	—	—	·48	—
24	—	—	·43	·21
25	·44	—	·72	·23
26	—	—	·06	·14
27	-	·37	·01	·24
28	·2	·26	·98	·55
29	--	·53	·08	·42
30	--	·13		
31	—	·27		

IV. REPORT OF THE GEOLOGICAL SUB-COMMITTEE.

BY PERCY F. KENDALL, F.G.S. ; J. H. HOWARTH, F.G.S. ;
AND W. LOWER CARTER, M.A., F.G.S.

GEOLOGICAL STRUCTURE OF THE DISTRICT.

The Carboniferous rocks exposed in the area present the following vertical succession, according to Tiddeman :—

SOUTHERN OR BOWLAND TYPE.	FEET.		FEET.	NORTHERN OR YOREDALE TYPE.
Millstone Grits... ...	3,900			Millstone Grits.
Bowland Shales ...	300–1,000	The Great Craven Faults.	400–900	Yoredale Series.
Pendleside Grits (including Pendlesant) ..	0–250			
Pendleside Limestone (with Knoll-Reefs) ...	0–400		400–800	The Carboniferous Limestone (with conglomerates at the base).
Shales with Limestone	2,500			
Clitheroe Limestone (with Knoll-Reefs)..	3,250 No base.			

On the northern (upthrow) side of the North Craven Fault the Carboniferous Limestone is seen to rest upon the Silurian Slates and Grits. The recent paper of Mr. Marr (Quar. Journ. Geol. Soc., Vol. LV., p. 327) offers a correlation of the beds differing somewhat from his, but for the present purposes the classification offered above may be accepted.

The general dip of the rocks is to the northward, but there are minor undulations which will be mentioned. Restricting attention to a tract of country about two miles from east to west, by four miles from north to south, and having Malham approximately as its centre, we find three regions separated by the northern and middle branches of the Craven Faults. The northern area consists of the white limestones of the lower part of the Carboniferous series resting upon Silurian Slates and Grits which form a narrow outcrop down to the northern fault. The median belt consists wholly of the Carboniferous Limestone and forms a high plateau, cut off on the south by a great indented

carpment falling to the line of the middle fault. The southern
ea consists of sharply undulating Bowland Shales, through
hich appears, in two outcrops, the Pendleside Limestone, with
e characteristic knoll-reefs of Tiddeman.   In the northern
·lt there originate three streams which have been the subjects
  investigation.   They suffer total or partial absorption upon
tering the median zone of strongly-jointed limestone, and the
ater is given out in three principal springs, two of which, at
alham Cove and Gordale respectively, come out at the foot of
e great escarpment, while the third, Aire Head, emerges at the
uthern edge of a synclinal fold of the Bowland Shales.

SUMMARY OF RESULTS.

It may be convenient to briefly relate the results obtained
om the investigations of the Chemical and Engineering Sub-
ɔmmittees, and those previously recorded by Mr. Tate.

A. CHEMICAL.

1. Reagents introduced into the Smelt Mill Sink have
·en traced nowhere except at the Cove.

2· The ammonium sulphate supplied to the Tarn effluent
ream in June, 1899, though almost exclusively discharged by
e Aire Head Springs, yet was traceable in minute quantities
  the water flowing from Malham Cove on July 4th to 11th.

(It is important to note that it synchronised exactly with
e discharge of the same reagent at Aire Head.)

3. The ammonium sulphate placed in Gordale Beck
appeared in the springs at the foot of Gordale Scar, and
·where else.

4. The fluorescein poured into the bottom of Grey Gill
ive has not been traced.

5. The salt introduced into Cawden Burst came out at
ire's Barn Spring about a quarter of a mile to the southward·

B. ENGINEERING.

1. Messrs. Morrison and Tate's experiments show that at that time (the middle of May, 1879) a large flush of water from the Tarn produced a responsive flow from Aire Head Springs in one hour and twenty-five minutes, and that the outlet from the Cove was affected 38 minutes later, although so much nearer the source of supply.

2. The experiments in June, 1899, when flushes were sent down from Malham Tarn, proved that *under the conditions then prevailing*, the Aire Head Springs responded in two hours, but the discharge from Malham Cove was not affected by even the smallest measurable quantity.

On August 7th, 1899, a large volume of water was sent down from Malham Tarn, at the request of Mr. Cuttriss, and produced a rise in the Cove stream of two inches at Malham Bridge.

On August 26th another large flush was sent down, and the Cove stream again rose not less than two inches at Malham Bridge.

These varying results have a special significance when considered in conjunction with the appearance at Malham Cove of minute quantities of the ammonium sulphate put into the Malham Tarn stream in June, 1899.

3. The Smelt Mill stream in June, 1899, only contributed 19,800 gallons of the half million gallons flowing out at Malham Cove, i.e., not more than four per cent.

THE GEOLOGICAL PROBLEMS.

The problems which these data set before the Geological Sub-Committee for solution are the following:—

1. The determination of the *route* taken by each underground flow. We regard the use of the word "stream" to describe these movements as inaccurate, since it suggests the idea of a definite channel rather than a diffuse flow through many fissures large and small.

BROAD SCAR, LOOKING EAST.

l by Godfrey Bingley, Headingley, Leeds.

BROAD SCAR, LOOKING WEST.

2. The cause of the adoption of the several routes.

3. The conditions of underground flow of water, with special reference to the retention of reagents for long periods, and the apparently capricious effects of the Tarn flushes upon the outflow at Malham Cove.

4. The nature of the underground spaces in which the water is contained during its transmission.

The second is the main problem, viz.: What is the cause which has determined the route taken by each flow after its absorption through the swallow-holes? In dealing with this, the other points named will receive some explanation.

The history of the early speculations upon this subject has shown how prone observers have been to assume that, where a dry valley, obviously the result of stream erosion, connects a sink with the place of emergence of a stream, the flow will follow the same general course *under* ground that it used to do *above*. This very natural assumption has been made in the case of the Tarn Water Sinks and Malham Cove. It seemed in some measure supported by Messrs. Morrison and Tate's experiments, when *all* the Tarn flushes affected both Aire Head and Malham Cove. Moreover it seemed difficult at one time to account on any other hypothesis for two circumstances:—

1. That the water issuing at Aire Head had *crossed* the track of the Cove stream unless the two stream courses were coincident.

2. That the Cove stream enormously exceeded in volume the Smelt Mill stream, which was not large enough to account for the former, as suggested by Dr. Whitaker and local tradition; besides which the hardness of the Cove water was *less* than that of the Smelt Mill, notwithstanding the limestone strata between.

The experiments of the Committee, however, have shown that (at least in June, 1899) no appreciable quantity of water flowed in the direction usually assumed.

The explanation which here follows, suggested in the first instance by Mr. Kendall from certain *a priori* considerations, was confirmed by an inspection of the Ordnance six-inch maps (not geological), and has been raised to demonstration so far as one-half of the area is concerned by a close examination of the country.

The production of spaces capable of transmitting water through a dense and compact rock like the average unweathered Carboniferous Limestone is affected by the solvent action of rain-water, which is always more or less charged with carbonic acid. This action is further greatly aided by organic acids derived from vegetable matter undergoing decomposition in the soil; and the activity of plants must also augment the quantity of carbonic acid available.

Now in a compact limestone solvent action is limited in the main to attacks upon actual *surfaces*, and the effects will be essentially different from those produced on more porous rocks of similar composition.

Such attacks proceed against the upper exposed surface, resulting in the fantastically furrowed and weathered forms which are so much in favour for the construction of "rockeries"; but they also proceed downwards along the exceedingly narrow joint fissures which traverse all rocks in at least two directions. The joints are widened, especially near the surface of the ground where the first contact with the acid water takes place, and there is also a selective action, some beds resisting solution more than others. Flow also occurs along bedding planes, with the effect of producing openings following the inclination of the bedding.

These occasionally assume the dimensions of spacious chambers, and they are frequently enlarged by falls of the roof. The limestone caverns of Craven are sometimes of this character. The joint fissures are, however, far more numerous and important, and most frequently give rise to caverns.

Where solution is taking place with the greatest freedom all the rainfall which escapes re-evaporation is absorbed into the rock, with the frequent result of widening the joints so much at the surface as to cause the swallowing up of the soil and the production of the well-known "Clints" or "limestone-pavements" (See Plate VI.).

raphed by Godfrey Bingley, Headingley, Leeds.

GREY GILL.

Now in all stratified rocks which are jointed it is found
of the two dominant sets of joints usually decussating at
:s of 70° to 90° one set, the master-joints, is much stronger,
is much more persistent both in continuity and direction
the other set, which may be called for present purposes
:s-joints."

It is safe to assume that the master-joints from their greater
nuity will afford a freer flow to water, and this will secure
they will be proportionately more affected by the solvent
n of water, so that, in any district, unless the direction of
.t ultimate escape for underground water be directly in the
of the cross-joints, it will travel as far as possible along
naster-joints.

Now to apply these principles to the problem of the sources
e Aire. The study of the six-inch Ordnance map showed
the high limestone plateau, beneath which the subterannean
·s flow between the several sinks and springs, is traversed
nes of scars having a very marked parallelism; some of
·alleys, moreover, or portions of them run in straight courses
the same general orientation; and it was suggested that
features were determined by the master-joints, of which no
observations had, up to that time, been made.

Certain obvious reflections were also made upon the results
dy obtained by the Chemists, which lent an air of proba-
· to the hypothesis that the features in question were due
naster-joints which had also a preponderating effect upon
ments of the underground water. Thus a line drawn from
t Mill Sink to Malham Cove coincided with the general
tion of the scars and other features of the plateau, which
assumed to depend upon the master-joints; it was also
:t coincident with the long valley of the Watlowes and
.in of intermittent valleys and scars running nine-tenths of
listance; but, further, it joined together the actual points
ntry and emergence of water which had been traced by
Chemists.

This observation lent some significance to the fact that a line similarly drawn (parallel to the one from Smelt Mill to Cove) from Malham Tarn Water Sinks (Plate V.) in a south-easterly direction struck the edge of the plateau just where it is breached by a ravine, Grey Gill. Grey Gill thus bears the same topographical relation to the Tarn Sinks that Malham Cove does to Smelt Mill Sink; but with the difference, suggestive in view of the fact that the Tarn Sinks feed Aire Head Springs, that Grey Gill Gorge is dry.

The view thus arrived at inductively seemed sufficiently plausible to encourage work by the Geological Sub-Committee, and an examination of the district was undertaken by Messrs. J. H. Howarth, F.G.S., P. F. Kendall, F.G.S., and W. Simpson, F.G.S., with assistance from Mr. Herbert B. Muff.

For the purpose of and prior to this investigation the section illustrating this report (Plate XII.) was drawn by Mr. Kendall, and the subsequent examination of the district disclosed a series of facts supporting in a remarkable way, without completely proving, the assumed analogy between the relations of the sinks and gorges in the two cases.

At the very outset of the inquiry it was found that the speculations regarding the directions of the master-joints on the plateau were thoroughly in accordance with the facts. Local discrepancies occurred here and there, especially in close proximity to the North Craven Fault, but in general the master-joints showed a quite inconsiderable variation from a direction 160° magnetic (or 142° true), which is 7° to S. of true S.E. This direction is indicated on the map accompanying this report by the two strong lines to the west of the words "Malham Lings."

Mr. Bingley's photographs (Pl. VI.) show a portion of the limestone surface known as Broad Scar between the Tarn Water Sinks and Grey Gill. The first is looking towards the water sinks, which lie under the curve of the moorland in a line with Low Trenhouse farmstead. The second is looking towards Grey Gill. On these scars, as is so admirably shown in the photographs, the master-joints are very strongly developed and their

Photographed by Godfrey Bingley, Headingley, Leeds. GORDALE.

Proc. Yorks. Geol. and Polytec. Soc., Vol. XIV., Plate VIII.

inuity is remarkable. It is possible to select one at the
h-westerly edge of the plateau which bears directly for the
ı Water Sinks, and to follow it almost without interruption
he crags overlooking Grey Gill. Cross-joints run nearly at
t angles to these, but any single one can rarely be traced
more than four or five yards, though its main direction may
etimes be continued *en échelon* by other joints. In some
es a third set of joints inclined about 35° to the master-joints
· be observed.

In the course of this investigation it was ascertained that the
ı at the head of Grey Gill (Plate VII.) was exactly in the
of this section, which fact forms an interesting coincidence.
ı cave bears 340° mag. (the equivalent of the bearing of the
ter Sinks joints, viz., 160° before mentioned), and is obviously
ely an enlarged master-joint. It is well to state here that
cave slopes downward from the entrance towards the N.W.

As to the connection between the joints and the flow of
ır complete proof cannot in all cases be expected, but the
ı now brought to light seem to justify the inference that
· have a determining influence, and several flows may now
:onsidered seriatim.

1. SMELT MILL SINK TO MALHAM COVE.

The route from point to point here so closely approximates
he direction of master-jointing that it may, it is considered,
aken as fairly certain that the water is transmitted directly
ugh a series of enlarged master-joints.

2. GORDALE BECK TO GORDALE SPRINGS.

These are taken out of their geographical order of succession
use much light was thrown by the behaviour of the water
ıis instance upon the very difficult case of the Tarn to Aire
ı flow.

It was not until this investigation had directed attention to
problems of underground saturation-levels that it was more

than dimly suspected that Gordale Beck (Plate VIII.) suffered
amount of absorption in its passage across the limestone

Some details of this stream are needful to an und
of the problem.

The general level of the limestone plateau in this region
about 1,250 feet. The stream passes off the Silurian rocks on
the Carboniferous Limestone at 1,160 feet, and its valley q ·
begins to assume the character of a gorge, gradually deepening
until at the waterfall in little less than a mile it has become
a wild rocky ravine, 200 feet deep (Plate IX). The stream-level
has fallen in this distance to 975 feet, while in the next
300 yards it descends in a series of cascades to 800 feet. A little
lower down, on the east side, great springs break out from the
foot of the crags which rise very steeply for 400 feet. From this
description it will be seen that Gordale Beck flows over the
limestone at an altitude more than 200 feet above the level
escape of the springs. This fact prompted the inference
the stream must at all times undergo some amount of a
in passing over so permeable a bed as is furnished by the
fissured limestone. This inference it became possible (for th
time within living memory, perhaps) to put to a decisive tes
August, 1899, Gordale Waterfall for the first time in its
was absolutely dry. The members of the Committee
several examinations of the beck, and on 28th August
following note was made of a journey down stream from
point where the beck enters the limestone country :—

"Gordale Beck, flowing strongly at foot-bridge, d ·
and becomes slimy and offensive; finally bed is quite ɩ
a little above the sheep-fold and wash-dub, where path
beck." (This point is just above the 1,000 ft. contour.)

Gordale Beck, then, under conditions of extremest d
suffers the fate of the neighbouring streams, and is swal
up by the limestone. This does not definitely prove that
absorption takes place in normal seasons, but it raises a
strong presumption in favour of the view that it does, and that
Gordale Springs are always mainly supplied in this way.

GORDALE SCAR.

In such a drought as prevailed then it was easy to demonstrate
he experiments described by the Chemical Sub-Committee
the water absorbed in the stream bed came out in Gordale
gs, which suffered only partial failure and never actually
d to flow.

The route followed by the water scarcely admits of doubt.
master-joints run obliquely across the beck and evidently
· the water away into the body of the limestone on the
rn side.   At the waterfall the gorge makes a south-westerly
evidently determined by some structural feature of the
and it would appear that the same structural feature
ls the underground. water flowing for some distance into the
in a south-easterly direction to take a similar south-westerly
in rough parallelism to the beck, and to reappear in the
gs.

MALHAM TARN WATER SINKS TO AIRE HEAD SPRINGS.

This case has been reserved for the last because the difficul-
it presents are much diminished by the light obtained from
ale.

The introduction has described how Malham Tarn water after
ging from the Tarn flows off the Silurian area and for some
nce over the Carboniferous Limestone before suffering com-
absorption at the water sinks.

From the point of disappearance a valley runs in a straight
deepening steadily until, making a very sharp turn, it forms
great dry gorge which terminates above Malham Cove in
eer drop of 260 feet.

The point at which absorption of the stream is complete
es with the seasons.   In very dry seasons it can be seen
the stream begins to dwindle directly the limestone area is
hed, and at the first sink it wholly disappears; but in more
nal seasons two more sinks come into operation.   In very
times the surface stream continues down the valley and,
tated in the Introduction, it has been known to reach Comb
· and Malham Cove also.   Little doubt need be felt, therefore,

that at some more or less remote period there was a constant flow along the dry valley, and that the great recess at Malham Cove has been produced mainly by the agency of the plunging stream which has prevented the accumulation of protective talus at the Cove foot, besides drilling out a pool. Such pools are invariably found at the foot of waterfalls with a sheer drop, and by the recession of falls they commonly form ravines. That no such pool or ravine now exists at the foot of Malham Cove may be interpreted as proving that a sufficiently long interval has elapsed since the stream regularly flowed over the top to admit of the obliteration of the pool by the accumulation of fallen blocks, or by the action of the great spring which emerges there and still prevents the formation of any talus.

It is worthy of note that springs emerging upon steep hill-sides commonly show the same feature of a scarp at the back. and for the same reason that the outflowing water prevents the accumulation of talus.

The cause of the great sheer face at the Cove is one familiar to all geologists, viz.: the occurrence of a more durable bed to form the sill of the fall, as at Niagara, Hardraw Scar, &c. In this case the massive bed of limestone so well shown in the photograph has been a determining factor. (Pl. X.).

Upon the stage when the stream followed the surface channel only, and which may have been connected with the frozen state of the ground at the close of the Glacial Period, several phases of partial absorption and retreat must have supervened when successive sinks gradually developed. The surface flow would thus cease by a regular retrogression, and now it can be seen that the recession has reached very nearly to the North Craven Fault.

During the early stages of retreat the swallow-holes must, it would appear, have carried the water into the master-joints below the point named the Watlowes on the map, and so out at Cove; the emergence being not necessarily all at the present outlet but sometimes at some height above it. There is a cave opening on the face of the Scar which may be one of these

y outlets. This arrangement would subsist until the point
final absorption had receded from the direct line of the
ter-joints running down to the Cove (that is out of the
lowes), and to a spot from which master-joints would lead
water eastward of the Cove; and so on until the water
eventually turned towards Grey Gill.

Upon this supposition we may now consider the probable
equent course of events. A great spring would be likely to
rge in Grey Gill and produce a stream flowing down the dry
nel which runs to Gordale Bridge; but it is well to remark
that the character of the upper part of Grey Gill gorge
of the country to the north rather favours the view that
arface stream (though of smaller volume than that necessary
ccount for the lower part of the gorge) descended here.

From the level of the cave downwards the floor of Grey
is composed of loose blocks of limestone intermingled with
(the latter indicating plainly that water flowed in a surface
am), and no attempts on the part of the Committee to get
ugh or to probe to live rock were successful. Attempts
э repeatedly made to reach the rock, as inhabitants of the
rict testified that in normal seasons a sound of running
er could be heard, and it was considered that the Malham
1 water was probably passing here on its way to Aire Head.
Tarn stream was twice flushed for the purpose of listening
in Grey Gill Cave and the screes below, but no water
l be heard running then. Fluorescein was introduced into
bottom of the cave in August, 1899, and washed down
100 gallons of water, but no trace of it has since been
l.

Grey Gill changes rather abruptly from its gorge-like character
e the Mid Craven Fault comes across near Cawden Flats
, and this fault juxtaposes to the great mass of pure white
tone of the plateau the region of "reef-knolls," consisting
ark flaggy limestones or shales enclosing great dome-like
es of highly fossiliferous limestone, of which Cawden is a
example.

D

The fault, the change in the character of the rocks, and the steep undulations of the bedding might be expected to have a great effect upon the flow of underground water.

Assuming that underground channels were established along the master-joints, what causes the water to flow out at Aire Head? It should be borne in mind that the *direction* of flow of underground water is as much or more determined by the ease with which it can get *out* at a particular place as the freedom with which it can get *in* at some other place; indeed, paradoxical as it may seem, the former is much more the cause of the latter than *vice versa*.

Now in the district south of the fault the rocks are bent into an arch and trough, so that the limestone under Cawden comes up against the plateau limestone at and about Grey Gill, dips under the shales below Malham village, and re-emerges at Aire Head; and here the great springs break out (Plate XI.). Again assuming that the Tarn water goes by way of or near Grey Gill, what causes it to turn off at so sharp an angle to its then course? The answer appears to be (though several reasons might be suggested) that Aire Head Springs are situated at or within a few yards of the *nearest* and *lowest* point of re-emergence of the limestone of Cawden as it rises towards Kirkby Top.

Moreover, the joint systems, which showed such remarkable persistence and regularity in the gently inclined limestone of the plateau, are inconstant in direction in the more disturbed, folded, and perhaps crushed, limestones of the southern area: but determinations made between Scale Gill Mill and Aire Head give readings of

$$
\left.
\begin{array}{l}
\text{N. } 10^\circ \text{ W. mag.} \\
\text{N. } 60^\circ \text{ W. } \quad,,
\end{array}
\right\} \text{ two sets.}
$$

$$
\left.
\begin{array}{l}
\text{N. } 35^\circ \text{ E. } \quad,, \\
\text{N. } 35^\circ \text{ W. } \quad,,
\end{array}
\right\}
$$

$$
\left.
\begin{array}{l}
\text{N. } 15^\circ \text{ W. } \quad,, \\
\text{N. } 45^\circ \text{ E. } \quad,,
\end{array}
\right\} \quad,,
$$

N. 20° E. .. master-joint.
N. 22° E. ,,   ,,

(Declination 18° W.)

Photographed by Godfrey Bingley, Headingley, Leeds.   AIRE HEAD SPRINGS.

The last two readings were taken at the weir about 200
, south of Aire Head, and were probably unaffected by
all unmapped fault crossing the Aire near the foot of the
dam which had disturbed the joints in the preceding cases.
.s bearing in the same direction as these two master-joints
d connect Aire Head with Grey Gill.

It appeared to the Committee a fact of some significance
a " burst," of great volume and sufficient hydrostatic head
:opel the water to a height of several feet, breaks out every
years at a point on Cawden, exactly on the line between
· Gill and Aire Head. A trial made in September, 1899,
·ver, just as the drought broke up, resulted in the discharge
.he re-agent not at Aire Head but in a small spring at
·'s Barn which lies to the north-east on the direct line from
Head to Cawden " Burst."

The crossing of the two flows, Smelt Mill to Malham Cove-
·am Beck and Tarn Water Sinks to Aire Head Springs, with-
·ny intermingling presents no difficulty. The intercrossing
·t take place above Malham Cove, for no impervious stratum
· in the Carboniferous Limestone of the plateau to keep
wo layers of underground water apart: it must take place
·· Malham Beck crosses a syncline of Bowland Shales just
· the village of Malham. It will be seen from the section
·ded to the map (Plate XII.) that Aire Head Springs
·;e from beneath the southern edge of these shales.

Fluorescein put into Tranland's Beck, by the first flood-gate
·· the bridge at the foot of Kirkby Top, re-appeared in
·ing at Scale Gill Mill. This again is consistent with the
·al direction of master-joints in the vicinity, but there are
·int exposures at the particular spots.

The late Mr. Tate considered that the Smelt Mill Sike could
·eed Malham Cove Spring because (a) the volume of the
·r was inadequate, being only about $\frac{1}{20}$th that of the Cove
·n, and (b) the quality of the water was different. Thus :—

	SMELT MILL.	MALHAM COVE.
Total Hardness ...	13·1	10·8

Mr. Tate said "It is impossible to believe that water flowing undiluted through a mile and three-quarters of limestone strata should, during its voyage, lose between two and three degrees of hardness."

The fact appears to be that, as Mr. Tate said, the Smelt Mill Sike contributes only about $\frac{1}{20}$th of the water issuing from the Cove, the remaining $\frac{19}{20}$ths consisting of the rainfall absorbed on the two square miles or more of country to the westward of the Watlowes, and of the water which sinks in the potholes on Outside, and that this water is somewhat less hard than that from the Smelt Mill. But any argument drawn from so slight a difference of hardness must be inconclusive.

V.   CONCLUSIONS AND REMARKS OF THE UNITED SUB-COMMITTEES.
A.—MALHAM COVE SPRING.

That this discharges :—

1.   The water from Smelt Mill Sike.

2.   The surface water from the limestone area west of the Cove and the Dry Valley.

3.   Under certain conditions a portion of the Tarn water.

NOTE.—It is not quite certain, however, that these conditions have not to be artificial or exceptional. Flushes from the Tarn sometimes affect the Cove, but if the Smelt Mill Sink could be similarly flushed at the same time *all* the Tarn water might pass on to Aire Head. As a general rule it probably does so, unless there happens to be a marked difference in the rainfall between the Tarn and Smelt Mill gathering areas which renders the Tarn supply abnormal relatively to that of the Smelt Mill side.

It is only the Tarn stream which can be experimented upon by flushing, so that it is possible that if the Smelt Mill side, whether artificially or by rainfall, were flushed to a greater degree than the Tarn stream some Smelt Mill water might come out at Aire Head. It seems probable that an underground watershed exists between the route taken by the Tarn Sinks to Aire Head water and Malham Cove.

4. The water from Outside and Black Hill which disappears in the pot-holes above Ha Gate.

NOTE.—Re-agents have not been tried in these pot-holes.

B.—AIRE HEAD SPRINGS.

These discharge the main portion of the water disappearing at Malham Tarn Water Sinks. (See Malham Cove, note.)

C.—GORDALE BECK SPRINGS.

These discharge the water absorbed into the stream bed in Upper Gordale.

D.—CAWDEN BURST.

The Salt put in here reappeared at Mire's Barn.

E.—TRANLANDS BECK WATER SINKS.

These discharge in the principal spring at Scale Gill Mill.

F.—GREY GILL AND CAVE.

No connection has been established between this water-worn cave and any outlet spring.

The Geological Sub-Committee consider that the Tarn Sinks water passes near here, but this has not been proved.

G.—SPRINGS BELOW JANET'S FORCE.

These have not been experimented upon.

The spring on the left bank of Gordale Beck below Janet's Force is probably supplied from the slopes above it.

On the right-hand side of the main stream a strong spring breaks out in floods under a limestone escarpment. This is probably supplied from the Cawden area when the underground passages and fissures there are filled to high levels.

H.—UNDERGROUND FLOW.

The investigations show that within the area the main direction of underground flow is along the master-joints in the limestone.

J.—DELAY IN TRANSIT.

The investigations show that the flow of underground water is much slower than was generally supposed.

K.—UNDERGROUND CAVERNS AND POOLS.

The investigations have thrown no very definite light upon the question as to whether these exist in the Malham area.

The Committee, however, believe that both Malham Cove and Aire Head springs are *below* the general saturation level of the rocks.   If so, caverns are only likely to exist, if at all, in the upper part of the limestone and *above* such saturation level.   Having regard to the character of the rock-joints caverns are more likely to exist than large pools.

L.—MALHAM TARN FLUSHES AND MALHAM COVE.

Of the problem as to why Tarn flushes should affect Malham Cove spring sometimes and not always, Mr. Kendall contributes a highly interesting solution.   (See Appendix.)

———————

It is our pleasurable duty to mention that Mr. Morrison's aid and advice have been given throughout the investigations at Malham, and the Committee have had the active co-operation of his steward, Mr. Winskill.

———————

APPENDIX.

MALHAM TARN FLUSHES AND MALHAM COVE.

BY PERCY F. KENDALL, F.G.S.

The behaviour of the water sent down in flushes from Malham Tarn demands some attention.   In Mr. Tate's experiments the flushes affected Aire Head in about $1\frac{1}{2}$ hours and Malham Cove about 38 minutes later.   In the experiments of this Committee in June, 1899, before the commencement of the great drought, a gauge was established at the Cove and

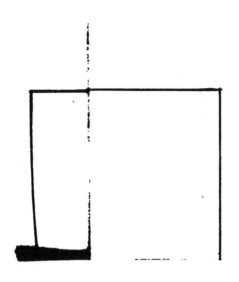

l
l
a

s
l

l
x
l
c

readings were taken every 15 minutes for many hours after the flush, and absolutely no rise occurred.   On the other hand, as has already been stated, in the beginning of August and again later in the month, on which occasions the drought was at its maximum severity, the Cove, or at least the stream, at Malham Bridge rose in response to each flush.   In the experiments in June, though no rise of the water-level took place at the Cove, traces of the ammonium sulphate introduced at the Tarn Water Sinks were found in the water issuing from the Cove eleven days later.   These apparent anomalies appear to be susceptible of a fairly simple explanation, and one which throws much light on the movement of underground water.   The upper limit of saturation of a pervious rock forms a somewhat irregular surface, whose altitude and slope vary according to (1) the freedom of escape of the water at various points, (2) the facility of percolation or flow in different directions, and (3) the interval which has elapsed since the last absorption of rainfall.

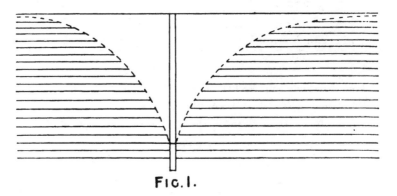

FIG. I.

1. The points of escape of underground water.   There will be a general slope of the saturation planes so as to produce a series of slopes converging upon each of the points of escape. In the case of a well the slopes will form an inverted cone (sometimes called the "cone of exhaustion") during pumping or in the recovery from pumping (Fig. 1).   When the escape is by a spring, the slopes will generally assume the form of a half cone

(Fig. 2). When the escape is along both sides of a valley the saturation planes will generally slope in the same direction as the sides of the valley but at a lower angle (Fig. 3).

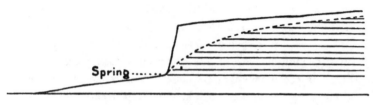

FIG. 2.

2. The greater facility of percolation or flow in a given direction, as compared with that in another direction, will have the effect of producing a saturation plane of lower gradient in

FIG. 3.

the direction of easiest movement. Thus we find that the saturation gradient is generally gentler in the direction of the dip of rocks than towards the rise, e.g., the cone of exhaustion

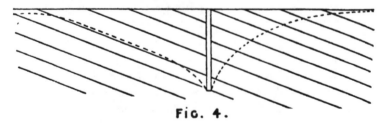

FIG. 4.

produced by pumping from a well or borehole in inclined beds has its axis inclined in the direction of the dip (Fig. 4); and on a given outcrop of inclined strata, the subterranean watershed,

formed by the meeting of two planes of saturation, one towards
the dip slope and the other towards the rise of the beds, will be
nearer to the foot of the escarpment or basset-edge than to the
edge which is covered by newer rocks (Fig. 5).

3. The influence of alternating wet and dry periods upon
the limits of saturation will be that during rainy periods the

FIG. 5.

saturation levels will rise and, the points of escape being
approximately constant (though new ones may come into operation at such times), the gradients will steepen.   In the simple
case of horizontal beds forming a plateau with free edges all
round, the saturated rock will assume the form of a dome.   In
the intervals between periods of rainfall the gradients will continuously diminish, and in the case of the plateau postulated
above, the dome representing the saturated rock will flatten
(Fig. 6).   All the gradients will tend to zero.

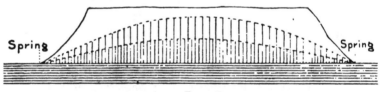

FIG. 6.

Applying these principles to the problem under consideration,
we will assume that the waters of Smelt Mill Sike are transmitted through a series of master-joints directly to Malham Cove,
and that the water entering at Malham Water Sinks flows
beneath the limestone plateau in a similar parallel series of joint-
fissures.

Each set of joints may be regarded as a sort of ground valley through which, with great retardation by fr a stream of water flows.   There will be between these two of flow an area of saturated rock receiving the direct perco of rainfall from the intervening tract of the limestone pl This saturated area will drain into the subterranean "va and produce saturation gradients following the rules already d

Suppose, now, a "flush" to be sent down from Malhan into valley No. 2 (Fig. 7).   Before any effect can be produce the flow in valley No. 1 the water must accumulate under to such an extent as to cause a rise of the saturation-lev the intervening watershed.   The readiness with whic

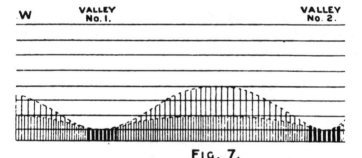

FIG. 7.

can be accomplished will depend upon two factors— magnitude and duration of the "flush"; (2) the height underground watershed. The height of the watershed is (a very material factor, for upon it will depend the vo water which must accumulate in the subterranean valle an overflow can take place. The magnitude of the flush for obvious reasons, of equal importance. When the sa level is high a large volume of water will be needed to it; when the level is low a smaller quantity will suffice.

Mr. Tate's experiments were made in a season of rainfall, but there appears to have been a very large v water sent down, for both sluices of the Tarn were ope it is stated that on the first day the sinks were overf

35 minutes from the opening of the sluices, and in little more than an hour the overflow was 150 yards long. On this occasion the Cove stream rose 2 inches. The next day the sluices (both be it observed) were opened and the Cove stream rose 2¾ inches. A very large volume of water in this case was employed, and, the season being normal, an overflow into the Cove drainage resulted. In June, 1899, also a season of normal rainfall, a flush was produced by opening *one sluice*; a photographic record shows that no water flowed beyond the Water Sinks. On this occasion not the slightest effect was produced upon the flow from Malham Cove; it is, however, worthy of note that a minute trace of the Ammonium Sulphate introduced into the Water Sinks in this experiment subsequently appeared at Malham Cove. This appears to indicate that the watershed of the saturation was actually to a small extent surmounted, but that the increased head was so small as to be practically nullified by friction. It may here be remarked that the 1899 experiments show decisively that a flush affects the springs at Aire Head or elsewhere not by the direct discharge of the water admitted to the sinks, but by increasing the water pressure in the rock and causing a forward thrust which drove other water before it, just as the water flowing from a reservoir thrusts water out of our domestic taps. The two flushes in August, 1899, both affected Malham Cove. They appear to have been of less volume than those recorded by Mr. Tate, yet they produced a very marked rise of the stream at Malham Bridge.

The great effect upon the stream issuing from Malham Cove may be due in part to a greater volume of water having been sent down, but it appears quite certain that during the long, unexampled drought which prevailed during July and August, 1899, the saturation levels in the limestone must have fallen to a very marked extent, producing a flattening and lowering of the subterranean watershed, which would greatly facilitate the overflow from one drainage system to the other.

Two subsidiary points need notice. It has been suggested, as stated in the Introduction, that the tardy response of Malham

Cove to a flush from the Tarn Water Sinks, as compared with more prompt outflow at Aire Head, may be due to the flow taking place along a horizontal fissure, while that to Aire Head was through a vertical fissure. It seems unnecessary to appeal to such a cause, the existence and adequacy of which are alike uncertain. The causes assigned, namely, the interposition of an area of saturated rock and the difference between master-joints and cross-joints seem fully to account for the phenomena.

The hypothetical cave which has been supposed to lie beneath the limestone plateau at Malham Cove rests upon no basis of observation, and, while all the facts can be explained without it, there are some which appear to be quite inconsistent with its existence.

THE COMPOSITION OF SOME MALHAM WATERS.

BY B. A. BURRELL, F.I.C.

(*Read November 2nd*, 1899.)

The waters analysed were those whose flow was investigated members of this Society on the 21st and 22nd June, 1899. ᵢ only published analyses of these waters are the partial minations (1) of the Malham Tarn Stream, Aire Head Spring, Malham Cove Spring,* made by the Rivers Pollution Com-ᵢioners in 1869, and (2) of the Smelt Mill Stream, Tarn am, and Cove Spring made by Rimington in 1878.†

A complete mineral analysis is now submitted of water n at the Tarn Sluice Gate, the principal Aire Head Spring, Smelt Mill Water Sink, and Malham Cove.

The waters were collected on the 5th and 6th August, 1899. ᵢg to the prolonged drought very little water was flowing, ᵢr from the Aire Head Spring or down the Smelt Mill er Sink. At the latter place, the sample was taken some yards above the spot where the common salt was added on ᵢ 22nd, 1899, so as to avoid any possible risk of contami-ᵢn. When this sample was taken two brilliant patches of escein were noticed, evidently due to some of the strong ᵢion having sunk to the bottom of the stream.‡

In the following tables the results are expressed in grains gallon. The atomic weights used are taken from the Sixth ual Report of the Committee on Atomic Weights, 1898 gen = 16).§

' Rivers Pollution Commission, Sixth Report, 1874, pages 43, 112.
‧ Proceedings Yorkshire Geological and Polytechnic Society, Vol. VII., 185. 1878.
‡ Six weeks had elapsed since this reagent was used.
¦ Chemical News, Vol. LXXIX., page 207. 1899.

In I. the quantities of the different constituents are given, in II. the acids and bases are combined together in the usual manner, and in III. the previous analyses are given for reference.

In II. a comparison of the Tarn water with that issuing from Aire Head Spring shows that the underground course has effected a considerable change in its composition : the calcium carbonate rising from 4 grains to 9·77 grains, or an increase of nearly two and a half times, the magnesium carbonate from 0·60 to 0·94 practically the same ratio, whilst there is a slight diminution in the calcium sulphate. No nitrates could be detected in the Tarn water, whereas the Aire Head water contains an appreciable quantity.

The experiments carried out by the Society during the summer of 1899 proved conclusively that the Smelt Mill stream issued at Malham Cove, but as Tate remarks, "the volume of the Smelt Mill Syke is not a twentieth that of the Cove stream,"[*] and it is therefore obvious that the Cove stream is fed from other sources. The comparison of the analysis of the Smelt Mill water with that from the Cove, leads to the supposition that in the latter there is a large volume of a much softer water.

In the Cove water the proportions of calcium and magnesium carbonates, the calcium sulphate and the total dissolved matter are all slightly less than in the Smelt Mill water. If the Smelt Mill water could be obtained free from admixture with other waters as it issues from its underground course, there is not the slightest doubt that there would be a considerable increase in the quantities of calcium and magnesium carbonate.

It will be noticed that the Cove water contains nitrates, though none are present in the Smelt Mill water. The conditions favourable for the development of the nitrifying organism are a base such as calcium carbonate, traces of phosphates, free oxygen and darkness, all of which are fulfilled during the underground courses of the Tarn and Smelt Mill waters.

* Proceedings Yorkshire Geological and Polytechnic Society, Vol. VII., page 184. 1878.

TABLE I.

)NSTITUENTS IN GRAINS PER GALLON (PARTS PER 70,000).

	Malham Tarn Sluice Gate.	Aire Head Spring.	Smelt Mill Water Sink.	Malham Cove.
.te when sample was taken {	August 6th, 1899, 12.35 p.m.	August 5th, 1899, 6.20 p.m.	August 6th, 1899, 12.5 p.m.	August 6th, 1899, 3.10 p.m.
·ature of air	65·0° F.	62·0° F.	67·0° F.	...
·ature of water	61·0° F.	48·5° F.	64·0° F.	46·0° F.
iO_2)	0·213	0·224	0·139	0·536
ric Anhydride (SO_3)	1·281	1·050	2·049	1·929
Acid (HNO_3)	None	0·100	None	0·150
ı Acid (HNO_2) ...	None	None	None	None
oric Acid (P_2O_5) }	None {	Minute trace	Minute trace	Minute trace
·e (Cl)	0·665	0·700	0·700	0·700
Oxide (Fe_2O_3) ...	0·030	0·030	0·009	0·023
?aO)	3·143	6·258	6·839	6·652
ıia (MgO)	0·169	0·403	0·841	0·608
ı (Na)	0·433	0·456	0·456	0·456
ıia (NH_3)	0·005	0·007	0·003	0·002
inoid Ammonia	0·006	0·006	0·018	0·005
our of solid residue on } ıition ... ... }	Blackens {	Does not blacken	Blackens	Does not blacken

TABLE II.

)ISSOLVED SALINE CONSTITUENTS IN GRAINS PER GALLON (PARTS PER 70,000).

	Malham Tarn Sluice Gate.	Aire Head Spring.	Smelt Mill Water Sink.	Malham Cove.
SiO_2)	0·213	0·224	0·139	0·536
n Nitrate ($Ca2NO_3$) ...	None	0·130	None	0·195
n Carbonate ($CaCO_3$) ...	4·008	9·778	9·646	9·344
n Sulphate ($CaSO_4$) ...	2·178	1·785	3·483	3·279
sium Carbonate ($MgCO_3$)	0·353	0·843	1·759	1·272
ı Chloride (NaCl) ...	1·097	1·155	1·155	1·155
s Carbonate ($FeCO_3$) ...	0·043	0·043	0·013	0·033
	7·892	13·958	16·195	15·814
lissolved matter by evap- ıtion dried at 110° C. ...	8·176	14·280	16·940	16·520

TABLE III.

RESULTS OF ANALYSES EXPRESSED IN GRAINS PER GALLON (PARTS PER 70,000).

NAME OF SAMPLE.	Stream from Malham Tarn, Sept. 30th, 1869.	Water Sinks, May, 1878.	Aire Head Spring, Sept. 30th, 1869.	Smelt Mill, May, 1878.	Spring in Malham Cove, Sept. 30th, 1869.	Malham Cove, May, 1878.
Analyst	Rivers Pollution Commissioners.	F. M. Rimington.	Rivers Pollution Commissioners.	F. M. Rimington.	Rivers Pollution Commissioners.	F. M. Rimington.
Total Solid Impurity	8·715	10·8	10·99	17·9	11·34	14·2
Organic Carbon	0·191	...	0·1155	...	0·200	...
Organic Nitrogen	0·021	...	0·0049	...	0·0098	...
Ammonia	0·001	...	0·0007	...	None	...
Nitrogen as Nitrates and Nitrites	None	...	0·0119	...	0·0084	...
Total Combined Nitrogen	0·022	...	0·017	...	0·0182	...
Previous Sewage or Animal Contamination	None	...	None	...	None	...
Chlorine	0·665	0·6	0·693	0·8	0·805	0·7
Hardness (Temporary)	6·5° (Clark's)	...	6·2° (Clark's)	...	8·1° (Clark's)	...
,, (Permanent)	3·0° ,,	...	2·3° ,,	...	3·1° ,,	...
,, (Total)	9·5° ,,	9·2° (Clark's)	8·5° ,,	13·1° (Clark's)	11·2° ,,	10·8° (Clark's)
Organic and Volatile Matter	...	1·2	...	2·3	...	2·0
Inorganic Matter	...	9·6	...	15.6	...	12·2
Temperature	...	...	10° C.	...	8·0° C.	...

NOTE.—The analyses published by the Rivers Pollution Commission were made to ascertain the suitability of the waters for drinking purposes. The analytical scheme, therefore, differs from that used in the present investigation.

A PEAT DEPOSIT AT STOKESLEY.

BY REV. JOHN HAWELL, M.A., F.G.S.

(Read August 4th, 1899.)

On this occasion of the meeting of the Yorkshire Geological nd Polytechnic Society at Stokesley, it has seemed to me to be itting to bring briefly before you a notice of a Post-glacial Peat)eposit or Forest Bed which occurs in the immediate vicinity f our place of meeting, and of which, so far as I know, no ecord has previously been made public.

In the autumn of 1892, Mr. Henry Fawcett, Head Master f the Preston Grammar School, Stokesley, called my attention o a section exposed in digging a tank in the garden adjoining iis house, some few yards east of the river Leven. After passing hrough 5 ft. 6 in. of surface soil and alluvial matter a thickness f 1 ft. 6 in. of fine clay was met with, and immediately below this ccurred a peaty deposit, the depth of which was not ascertained. ubsequently Mr. Fawcett and I gained further information egarding this bed of peat. Some years previously, when the ate Canon Bruce was rector of Stokesley, he made an attempt o sink a well near the rectory, which is on the same side of he stream. The same bed of peat was then reached. The vater which came up smelled so offensively that the well had to e filled in again immediately. But before doing this the vorkmen thrust an iron rod into it to the depth of 12 ft. At ı depth of 9 ft. from the surface they met with leaves and wigs of trees. There is a mill at a distance of two fields from Mr. Fawcett's house, and when the foundations of this mill were being dug a tree of black oak was found embedded in the clay at a depth of about 7 ft., and below it occurred hazel bushes with nuts upon them at a depth of 10 ft. from the surface of the ground. At a greater depth gravel was come upon. In a garden adjoining the mill some large horns were found at

E

some distance from the surface. These horns are described as having been "very large and curved and similar to those found in the railway cutting near Kildale Church." The horns found at Kildale were the antlers of *Cervus elaphus* and *C. tarandus,* but the similarity to those of the Stokesley horns, which crumbled away shortly after being exhumed, must not be too much insisted on.

The deposit of peaty matter appears to extend over a considerable area on the eastern side of the Leven at Stokesley. I am informed by the local plumber that in sinking wells on the western side gravel is usually met with. In the Appendix to the Geological Survey Memoir on "the Geology of the country around Northallerton and Thirsk" the following section is given of a well at Stokesley Brewery :—

	ft.	in.
Made ground	1	2
Beck silt	2	0
Sand and gravel, with many pebbles of Magnesian Limestone	30	0
Clay	30	0
Sump and sand	8	0
Brown clay	16	0
Sand	3	0

This Stokesley Peat Deposit is evidently Post-glacial in date. These Post-glacial peats or forest beds may be traced at many points in the Cleveland and adjoining districts. There are large tracts of them on both sides of the Tees estuary, extending from Hartlepool to Redcar. Near Hartlepool the bed is in one place 40 ft. thick ! (See Proc. of Yorks. Geological and Polytechnic Society, 1883, page 224.) When the railway-cutting between Ingleby Station and Battersby Junction was being made one of these deposits was cut into, and a large quantity of hedge-cuttings, sleepers, and similar material had to be thrown in in order to obtain solid ground for the railway.

The peat bed which will be seen to-morrow near Kildale Station would appear to be of later date. In fact, I should regard the Stokesley bed as more likely to have been contemporaneous with the *shell* deposit at Kildale than with the overlying peat.   Both were at this epoch localities of arrested drainage, but at the Kildale tarn the forest growth was less dense, and the sun's rays got through with some degree of freedom, and decaying vegetable matter was transferred to the atmosphere, whilst molluscs lived happily and died peacefully, and their shells fell to the bottom of the water, forming in time a thick deposit.   At Stokesley, on the other hand, the forest-growth was probably dense, and the vegetable matter accumulated freely there, while the circumstances were unsuited to the life of molluscs.

It is practically certain that this part of Yorkshire has undergone some amount of elevation since the Stokesley peat was deposited.   This elevation, which probably amounted to 20 ft. or 30 ft., would give origin to an improved drainage system for the locality.   There was, however, perhaps first a temporary subsidence during which the fine clay was laid down over the peat.

ON THE GENUS MEGALICHTHYS, AGASSIZ: ITS HISTORY, SYSTEMATIC

POSITION, AND STRUCTURE.

BY EDGAR D. WELLBURN, L.R.C.P. AND S.E., F.R.I.P.H., F.G.S., ETC.

(*Read November 2nd,* 1899.)

INTRODUCTION.

AT the meeting of the British Association, held at Edinburgh
in 1834, Dr. Hibbert read a paper before the Geological Section
on a series of fossil remains found in the Burdiehouse limestone,
near Edinburgh.   These contained a series of fish remains, among
which, besides · *Gyracanthus, Palæoniscus, Erynotus, Pygopterus,*
were some bones, scales, and teeth, remarkable for their great
size, and also some smaller rhombic enamelled scales.*

Prof. Agassiz being present the remains were submitted to
him for his opinion.   They proving new and strange to him,
he, Drs. Hibbert and Buckland formed a committee to report
on them.   About this time Agassiz, whilst on a visit to Leeds,
saw in the Museum there a fine and well-preserved head and
part of the trunk of a fish, which he seems to have considered
of the same species as the Burdiehouse remains.   This new find
having relieved his doubts concerning the Burdiehouse fish, he
took the Leeds specimen as the type of his genus *Megalichthys,*
and at that time included the large rounded scales and gigantic
teeth, as well as the smaller rhombic enamelled scales, in this
genus.   Later, however, he separated the large rounded scales
and the teeth, placing them in a new genus *Holoptychius.*†

---

* See Poissons Fossils (Agassiz), Vol. 2, Pt. I., pp. 89 and 90.

† Poissons Foss., Vol. 2, Part I., p. 90.

It is very unfortunate that Agassiz made the Leeds fish his type, as undoubtedly the name Megalichthys was suggested to him by the great size of the Burdiehouse remains, for which in 1840* Prof. Owen instituted the genus *Rhizodus.*

SYSTEMATIC POSITION.—Agassiz classed *Megalichthys* in his heterogeneous group of "Sauroides."†

Sir P. Edgerton‡ next proposed its inclusion in the family *Sauroidei-dipterini* (*Sauroides-dipteriens* of Agassiz); its position in the *Saurio-dipterini* was also indicated by Pander § and Huxley‖ on account of the close relationship of its head bones, &c., to those of *Osteolepis*, though they both seemed to hesitate for want of knowledge of the conformation and position of the fins.

In 1861 Prof. Young, in Dec. X. Geol. Survey, mentioned specimens in the Jermyn Street Museum, showing the form of fins, but unfortunately gave no description or figures.

In 1875 Mr. J. Ward, F.G.S. (Fossil Fish of North Staffordshire Coalfields), classed *Megalichthys* in this same family (*Saurio-dipterini*), and also stated that the pectoral fin is lobate.

Dr. R. H. Traquair, F.R.S., in a paper read before the Royal Physical Society, Edinburgh, on Feb. 20th, 1894, says that there can be no doubt that the true position of *Megalichthys* is in the family *Saurio-dipterini* as defined by Pander, Huxley, and others. In every matter of "Family" importance its structure closely conforms to that of *Osteolepis.*

In 1890 Mr. J. Ward, F.G.S., in his "Geology of the North Staffordshire Coalfields," classifies it in the same family; but in 1891 Mr. A. Smith Woodward, F.G.S., in vol. ii. of his Catalogue Fos. Fishes in the British Museum, places the genus *Megalichthys*

---

* Odontography, 1840, p. 75.

† Poissons Foss., Vol. II., Pt. II., p. 152.

‡ Morris's Catalogue Brit. Fossils.

§ Die Saurodipterinew, &c., devon Syst., p. 5.

‖ Dec. Geol. Survey X., 1861, p. 12.

along with *Osteolepis, Thursius, Dipterus,* and *Glyptopomus* in the family Osteolepidæ, and in this view Dr. Traquair seems to concur.*

STRUCTURE.—The body is much elongated, being about five times the length of the head, rounded and covered with rhomboidal scales which run in obliquely sigmoidal parallel lines from before backwards, the greatest obliquity being on the dorsal and ventral surfaces, the scales becoming smaller on the latter surface. Well marked ridge scales present (at least) in the posterior half of the fish, where they pass some little distance on to and strengthen the anterior basal portions of the unpaired fins ; they also pass for some distance on to and strengthen the upper lobe of the caudal fin.

SCALES.—The superior surface is divisible into an anterior or covered, and a posterior or exposed, portion (Pl. VIII., Fig. G). The *anterior covered* area is smooth and covered with a thin layer of non-corpusculate bone or kosmin, and is crossed by a grove which runs more or less parallel with the anterior and superior edges of the posterior or exposed portion. The "overlap" of the scales is from above downwards and backwards. The posterior exposed area is rhomboidal in form and is covered with a glittering layer of ganoine which ceases on the sides with abrupt rounded margins which dip down to and slope to the surface of the scale. This part of the scale is deepest at the centre, and on section is seen to be composed of non-corpusculate bone, tufts, capillary tubes and the upper series of the haversian canals (Williamson). The *internal* surface is smooth with the exception of an elongated ridge or boss (not present in *M. læris* Traquair) which runs more or less vertical to the axis of the body of the fish, and is situated between the anterior border and the centre of the scale.

The Haversian system is in direct communication with the scale surface giving rise to the wide pores.

---

* Geo. Mag., Dec. III., Vol. VIII., No. 321, p. 123, Mar. 1891.

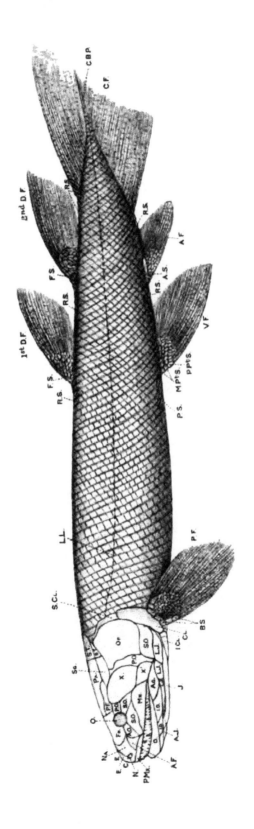

LATERAL LINE.—I have not seen any evidence of this "sense organ," but it probably arises at a point on a level with the upper border of the operculum and traverses a longitudinal series of scales to an indetermined point on the caudal pedicle.

SHOULDER GIRDLE.—The pectoral arch exhibits well developed membrane bones, there being a large clavicle (Fig. C, Pl. XIX.) and a smaller infra clavicle (Fig. D, Pl. XIX.). A supra clavicular element was in all probability also present (as is the case in some other members of the order), but not having seen the bone I am unable to offer any opinion as to its characters.

FINS.—(a) *Paired fins*. These were represented by the " Pectoral" and "Ventral" fins, the latter being abdominal in position.

(1) *Pectoral fins*. These fins are obtusely lobate, and their superficial characters are beautifully shown in a specimen in the Science and Art Museum, Edinburgh (Pl. XVII., Fig. B), and also in the fine fish in the Leeds Museum (see Fig. C, Pl. XVII.).

CHARACTERS.—(a) Superficial. At the base of the fins are a series of large scales, which are continued along the post-axial and preaxial borders, the space between these being occupied by smaller scales arranged in many parallel rows.   (b) The internal skeleton is, as pointed out by Prof. Miall,* indicated in *Magalichthys*, as in other fishes with lobate fins, by the external characters, the larger or fulcral scales covering the more rigid, and the smaller the more flexible parts of the internal structure (in the nearly allied family Rhizodontidæ this character is clearly indicated in a pectoral fin of *Strepsodus*, which is in the Science and Art Museum, Edinburgh).   The lobe of the fins seems to be, as pointed out by Mr. A. Smith Woodward, supported by an endoskeletal cartilage (covered with a thin layer of dense bone, Cope), arranged on the plan termed archipterygial by Gegenbaur; the axis being shortened, whilst the parameres of the one side are atrophied, those of the other border enlarged.   There is thus no di- or tri-basal arrangement of the cartilages as in

---

* Quart. Journ. Geo. Soc., Vol. XL., p. 347.

Polypterus, the skeleton being more like that found in *Ceratodus*, with the difference that the basal cartilage (*Metapterygium*) is somewhat shortened, the radials on its anterior border atrophied, those on the posterior border enlarged, and the cartilage seen along the post axial border of the fin being elongated to form a propterygium, this giving a structure similar to that shown in Fig. A, Pl. XVII.

Prof. Cope gives a section (Proc. W. S. Natl. Museum, Vol. XIV., p. 457) of the lobe of the pectoral fin of *M. nitidus* Cope, which shows a well-marked metapterygium, with radials springing from its tip and outer or posterior edge.

The dermal fin rays form a fringe around the lobe, they are closely articulated, the articulations being rather longer than broad, and covered with ganoine similar to that on the scales, distally they increase in number by dichotomisation and become much finer. The anterior rays are much more robust than those situated further back.

Ventral fins.—The fins are abdominal and their position and character are well shown in a fish in the Hugh Millar Collection in the Science and Art Museum, Edinburgh. Their basal characters are also well shown in the Leeds fish, the right one being the better preserved.

The fin is obtusely lobate, the base being invested with large scales which are continued along the internal or post-axial border; along the outer or preaxial border is a shorter series of large scales which meet the others (post-axial) at an acute angle. The space between these rows is occupied by a close series of smaller scales arranged in many parallel rows. Here, again, the external characters probably indicate the internal skeleton which is thus described by Prof. Miall: "The larger scales conceal a strong pro- and a metapterygium, whilst the smaller scales cover numerous radials which spring from the outer edge of the metapterygium."

The section of the basal portion of this fin given by Cope (op. cit.) goes to prove that a strong, well-marked axial rod or metapterygium was present, with well-marked radials springing from its tip and posterior border, but none are shown on the anterior margin. I don't take this as proving that they were absent from the anterior border, as they might easily have been missed owing to the direction of the section, and it is highly probable, considering that the dermal rays not only spring from the tip and posterior border, but also from at least the distal portion of the anterior border of the fin lobe, that there were short radials on the anterior margin of the axial support distally; and considering the fact that the dermal rays, springing from this portion of the lobe, are much stronger and more robust than the others, it is very probable that their supporting ossicles were, although short, strong and robust, and from the above the conclusion seems to be that the skeleton of this fin was of a nature similar to that shown in Fig. E, Pl. XVII.

The dermal fin rays are similar to those of the pectoral fin in character and arrangement.

The pelvis is probably represented by an elongated cartilaginous element, covered with a layer of dense bone and having the distal end concavo-truncate (see Cope, op. cit., p. 458).

In the Leeds fish, between these fins are three large, elongated scales, one median and two lateral, which may be called "pelvic scales." On the left side of the median one the anus is well shown. The anus is not always in this position as is shown by other specimens. The difference is probably connected with the sex of the fish. (Pl. XIV., L P S and M P S, also Pl. XIII., P S.)

Unpaired fins.—There are two dorsal fins situated far back, the first being opposed to the ventral and the second to the anal fin, which arises close to the root of the tail. All the fins are lobate, the lobe being more acute than that of the paired fins.

Anal fin.—The superficial characters are well shown in several specimens, viz., in the Leeds fish, in the specimen in the Science and Art Museum, Edinburgh, described by Dr. Traquair (Proc. Roy. Phys. Soc. Edinb., Vol. VIII., p. 67), and in a specimen in the Lister Collection, Brighouse, &c. (Pl. XIV., A F, and Pl. XVI., A F).

CHARACTERS.—On each side of the lobe is a large "basal" scale, the function of which seems doubtful. The dermal rays are of similar character and arrangement to that of the pectoral fin.

Internal Skeleton consisted of a single club-shaped axonost, its broad basal portion bearing several rod-like baseosts, which were jointed at intervals and bifurcating, the more anterior ones being the most robust. The dermal rays are much more numerous than the supporting ossicles, and were of a character similar to those of the pectoral fin.

Dorsal fins.—These fins were lobate, the lobe being more acute than that of the paired fins, the posterior fin is more strongly developed than the anterior, the dermal rays of both fins are similar in nature to those of the other fins, and the supporting skeleton is similar to that of the anal.

NOTE.—A specimen in the British Museum (No. 38,007) shows that the supporting (or internal) skeleton of the unpaired fins was of a similar nature to that described above.

Caudal fin.—(Pl. XVI. and Pl. XVII., Fig. H.) The structure (superficial) of this fin is well shown in several specimens in the Science and Art Museum, Edinburgh ; Lister Collection, Brighouse ; and Owens College, Manchester. The fin is intermediate in type between the diphycercal and heterocercal stages, and in general form reminds one of that of *Tristichapterus*, as pointed out by Dr. Traquair.

The rays arise from both the upper and lower margins of the body prolongation, those of the lower side commencing in advance of those of the upper (see Pl. XVI.). After the commencement of the rays the upper margin of the body slopes a little downwards, whilst the lower one first slopes somewhat rapidly upwards and backwards, then more gradually to meet the upper in a fine point, which is finally lost among the dermal rays, the scaly covering being continued to this point (Pl. XVI., and Pl. XVII., Fig. H).

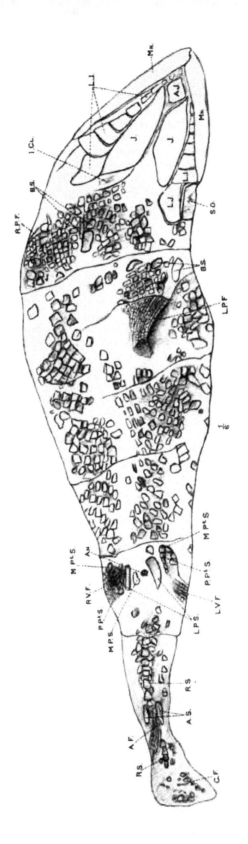

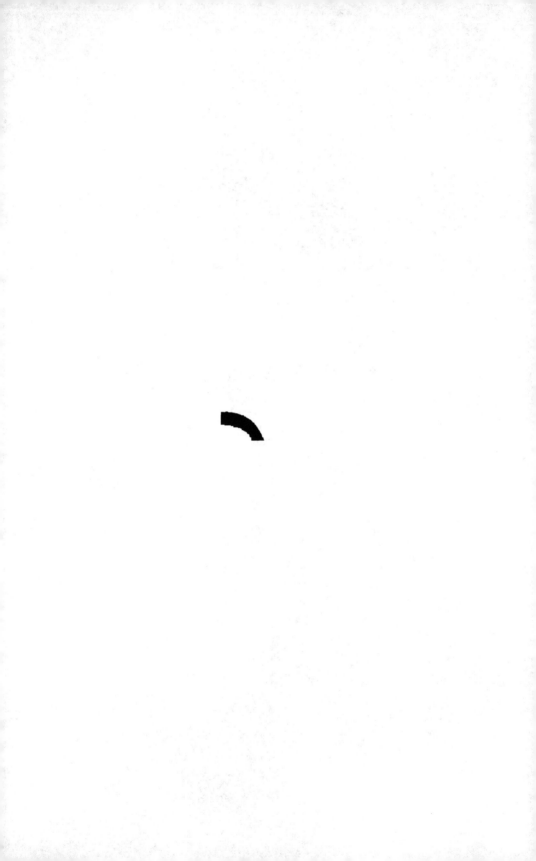

The greater number of the dermal rays arise from the lower aspect of the body prolongation, whilst the apex is formed by those arising from the dorsal side of the axis. (This is clearly shown in a specimen in Owens College Museum.) (Pl. XVII., Fig. H.) The posterior margin of the fin slopes obliquely upwards and backwards, and the dermal rays are articulated, covered with ganoine, increased by dichotomisation distally, the uppermost ones are the most robust and the proximal part of the upper border of the fin is strengthened by well-marked "ridge scales," which are continuous with those on the dorsal ridge of body.

Internal Skeleton. (Pl. XVII., Fig. F.) The specimen of *Megalichthys Hibberti* Ag., No. 38,007, already mentioned, shows that the internal structure was of a similar nature to that of *Tristichopterus alatus* Egerton,* viz., the more anterior dermal rays are supported on a series of "hour-glass" shaped interapophyscal osselets, each osselet having several rays opposed to their distal end, their proximal ends uniting with the distal extremities of elongated and thickened neural and hœmal spines of the vertebral column. More posteriorly the dermal rays seem to abut on the vertebral axis.

Vertebral Column.—In the Andersonian Museum, Glasgow, there is a slab which contains, besides the upper surface of the head, a good display of the vertebral column. About fifty vertebræ are shown, of which the anterior are the shortest and broadest, the caudal being the longer and narrower.

The notochord is partially persistent, the cartilages of the arches are superficially calcified, there are robust ring-shaped vertebræ, and several specimens (in the Science and Art Museum, Edinburgh ; British Museum ; in the Author's Collection, &c.) show well-marked neural spines, which have a cylindrical shaft, articular head, and are somewhat flattened distally. Hœmal spines are also shown in the caudal region in several specimens.

---

* See Mem. Geol. Survey (Figures and descriptions organic remains), Dec. X, Pl. 4 and 5, pp. 50-53.

HEAD.—*Internal anatomy* in *M.* (*ectosteorhachis*) *ciceronius* Cope,[*] the chondrocranium is in some degree ossified, and the parachordal cartilages are ossified to form two subtriangular bones which present one angle forwards, and having the internal side which bounds the chordal groove straight and longitudinally grooved. The antero-external side is oblique and nearly straight, and is overhung by the cranial roof. These ossifications embrace the chorda dorsalis posteriorily, and are continued a short distance posteriorily as a tube. Anteriorily the chordal grove is open, and we here get a good illustration of a permanent embryonic type (Cope, opus cit.).

According to Dr. Young (see Quart. Journ. Geol. Soc., Vol. 22 (1866), p. 605) the basilar region is well ossified and includes a massive basioccipital which projects behind the vertical posterior wall of the cranium and sometimes has its length increased by coalescence with at least the first vertebral ring, whose neural process remains distinct. In a lateral view, the aliophenoides (?) and an incomplete interorbital septum (?) are well shown.

The hyomandibular is not shown in any of the specimens I have seen, but it is probably (as in *Rhizodopsis*)[†] covered by the preoperculum, and extends from the squamosal above downwards and slightly backwards to the articular extremity of the mandible below.

Cranial Anatomy (Pl. XV., Figs. A, B and C).—The whole of the cranium is covered with thick dermal plates, which exhibit a definite arrangement, and there is a considerable development of membrane bones on the roof of the mouth. The shield of the cranial roof is divided by a much-pronounced transverse suture into two parts. The posterior portion consists chiefly of a pair of long narrow bones (Pa.), the parietals, which are divided down the middle line by an irregular suture, the bones are twice as wide behind as in front, their external margins first run nearly straight

---

[*] Proc. Amer. Phil. Soc., Vol. 20, page 628. 1883.

[†] Trans. Roy. Soc. of Edinburgh, Vol. XXX., page 171 (Traquair).

forwards to a point a little behind their middle, then forwards and inwards for a short distance, and then nearly straight forwards to meet the posterior boundary of the bones of the anterior division. Along the outer edge of each parietal are two smaller bones (P.F. and Sq.), the anterior ones being narrow, elongated bones which in front meet the posterior extremity of the bones (Fr.) of the anterior division; from this point, where the bones are the broadest, the external margins run backwards to a point, a little in front of the centre of the parietal bones, where they join the posterior pair of bones (Sq.) by a suture which runs from without, inwards, and backwards. The posterior border of the bones (Sq.) is straight, and they are wider here than in front. Their outer border at first runs forwards and slightly outwards, then forwards, and then inwards and forwards to meet the external border of the anterior pair of bones (P.F.) at the junction of their posterior and middle thirds.

The anterior division of the cranial shield is divisible into a posterior (Fr.) and an anterior moiety (C.E.).

The posterior division is composed of two bones (Fr.), the Frontals, which are divided down the middle line by an irregular suture; their inner sides are longer than their outer, which are notched to form the upper boundary of the orbit. From the anterior edge of this notch, where the bones are the widest, they gradually narrow to form an obliquely truncated anterior extremity, which indents the posterior border of the bones (C.E.) in front, the union being by a semi-lunar suture, with the convexity forwards.

The anterior division (*Moignon inter maxillaire* Agassiz) is a cresentic shield which terminates the head anteriorly, and presents distinct indications of a division into a number of pieces, viz., Ethmoids (E.), Pre-frontals (P.F.), Nasals (N.), and pre-maxillary (P.Mx.) bones. The bones are usually firmly united, and form the "Compound Ethmoid." The pre-maxillary portions are separated by a median suture, and form the lower and anterior boundary of the shield. Above these in the centre is the

ethmoidal (E.), with the Nasals (N.) on each side, and more
external still the bone probably represents the Pre- or Anterior
frontals (A.F.). The nasals were perforated by and contained
the olfactory organs, and the nares (Na.) were anteriorly placed
on each side of the snout, as is the case in *Osteolepis* and in the
recent fish *Polypterus* (see Pl. XV., Fig. B).

Behind the cranial shield, in the occipital region, are three
bones, one median and two lateral. The central one is in the
form of a narrow isosceles triangle, and the lateral ones form a
pair on each side of it. These bones are regarded as the supra
temporals by Dr. Traquair, Mr. Smith Woodward, and others.

Side of the Head. (Pl. XIII. and XV.)—This region is
entirely covered with loose dermal plates.

The Orbit is anteriorly placed, being situated at the junction
of the anterior and middle thirds of the head. Above it is
bounded by the frontal bones, in front by a bone (A.O.) which
probably represents the Anterior Orbital of *Polypterus*, below by
an elongated triangular bone (S.O.), the Sub-Orbital, which rests
on the anterior portion of the upper edge of the maxilla (Mx.)
below, and extends from the pre-maxilla in front to a third bone
(S.O¹)., which is triangular in shape, and forms the posterior
inferior boundary of the orbit. This bone is the supra-orbital
of Prof. Huxley. Behind the orbit is a square - shaped bone
(P.O.), the post orbital.

Behind the orbital region is a large plate (X) which is
somewhat oval and obliquely placed, it covers a large portion
of the cheek, and is bounded in front by the post- and supra-
orbitals, behind by an elongate bone (P.Op.), above by parts of
the posterior frontals and synamosals, and below by another plate
(X¹) which is somewhat rhomboidal in shape and fills the space
between the larger cheek plate (X) above, the articular extremity
of the mandible below, the hinder border of the maxilla in front,
and the lower third of the anterior border of the bone (P.Op.)
behind. This latter bone (P.Op.) may be considered as the
preoperculum. It is a narrow, elongated, somewhat arched bone.

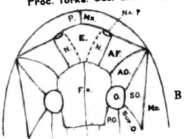

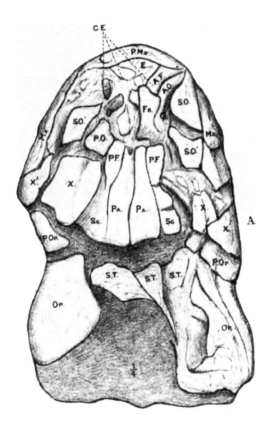

)nvexity being backwards; its direction is from above down-
, and slightly backwards.   Above it articulates with the
1osals (Sq.), and from this point it passes down behind the
·ior border of the bones X and X¹ to meet the articular
nity of the mandible below, and behind are two bones (Op.
\.Op.).   The former of these bones (Op.) represents the Oper-
, and the latter (S.Op.) the Suboperculum.

'he two cheek plates X and X¹ are probably equivalent to
1eek cuirass of *Lepidosteus* (or as Dr. Traquair remarks of
ιme bones in *Rhizodopsis*),* to the posterior set of sub-
's in *Lepidotus* and in the Palæoniscidæ.   By Agassiz † the
X and P.Op. were considered to be the equivalent of
-called pro-opercalum of *Polypterus*, while the lower one X¹
npared to the little bone fixed above the posterior edge
; maxilla in Salmonidæ, &c., and which Mr. Parker con-
to be the homologue of the malar bone of other vertebrata.

he opercular bones (Op. and S.Op.) were largely developed.
perculum is a large square-shaped bone with its posterior
or angle much rounded, it is broader above than below and
l than in front.   It is bounded above by the posterior half
squamosals in front, and behind these by the lateral pair
prateuporals; in front is the preoperculum; behind the
of the shoulder girdle and below by a plate (S.O.), which
rlaps, the suboperculum.   This latter bone is much narrower
as the anterior superior and posterior inferior angles much
əd.   The bone is bounded above by the operculum, in front
ə lower part of the preoperculum, behind by the clavicle,
elow by a bone (L.J.) which is to be regarded as the most
ior of the lateral jugulars.   (Plates XIII., XIV., and XV.,
Σ.)

ʌws.—The maxillæ are of an elongated triangular shape,
lveolar border being the longest and the posterior one the
·st.   The greatest depth of the posterior expansion varies in

---

* Trans. Roy. Soc. Edinburgh, Vol. XXX., p. 177.

† "Poisson's Fossiles," Vol. II., part 2, p. 92.

the different species, being in *M. Hibberti* Agassiz about a third the greatest length. The anterior angle is pointed, the others somewhat blunt. A specimen in the Author's Collection shows on the upper edge of the bone, and a short distance from the anterior extremity, a short, blunt projection, similar to the one mentioned by Dr. Traquair on the maxilla of *Rhizodopsis* (op. cit., p. 172). (Pl. XIX., Fig. A.)

The pre-maxillæ are separated by a median suture, and form the lower and anterior boundary of the cranial shield. When seen from the palatal surface the bone is spatulate, with a rounded fore edge.

The mandible is of a very complex structure, but the component parts are, in the older fishes, firmly united together. Behind there is a distinctly ossified articular element. The upper and outer border, in front of the angular bone (Ag), is formed by an elongated element (D), the dentary bone, which is deep and thick at the symphysis, but from this point it gradually tapers backwards to a fine point. Its lower border is bounded by a series of three plate-like, lenticular bones, which form a series in front of the angular element (Ag), and are termed infra dentaries. The inner wall of the ramus is formed by a thin sphenial lamina, and between this and the dentary is a series of three or four stout lenticular bones, the laniaries. (Pl. XVIII. and Pl. XIX., Fig. B.)

DENTITION. — *Upper*. The pre-maxillæ and vomerine bones bear within and close to either outer extremity a large tooth, and on each side of the middle line in front is a similarly socketed large tooth. The small marginal teeth are continuous with two curved rows of equally small teeth which pass in front of the outer tusks, and curving inwards, meet in the middle line anterior to the basilar bar, whose surface is closely set with fine denticles.[*] Behind, on either side, are two palatine bones, which seem to be wedged in between the maxillæ and pterygoid bones. Each plate bears a marginal row of short, stout, conical teeth,

---

Young, op. cit., p. 605.

he rest of the surface being set with similar but smaller teeth, 'hich are more distant over the anterior portion of the bone, ut posteriorly pass into a dense rasp of minute denticles. Outide these the edges of the maxillæ are set with small, conical eth, continuous in front with those on the edge of the preaxillæ.

Lower dentition.—The outer edge of the dentary bone of the andible bears a row of small, conical teeth. Within these are ur large, strong, conical teeth, which are distantly placed. The nterior one is the largest, and is firmly socketed in the thickened ymphisial extremity of the dentary bone, the others lie *within* he edge of the dentary bone, and are attached to the series of niary bones (Pl. XVIII.). The edge of the sphenial bone also ears numerous rasp-like teeth.

From the above it will be seen that the dentition is that of predatory fish.

TEETH.—*External characters.* They are round in transverse ection, conical, more or less curved, bases plicated, and many re covered with very fine striæ, which merely involves the outer ortion of the enamel (Young, Davis). These lines are someimes parallel or anastomose to form a fine reticulation.

Internal characters.—The walls of the teeth are infolded, he folding being simple at the commencement of the external luting, but as we pass towards the root the folds become wonderully beautiful and complex, but the vertical tubes formed by the nfolding never form such an interlacing network as in the Dendrodont type (Dr. Traquair).*

The gill flap, anteriorly and inferiorly, is completed by a eries of bony plates, the jugulars, which lie between the mandiular rami, and which, together with the infraclavicular bones which lie along their posterior border, cover in and protect the nderlying branchial arches.

In the centre are two elongated plates, the principal jugulars P.J.). They are broader behind than in front; their posterior

---

* Geo. Mag., Dec. III., Vol. VIII., No. 321, p. 123. 1891.

F

end is rounded, the anterior truncated. Behind the symphysis of the mandible, and between, and partly overlapped by, the anterior ends of the principal jugulars, is a well-marked azygos plate, the medial jugular (M.J.). On each side are a series of well-marked lateral jugulars (L.J.), which increase in size from before backwards ; they run along the inner sides of the mandibular rami, and extend some little distance behind the posterior border of the principal jugulars. There are probably nine of these plates on each side. (Pl. XIV.)

CONCLUSION.—(Pl. XIII.) The body of *Megalichthys* was much elongated, somewhat rounded, and covered with rhomboidal ganoine-covered scales, arranged in sigmoidal rows, running from above downwards and slightly backwards.

The head is long (about one-fifth length of body), broad, and depressed, all the external bones being covered with a dense layer of ganoine. Cranial roof is covered with well-developed bony plates, which form a compact shield divisible into an anterior and a posterior moiety. The nares are placed on each side of the rounded, depressed snout. The orbit is anteriorly placed, being at the junction of the anterior and posterior third of the head, and is bounded in front, below, and behind by well-developed bones. Behind, the cheek is covered by a series of loose dermal bones. The opercular bones are well marked. The jaws are powerful and well developed, the gape extending far back. There are teeth of two sizes on the jaws, the large ones being internal. Small, numerous, rasp-like denticular teeth were also present on the well-developed membrane bones of the mouth and on the edge of the sphenial. Between the mandibular rami the branchial apparatus is defended by a series of jugular plates, there being two principal, one median, and nine lateral plates on each side. The vertebral column is well developed, there being well-ossified ring-shaped centra, neural and hæmal arches. Neural spines were present, and in the posterior part of the fish hæmal also.

The shoulder girdle well developed ; paired fins obtusely lobate and fulcrate, the ventral being abdominal in position.

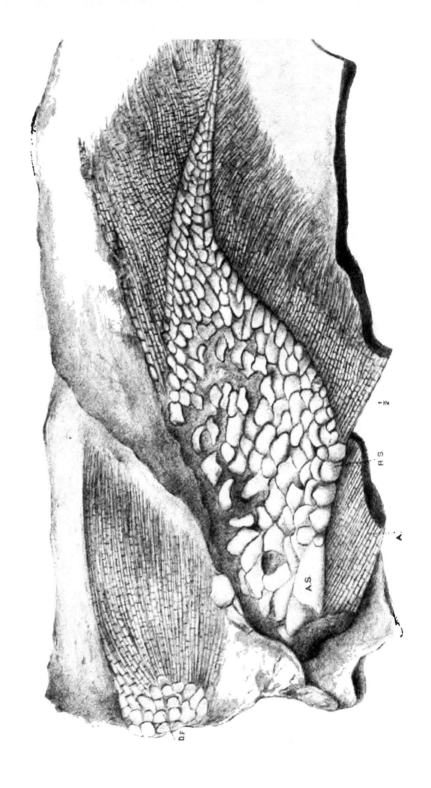

There are two dorsal fins, the second being the larger and more powerful; the first rises at a point about equal to five-eighths the total length of the fish, measuring from the snout, the second being about three-sixteenths further back; the first is opposed to the ventral and the second to the anal fin, which is smaller than the others and more lanceolate in shape, and is situated close to the origin of the caudal fin. All these fins are lobate, the lobe being more acute than that of the paired fins, and well-marked fulcral scales are also present. The caudal fin is powerful, and intermediate in type between the diphycercal and heterocercal stages; its posterior margin slopes obliquely upwards and backwards, and the majority of the dermal rays spring from the under side of the body prolongation. The fish was evidently very powerful, and of predaceous habits.

NOTE.—The proportions of certain of the bones of the head vary, as shown in the following table:—

	Parietal Division of Cranial Roof.	Maxilla.	Mandible.
M. Hibberti Ag.	Longer than the Fronto-Ethmoidal division.	Three times as long as greatest depth.	Five times as long as deep.
,, intermedius A.S.W.	Do.	Posterior expansion deep.	Do.
,, pygmæus Tr. ...		?	Three and a half times as long as deep.
,, laticeps Tr. ...	Shorter than the Fronto-Ethmoidal division.	More than four times as long as deep.	Do.

Before concluding this paper I must acknowledge, with warmest thanks, the great obligation I am under to the following gentlemen for the privilege of examining the specimens under their care, viz., Dr. R. H. Traquair, F.R.S., Mr. A. Smith Woodward, F.G.S., Dr. C. B. Crampton (Owens College), Mr. Crowther (Leeds Museum), and Mr. Rowe (Brighouse).

EXPLANATION OF PLATES.

Throughout the plates the same letters apply to the same bones or parts of the fish.

HEAD BONES, &c.

ST.	Supratemporals.	P.Op.	Pre-operculum.
Pa.	Parietals.	Op.	Operculum.
Fr.	Frontals.	S.Op.	Sub-operculum.
Sq.	Squamosal.	PMx.	Premaxilla.
P.F.	Posterior Frontal.	Mx.	Maxilla.
E.	Ethmoidal.	D.	Dentary.
N.	Nasal.	ID.	Infradentaries.
Na.	Nares.	Ag.	Angular.
A.F.	Anterior or Prefrontal.	L.J.	Lateral Jugulars.
A.O.	Anterior orbital.	J.	Jugulars (principal).
S.O.	Sub-orbital.	S.Cl.	Supra-clavicular,
S.O^1.	Supra orbital.	Cl.	Clavicle.
P.O.	Post orbital.	I.Cl.	Infra-clavicular.
O.	Orbit.	Az.J.	Azygos Jugular.
X & X^1.	Cheek plates.		

PARTS OF BODY, &c.

L.L.	Lateral line.
P.F.	Pectoral fin.
V.F.	Ventral fin.
A.F.	Anal fin.
1st D.F.	1st Dorsal fin.
2nd D.F.	2nd Dorsal fin.
C.F.	Caudal fin.
B.S.	Basal scales.
R.S.	Ridge scales.
F.S.	Fulcral scales.
P.S.	Pelvic scales. { L.P.S. Lateral pelvic scales. M.P.S. Medial pelvic scales.
A.S.	Anal scales.

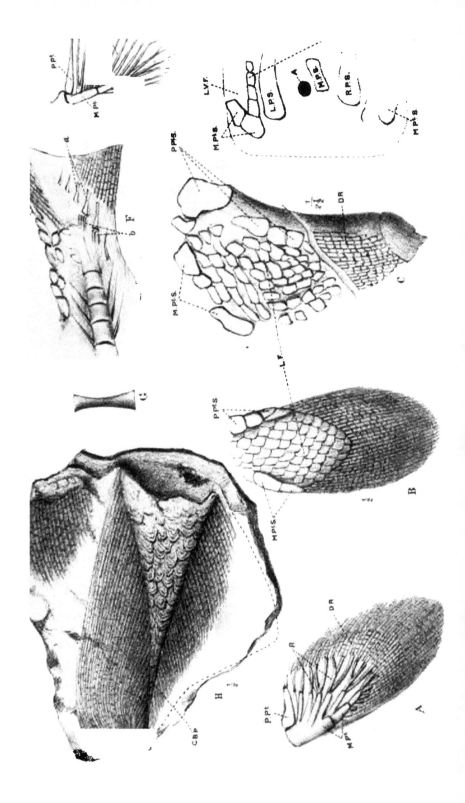

A.   Anus.
MPt. S.   Metapterygial or preaxial scales.
MPt.   Metapterygium.
P.Pt. S.   Propterygial or post-axial scales.
PPt.   Propterygium.
C.B.P.   Caudal body prolongation.
L.F.   Lobe of fin.
R.   Radials.
D.R.   Dermal rays.
Pel.   Pelvis.

PLATE XIII.

ɔred figure of *Megalichthys* Ag. From specimens in the
    Leeds, Science and Art, Edinburgh, Owens College, and
    Brighouse Museums, the author's, and other collections.

PLATE XIV.

ꞃen of *Megalichthys Hibberti* Agassiz. Ventral surface
    shown, one-sixth natural size. Leeds Museum (after
    Prof. Miall, F.R.S., slightly altered).

PLATE XV.

Megalichthys Hibberti Agassiz (type). Skull seen from the
    upper surface. Leeds Museum.

M. Hibberti Ag. Diagrammatic representation of the bones
    of the fronto-ethmoidal and orbital regions, and also
    premaxillæ and maxillæ.

M. (maxillaris) Hibberti Ag. Skull, lateral view. Leeds
    Museum.

PLATE XVI.

Hibberti Agassiz. Caudal region, the body prolongation into
    the caudal fins being beautifully shown. Second dorsal
    and basal portion of anal fins also seen. Brighouse
    Museum.

PLATE XVII.

A. Diagrammatic representation of the internal skeleton of the
   pectoral fin of *Megalichthys*.

B. *M. Hibberti* Ag. Pectoral fin. Science and Art Museum,
   Edinburgh.

C. *M. Hibberti* Ag. Left pectoral fin of the specimen figured
   on Plate II. Leeds Museum.

D. *M. Hibberti* Ag. Anal region, showing anus, pelvic scales,
   basal portions of ventral fins.

E. *Megalichthys*. Diagrammatic representation of the skeleton
   of the lobe of the ventral fin. The distal end of the
   pelvis also shown (Pel.).

F. *Megalichthys Hibberti* Agassiz (No. 38,007, type, British
   Museum). Caudal fin showing portions of the internal
   skeleton. S.O.—Supporting ossicles (axonosts).

G. Supporting ossicle of caudal fin of *Megalichthys Hibberti* Ag.

H. *M. (Rhomboptichius) intermedius* A. S. Woodward. Caudal
   fin, Owens College Museum, Manchester.

PLATE XVIII.

M. intermedius? A. S. Woodward. Mandible seen from the inner
   side the sphenial bone being absent. S.Mn.—Symphysis
   of mandible (dentary bone of). S.L.—Symphysical
   laniary tooth. L.T. 2 and L.T. 3.—Second and third
   laniary teeth. S.T.—Small tooth from edge of dentary
   bone. L.B. 1 and L.B. 2.—First and second laniary
   bones. ID.—Infradentary. Specimen in the Brighouse
   Museum.

PLATE XIX.

A. *M. Hibberti* Ag. Maxilla (Author's Collection), showing the
                     anterior articular projection (*a*).

B.   „   „   Mandible and maxilla. Author's collection.

C.   „   „   Clavicle. British Museum.

D.   „   „   Infra-clavicular. Author's collection.

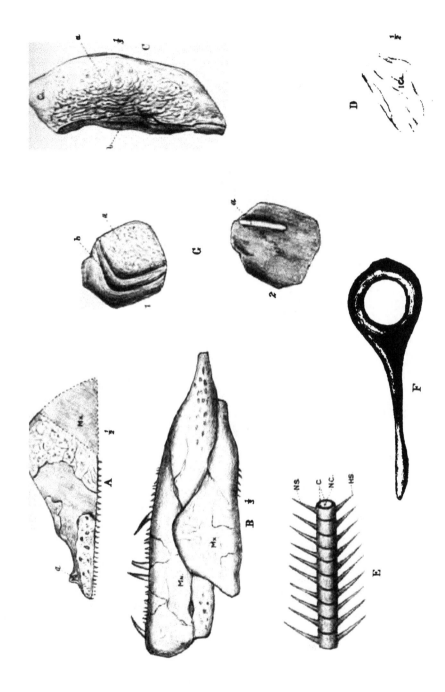

. Hibberti Ag.   Portion of vertebral column.   N.C.—Noto-
chordal interspace.   C. — Centra.   N.S. — Neural, and
H.S., hæmal spines.   Science and Art Museum, Edinburgh.

. (Rhomboptychius) intermedius A. S. W.   Vertebra with
spinous proces.   Author's collection.

. Hibberti Ag.   Scales.   1. Superior surface.   (*a*) Ganoine-
covered exposed portion.   (*b*) Covered area showing
groove.   2. Inferior surface.   (*a*) Bos.   Author's collection.

 OTE.—When not mentioned otherwise, the specimens are
natural size.

CONTRIBUTIONS TO A HISTORY OF THE MESOZOIC CORALS OF THE
COUNTY OF YORK.

BY ROBERT F. TOMES, F.G.S.

A great many Oolitic corals from the counties of Oxford, Gloucester, Wilts, Somerset, Dorset, Northampton, and Cambridge, have of late years been examined, and a considerable number of new species and some new genera from those counties made known. But while an abundance of material has been obtained, it has been a matter of some surprise that an extremely small number of specimens from the Yorkshire and Lincolnshire Oolites have come under notice. Almost equally meagre are the records of madreporaria from the Jurassic formations of those counties, and it is the purpose of the present communication to bring together such notices of Yorkshire species as have already appeared in print, and to add some others which have not hitherto been made known. I propose also to enumerate the Liassic species, and the few Cretaceous ones which have been found in the Speeton Clay.

In the Illustrations of the Geology of Yorkshire, by Professor Phillips, ten supposed Oolitic corals are mentioned, but they are not sufficiently defined for satisfactory reference. MM. Milne Edwards and Haime, in their History of British Fossil Corals, have, indeed, referred them to certain recognised species, as I shall now show, but it is most desirable that an actual comparison should be made between Yorkshire specimens and undoubted examples of the species to which they have been referred by those celebrated zoophytologists.

The following is the list of Oolitic corals given by Professor Phillips, as revised by MM. Milne Edwards and Haime:—

STYLINA TUBULIFERA.

> *Astræa tubulifera* Phill. Ill., Vol. I., p. 126, pl. iii., fig. 6.
> 1829. Coral Rag, Malton.

MONTLIVALTIA DISPAR.

> *Turbinolia dispar* Phill. Ill., Vol. I., p. 126, pl. iv. 1829.
> Coral Rag, Malton.

THECOSMILIA ANNULARIS.

> *Caryophyllia cylindrica* Phill. Ill., Vol. I., p. 126, pl. iii.,
> fig. 5. 1829. Coral Rag, Malton.

RHABDOPHYLLIA PHILLIPSI.

> *Caryophyllia* Phill. Ill., Vol. I., p. 126. 1829. Coral
> Rag, Malton.

CLADOPHYLLIA CONYBEARI.

> Coral like *Caryophyllia cespitosa* Phill. Ill., Vol I., p. 126.
> 1829.

ISASTRÆA EXPLANATA.

> *Astræa favosioides* Phill. Ill., Vol. I., p. 126, pl. iii., fig, 7.
> 1829. Coral Rag, Malton.

THAMNASTRÆA ARACHNOIDES.

> *Astræa arachnoides* Phill. Ill., Vol. I., p. 126. 1829.
> Coral Rag, Malton.

THAMNASTRÆA CONCINNA.

> *Astræa micraston* Phill. Ill., Vol. I., p. 126. 1829. Coral
> Rag, Malton.

Of *Thamnastræa arachnoides* I would observe that as two species have certainly been confounded under that name, the determination of the Yorkshire specimens is by no means satisfactory.

The following are also given by Professor Phillips as occurring in the Yorkshire Oolite, but they have not at present been referred to any acknowledged species :—

Astræa inequalis occurs according to Phillips at Malton. Edwards and Haime make no remarks on this species.

Astræa with "calices circumscribed" is also given by Phillips as a Malton coral.

Caryophyllia convexa is stated by the same authority to have been found in the Inferior Oolite at Cold Moor. It was supposed by Edwards and Haime to be a *Montlivaltia*.

Professor Phillips also figured without any description a coral from the Speeton Clay, to which he gave the name of *Caryophyllia conulus*.

In 1848* McCoy described a coral in the collection of the University of Cambridge by the name of *Dentipora glomerata*, from the Coralline Oolite of Malton, where it was said to be common.

Professor Duncan, in the Supplement to the History of British Fossil Corals,† added two species to the Oolitic coral fauna of Yorkshire from specimens which were in the collection of Mrs. Leckenby, of Scarborough, to which he gave the name of *Gonioseris angulata* and *G. Leckenbyi*, both of which had been obtained from the Millepore bed of the Inferior Oolite at Claughton Wyke.

The Rev. J. F. Blake, at page 447 of the "Yorkshire Lias," mentions four species of corals, to which I shall refer.

An addition was made by me in 1888, when I described and figured a species under the name of *Latimaeandraria decorata*,‡ which I had received from Mr. Bean, of Scarborough.

In the catalogue of type specimens in the Woodwardian Museum, Cambridge, dated 1891, *Stylina tubulifera*, from the Coral Rag of Malton, is mentioned as the type of *Dentipora glomerata* of McCoy, so named in 1848.

Finally, the present writer, in a recent communication to the Geological Magazine,§ indicated the presence of three species of corals in the Speeton Clay.

Having now given an outline of the literature relating to the Yorkshire Mesozoic Madreporaria, I will proceed to a critical consideration of such species as I have had the opportunity of

* Ann. and Mag. Nat. Hist., Ser. 2, Vol. II., p. 399. 1848.
† Sup. Brit. Fos. Cor., pt. iii., p. 21, pl. vii., figs. 1-11. 1872.
‡ Quart. Journ. Geol. Soc., Vol. XXXIX., p. 562, pl. xxii., figs. 7, 8, 9,10, and 15. 1883.
§ Decade IV., Vol. VI., p. 302. 1899.

ıining, commencing with the Liassic species, and proceeding
ırds to those of the later formations.

I.—LIASSIC SPECIES.

.ETERASTRÆA EXCAVATA Fromentel sp.

> *Septastrœa excavata* E. de From., in Martin Infra Lias,
> Côte d'Or, 1860.

> *Septastrœa excavata* Blake.   Yorkshire Lias, p. 447, 1876.

The genus *Septastrœa* of D'Orbigny having undergone a very
ching examination by Dr. Hinde,[*] can no longer be admitted
ı British or even a European genus.   It is an American
ı, and confined to the Tertiary formation.   In the Geological
ıazine of May, 1888,[†] the present writer showed that all the
ish species of *Septastrœa* possessed characters which were
lly inconsistent with those of that genus, and created for
r reception a new genus, which was designated *Heterastrœa*.
necessity for such a genus was obvious, and has been con-
ed by the discovery of a considerable number of new species.

ONTLIVALTIA POLYMORPHA Terquer et Piette.

> *Montlivaltia polymorpha* Terq. et Piette.   Lias Inf. de
> l'Este de la France.   Pl. XVI., Figs. 17-21.   1865.

As a Yorkshire species the present appears to be rare, only
t examples being recorded by the Rev. J. F. Blake in 1876.[‡]
vas included by the late Professor Duncan in the list of
reporaria from the South Wales conglomerates, but, as I have
vhere shown,[§] the specimens from that locality were really
ing more than fragments of a species of *Thecosmilia*.

Up to the present time I believe *Montlivaltia polymorpha*,
British species, is confined to Yorkshire, from which locality
ıve examined specimens.

---

[*] Quart. Journ. Geol. Soc., Vol. XLIV., p. 200, 1888.
[†] Decade III., Vol. V., p. 215. 1888.
[‡] Yorkshire Lias, page 447. 1876.
[§] Quart. Journ. Geol. Soc., Vol. XL., p. 365. 1884.

MONTLIVALTIA GUETTARDI Blainv.

> *Montlivaltia guettardi* Blainv.   Dict. des Sci. Nat., T. LI,
> p. 302. 1830.

> *Montlivaltia guettardi* Chap. et Dewal.   Mem. Cour. Acad.
> Belge, T. XXV., p. 264, pl. 33, fig. 6. 1854.

Considerable doubt exists respecting this as a Yorkshire, and indeed as a British species.   Specimens which were in the collection of Professor Tate, and said to have been collected at Redcar, which, with *Montlivaltia haimei*, came into the hands of the present writer, prove on examination to be identical with *Montlivaltia mucronata*.   They have so much the colour and general aspect of Warwickshire specimens of the latter species as to suggest that some mistake has been made as to locality. If, however, they are really Yorkshire examples, it is more than probable that *Montlivaltia guettardi* is not found in that county. The examination of undoubted Yorkshire specimens is most desirable.

MONTLIVALTIA HAIMEI Chap. et Dewal.

> *Montlivaltia haimei* Chapius et Dewalque.   Mém. Cour.
> par l'Acad. Belgique, T. XXV., p. 262, pl. 38, fig. 3.
> 1854.

A great many specimens of this common and well-marked species have been taken from the Lower Lias of Yorkshire, and may be seen in the York and other museums, as well as in private collections.   A considerable number from Redcar, which were in the collection of Professor Tate, are now before me, all of which have the regular and numerous denticulations on the septal edge which characterise typical examples.

II.—OOLITIC SPECIES.

Genus GONIOSERIS Duncan.

The corallum is simple, dome shaped or pyramidal, and has six very prominent angles extending from the apex to the base, and formed by a great development of the primary septa.

The basal wall is nearly horizontal, star shaped, and has
points, which are the lower terminations of the prominent
es. It is naked and costulated, and the costæ are continuous
ι the later cycles of septa, but not with the primary ones.
re are dissepiments connecting the costæ.

There is a small apical fossula. The margins of the septa
entire, and their sides are smooth. There is no synapticular
rth of any kind.

In the course of the investigations which have led to the
re definition of the genus, considerable doubt arose as to
ther the whole of the upper surface is calicular, or whether
; part is confined to the apical fossula. I conclude, however,
; Professor Duncan very rightly regarded the whole of the
er part of the corallum as calicular, and for the following
on :—The prominent angles have between them what must
taken as true septa, because in relative height they correspond
ι the normal development of the cycles in certain of the
-ceidæ. Sometimes *Montlivaltia lens* is so much elevated as to
almost dome shaped, and the cycles very closely resemble the
les in *Gonioseris*. I am entirely, therefore, in accordance
h the original describer in believing that the whole of the
er surface of the corallum in *Gonioseris* is calicular, though
iffer widely from him as regards the affinities of the genus.

It is in the comparative lateral prominence of the cycles that
form presents such an anomaly, the septa of the second cycle
ng in the receding angles have much less prominence than
se of the third and fourth. *Gonioseris* is certainly a genus
the *Astræidæ*.

GONIOSERIS LECKENBYI Dunc.

Gonioseris Leckenbyi Dunc.   Sup. Brit. Fos. Cor., pt. iii.,
p. 22, pl. vii., figs. 6-9. 1872.

Of the two species of *Gonioseris* described by Prof. Duncan
present is much the more typical, and a somewhat detailed
cription of it is desirable. There are some specific peculiarities
ich escaped the notice of the original describer, namely, the

very distinct apical fossula, the well-defined cycles of septa, and the subcristiform superior termination of the primary septa, as well as the perfectly smooth sides and margin of all the septa.

The septal fossula requires especial notice, and distinct definition. It is as follows:—The septa forming the first cycle extend to the apex of the corallum and have great prominence outwardly as well as superiorly. Those of the second cycle have no outward prominence, being in the receding angle, but they also extend the whole height of the corallum. The third cycle consists of septa which are three-fourths the height of the primary ones, and the septa of the fourth cycle are a little shorter than those of number three. There are septa of a fifth cycle, which extend for a very short distance up the side of the corallum.

GONIOSERIS ANGULATA Duncan.

> *Gonioseris angulata* Dunc.   Supp. Brit. Fos. Cor., pt. iii.,
> p. 22, pl. vii., figs. 1-5. 1872.

After so full an account of the characteristics of the genus *Gonioseris*, as well as of the preceding species, I need only say of the present one that it has a much less elevated form, and that all its details are much less strongly made out than in *Gonioseris Leckenbyi*, more especially in the shallowness of the apical fossula, and in the want of prominence of the upper and inner ends of the primary septa which form it.

Numerically it is much rarer than the preceding, one only having been received by the present writer to fourteen of *Gonioseris Leckenbyi*. Both species, I am informed, occur together in the Millepore bed of the Inferior Oolite at Claughton Wyke.

Genus DIMORPHOSMILIA gen. nov.

The corallum is circular, depressed, with the upper surface convex, and the lower flat or concave. There is a basal wall and epitheca showing a central point of former attachment. The whole of the upper surface is calicular, and there is a central calice surrounded by others which are in lines in broad, open valleys radiating from the parent calice. The lines of calice divided by much developed and prominent ridges, formed by union of costal prolongations from the septa, and the latter connected by septal costæ.

The sides of the septa have vertical ridges, ending in denti- ations on the septal edge. Increase takes place by gemmation.

The genus is one of the *Astræidæ*, and of the *Astræinæ*, and some affinity with with *Dimorphastræa*, but differs from it terially by having the calices arranged in radiating lines, arated by strongly-marked cristiform ridges around the parent ice instead of in circles.

DIMORPHOSMILIA EBORACENSIS sp. nov. Pl. xx., figs. 1, 2, 3, 4.

The corallum is subcircular or ovoid, and the outer margin somewhat lobular. The calicular surface varies considerably its degree of convexity, and the outer margin is thick or thin ording to the elevation of the corallum.

The basal wall is sometimes nearly flat, but more frequently a central circular depression with a point indicating former achment. It is clothed with a thick epitheca, which is con- trically wrinkled, between the rings of which there are some- nes costæ, which are straight and uniform in size.

The valleys are wide and open, but the ridges between them e prominent, sub-acute, and the costæ of which they are com- sed meet, blend, and form what has the appearance of a wall, ut in which no true wall has been determined. There are ually five or six of these ridges in a small specimen, increasing number up to eight or nine in a large one.

The calices are few in number, but there is often more than e row in a valley, more especially in specimens of intermediate ze. When of greater dimensions the rows of calices are nearly ways single. They are open and superficial, but the fossula all and well defined.

There are four cycles of septa, and a few short ones of fifth. The primary septa nearly meet in the centre of the sceral cavity, where they are sometimes a little curved. The ondary ones are a little shorter than the primaries, and those of the third cycle are two-thirds the length of the first cycle, and of the fourth cycle one-third the length of the first. All the septa are of nearly uniform and medium thickness, and when unworn have prominent denticulations on their edge.

	in. line.
Diameter of the figured specimen... ...	1 6
Height of the same...	0 7
Breadth of the widest valley	0 6

Number of rows of calices, 9.

HAB.—The Millepore bed of the Inferior Oolite, Claughton Wyke, Yorkshire.

With specimens of *Gonioseris* and *Dimorphosmilia* from the Millepore of the Inferior Oolite at Claughton Wyke, I have received a single specimen of a coral, which, though too ill preserved for specific determination, has characters which are deserving of particular notice, as they certainly indicate specific, if not generic, differences from any of the Jurassic corals at present known in Yorkshire.

It is small, tuber shaped, and the whole of its surface, with the exception of a small area of attachment, is made up by a few large shallow calices, which are united by costal prolongations of the septa. The edges of the septa are denticulated.

THECOSMILIA ANNULARIS Fleming sp.

> *Caryophyllia annularis* Fleming, Brit. Amm. p. 509. 1828.
> *Thecosmilia annularis* Edw. and Haime, Polyp. Terr.
> Palæoz., p. 77. 1851.

The Malton specimens of this well-known and common species need no further remark than to say that they do not differ in any way from those found in other localities in England.

THECOSMILIA COSTATA Fromentel.

> *Thecosmilia costata* E. de From. Intro. Etud. Polyp. Foss
> p. 143. 1858-1861. Polyp. Cor. Environ. de Gray, p. 1
> pl. 6, fig. 1. 1864.

One specimen only of this coral has been examined, and it ▰▰ very closely with the description and figure given by . de Fromentel.   The branches are small, sub-cylindrical, and ▮ ramifications are strictly dichotomous.   The epitheca is thin ▮ pellicular, and the costæ are thin, rather prominent, and ▰nt, the spaces between them being nearly three times the ▰dth of the corresponding spaces in *Thecosmilia annularis*. ▰ the breadth of the intervals is the result of an almost ▮imentary condition of the alternate costæ, which, indeed, are ▰etimes scarcely observable.

The calices are not well preserved, and the cycles of septa ▰ difficult to trace, but there are certainly four cycles, which ▰mber corresponds very nearly with the formula given by the ▰ginal describer of the species.

ISASTRÆA EXPLANATA Goldfuss sp.

> *Astræa explanata* Goldf., Petrif. Germ., V. I., p. 112, tab. xxxviii., fig. 14. 1829.
>
> *Isastræa explanata* Edw. and Haime, Brit. Fos. Cor., p. 94, pl. xvii., fig. 1. 1851.

I have seen and examined some casts of an *Isastræa* from e Coral Rag of Malton, which I do not hesitate to refer to ▮is species.

LATIMÆANDRARÆA DECORATA Bean sp.

> *Meandrina decorata* Bean MS.
>
> *Latimæandraræa decorata* Tomes, Quart. Joun. Geol. Soc., Vol. XXXIX., p. 562, pl. xxii., figs. 7, 8, 9, 10. 1888.

Two specimens of this well-marked species which were given ▮ me by Mr. Bean, of Scarborough, labelled "*Meandrina* ▮corata Bean, Coralline Oolite, Malton," were described and ▮ured by me in the thirty-ninth volume of the Quarterly ▮urnal of the Geological Society. Specimens of this species om the Coralline Oolite of Malton and Langton Wold are in e York Museum, which, having been submitted to the author, dicate great variation in the general form, while they show eat uniformity of structure.

G

PROTOSERIS WALTONI Edwards and Haime.

Protoseris Waltoni Edw. and Haime, Brit. Fos. Cor., p. 104,
pl. xx., fig. 1. 1851.

Protoseris, as a genus, was first characterised by MM. Milne
Edwards and Haime in 1851* in the following words:—

"Polypier fixé, en lames foliacées, lobées et pliées en cornet;
les faces extérieures nues et présentants des stries costales fines,
et les intérieures des calices superficiels à cloisons flexueuses et
confluentes qui ne sont jamais séparées par des collines ou crêtes;
columelle papilleuse."

A single specimen only from the Corallien near Weymouth
had been examined by the original describers and furnished the
above particulars, but subsequently a good many more have been
obtained from the same locality, some of which exhibit considerable variation in general form, but correspond with great exactness
in some very important characters. All are attached by a narrow
foot, generally a mere point, and all are equally destitute of even
a trace of epitheca on the outer wall, which is very distinctly
costulated. Also the calicular details are constantly as in the
type. The following will give some idea of the variation in
external form of *Protoseris Waltoni* from specimens taken from
the type locality :—

1. Subcrateriform, more or less lobate, and "invaginated:"
   generally having one side higher than the other, and
   the point of attachment at the lower side.

2. Irregularly saucer-shaped, lobate, and the point of attachment ex-central.

3. In the form of a thin, irregular disc, with a very thin
   and lobate margin, slightly curled upwards.

The examination of a considerable number of specimens
leads to the conclusion that in general outline the type specimen
does not correspond with the greater number, but has a more
elevated form than is usual.

---

* Polyp. Foss. Palæoz., p. 129.

Two species of *Protoseris* have been described from the Oolite of Nattheim, in Wurtemburg, one only of which has been examined by the present writer, namely, *Protoseris foliosa* Becker. It is a less typical species than *Protoseris Waltoni*. Three species have also been described and figured by Professor Koby in his work on Swiss Jurassic Corals,* namely, *Protoseris Gresleyi*, *P. plicata* and *P. Jaccardi*, but an expression of doubt as to the genus accompanies the description of the two last named. They are all from the Corallien.

Two specimens of *Protoseris* from Settrington, Yorkshire, which have been forwarded to me by the Rev. W. Lower Carter, prove on examination to be referable to *P. Waltoni*.

PROTOSERIS sp.

A single, very ill-preserved coral from the Corallien of Malton differs from *Protoseris Waltoni* in having crowded calices which are very small. It is not sufficiently preserved for description, but is certainly undescribed.

III.—CRETACEOUS SPECIES

The small turbinate corals found in the Speeton Clay have been so little understood that Mr. Judd in his paper on that peculiar deposit, published in 1868, has the following:—†

"*Caryophyllia conulus* Phil.—I have long doubted the identity of the minute Yorkshire coral with the large and well-marked species from the Gault, figured and described by Milne-Edwards. Mr. Dallas, who kindly made a comparison for me, found it impossible to come to any certain conclusion on the subject owing to the imperfect state of preservation of the type specimens."

The opportunity of comparing some pretty well-preserved specimens from Speeton with a considerable number from the Folkestone Gault, has enabled me to determine two well-defined species, probably a third.

---

* Monogr. Polyp. Jurass. Suisse, p. 350, &c., pl. xcvi., figs. 2, 3, 4, 5, and 6. 1885.

† Quart. Journ. Geol. Soc., Vol. XXIV., p. 225. 1868.

TROCHOCYATHUS CONULUS Phillips sp.

> *Caryophyllia conulus* Phill.   Ill. Geo. Yorkshire, 2nd edit.,
> pl. ii., fig. 1.   1835.
>
> *Turbinolia conulus* Mich.   Icon. Zooph., pl. i., fig. 12a (not
> 12b on the same plate).   1840-1847,
>
> *Trochocyathus conulus* Edw. and Haime.   Brit. Fos. Cor.
> p. 63, pl. ii., fig. 6.   1850.

This species occurs in the Speeton Clay, Yorkshire, and in the Gault at Folkstone.   An elongated variety is also found at both localities, which was described and figured by Professor Duncan as *Smilotrochus cylindricus.*

TROCHOCYATHUS? CALCARATUS Tomes.

> *Smilotrochus calcaratus* Tomes.   Geol. Mag., 1815, p. 543.

A well-preserved specimen of this species, obtained from the Speeton Clay, is in the hands of the present writer, and has been compared with others from the Folkestone Gault.   It is a peculiar species, the genuine position of which is by no means clear.

TROCHOCYATHUS WILTSHIREI Duncan?

> *Turbinolia conulus* Mich. Icon.   Zooph., pl. 1, fig. 12b
> (not fig. 12a), p. 1, 1840-1847.

Michelin figures two distinct species of corals, presumably from Speeton, one of which is free and now known as *Trocho-cyathus conulus*, and the other (figure 12b) an attached form. Of the latter I can only say that it does not represent *T. conulus*. but that it has considerable resemblance to *T. Wiltshirei*, as I have already stated in a recent communication to the Geological Magazine.*

The foregoing list of Yorkshire madreporaria, though a very short one, is remarkable for the number of interesting, and I might say anomalous, forms.   In the Millepore bed of the Inferior Oolite there are three not found elsewhere, and the Corallien is distinguished by the presence of the remarkable genus *Protoseris*, and by a species of *Latimæandraria* and one of *Thecosmilia* not

---

* Decade IV., Vol. VI., p. 302.   1899.

3 .

1

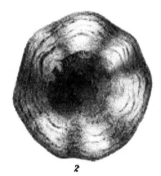

2

5

7

6

:t with in any other part of England.   Doubtless the
 of collectors would be rewarded by the discovery of other
:ing madreporaria were diligent search made. .

DESCRIPTION OF PLATE XX.

Dimorphosmilia eboracensis, natural size, seen from above.
The denticulations on the edges of the septa have been
worn off.

Dimorphosmilia eboracensis, the under surface of the
same species, showing the epitheca, and point of
former attachment.

Dimorphosmilia eboracensis, the side view of the same
specimen.

Dimorphosmilia eboracensis, a septum magnified showing
the strongly-developed denticulations.

Gonioseris Leckenbyi, a tall specimen, natural size, having
the apical fossula well defined.

Gonioseris Leckenbyi, a magnified figure of the apical
fossula showing the sub-cristiform development of the
upper termination of the primary septa.

A plan of the septa in *Gonioseris* showing the primary
septa forming the salient angles, and the secondary
ones in the receding angles.   A little magnified.

ON THE OCCURRENCE OF STREPSODUS SULCIDENS, HANDCOCK AND ATTHEY, IN THE YORKSHIRE COAL MEASURES.

BY EDGAR D. WELLBURN, F.G.S.

INTRODUCTION.

Whilst looking through some fish-remains in the Museum, Brighouse, Yorkshire, I found a fine mandibular ramus which was labelled *Megalichthys*, but which on examination proved to belong to the fish *Strepsodus sulcidens*. As it is the first time this fish has been found in the Yorkshire Coal Measures, and as the specimen shows some points of great interest, I judged it worthy of a brief description.

DESCRIPTION OF THE SPECIMEN.

The specimen is that of a mandibular ramus—imperfect posteriorly—with three fine laniary teeth, seen from the inner side. In order to understand aright the points shown in the specimen as figured on Plate XVIII., the author considers it advisable to give a brief description of the structure of the mandible of the Rhizodontidæ. In that family the mandible is of a very complex structure, and as shown by Dr. Traquair, F.R.S., is built up in the following manner, viz. :—There is first a dentary bone which is deep and thick at the symphysis, from which point it tapers backwards, and bears a series of small teeth, with one large laniary tooth in front. Below this bone there is a series of three or four plate-like lenticular bones, the hindermost of which corresponds to the angular bone, whilst the others are termed the infradentaries. A thin, splenial lamina forms the inner wall, and between these and the dentary are a series of three or four stout lenticular bones, the laniaries, each of which bears a strong laniary tooth.

On again turning to the specimen, it is at once apparent that the inner wall of the jaw, or splenial, is absent, but the outer and inner segments are present.

Of the outer wall, or segment, the inner side of the first and part of the second infradentaries are shown, the dividing

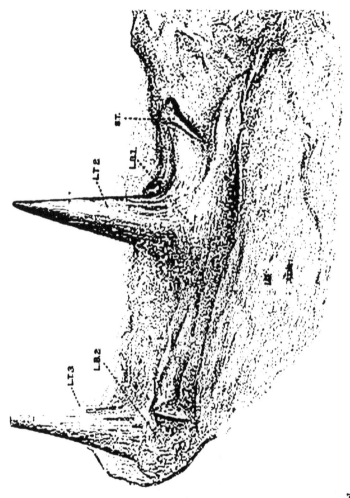

W/Allburn del.

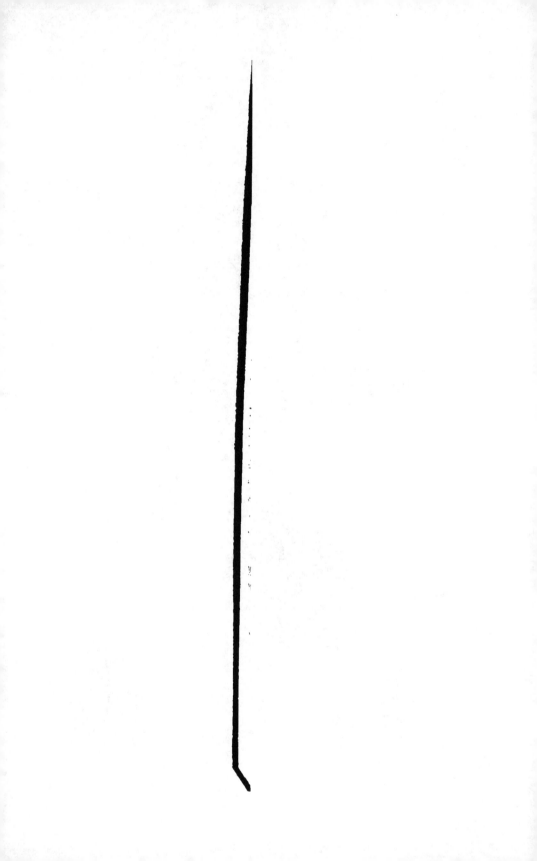

etween the two being indicated by a sutural line which
ownwards and backwards from a point immediately behind
st laniary tooth to cut the inferior edge of the jaw at an
angle.   Above these runs the dentary bone, which—with
ception of a portion of the deep fore-end or symphysis—
stly covered by the strong, thick laniary bones, which
· to be firmly anchylosed to it, although there is plain
ce at the posterior end—where the bone is broken across
, the bones were originally developed in separate centres of
ition.   Three of these laniary bones are shown, that at the
ysis being the largest ; in position they run in a line, from
backwards, a short distance below the dentary edge of the
·y bone, and each bears a strong laniary tooth, of which
cond and third show the usual characters, viz. :—They are
, elongated, somewhat flattened, apex straight, fine striæ
ier surface, base broad and showing vertically elongated
sions extending a short distance beyond the basal furrows.
nterior laniary at the symphysis is a larger and more
ul tooth than the others, it shows characters similar to
of the other teeth with the exception that they (the
ers) are more strongly marked, especially the vertically-
ed depressions at the base, and the apex, instead of being
t, is strongly and evenly bent backwards and slightly
s, this being a point of interest, as in the specific diagnosis
ex of the teeth are given as being straight.

orm and locality : Better Bed Coal Shale, Lower Coal
es, Low Moor ; and Barnsley Thick Bed Coal Shale,
: Coal Measures, Barnsley ; both in Yorkshire.

EXPLANATION OF PLATE XVIII.

odus sulcidens Handcock and Atthey. Mandible seen from
  inner side. S.Mn.—Symphysis of dentary bone. S.L.
  —Symphysical laniary tooth.   L.T. 2 and L.T. 3.—
  Second and third laniary teeth.   L.B. 1 and L.B. 2.—
  Laniary bones. ID.—Infradentaries. Specimen in the
  Brighouse Museum, Yorkshire.

NOTES ON THE HISTORY OF THE DRIFFIELD MUSEUM OF ANTIQUITIES AND GEOLOGICAL SPECIMENS.

BY J. R. MORTIMER.

(Read May 25th, 1900.)

It was the Great London Exhibition of 1851 that first decided my taste for scientific inquiry. Afterwards, Mr. Edward Tindall's geological and archæological collections at Bridlington fired me with a strong desire to make a similar collection. A curious chalk cast was the first specimen I obtained, whilst a small Ammonite, which I bought of Mr. Tindall, was the first of its kind I possessed.

My brother, the late Robert Mortimer, of Fimber, had a like love for collecting.

For the first ten or twelve years the late Edward Tindall, of Bridlington, and the late George Pycock, of Malton, were almost our only rivals; yet we accumulated specimens but slowly. We had, however, during this period, trained many of the farm servants in this neighbourhood to distinguish and keep for us any geological and archæological specimens they could find. The small collection we then made mainly consisted of chalk fossils and a very few stone and flint tools. These we exhibited in cases in my offices at Fimber. Small though this display was it seemed to stimulate others to indulge in the same hobby, and soon our neighbourhood was more or less periodically visited by the thirteen competitors hereinafter named, and their agents during a period of about 35 years, ranging from 1861 to 1896. None, however, of these enthusiasts, except Mr. Tindall and Mr. Chadwick, collected geological specimens, though all of them were active competitors for stone, flint, and bronze weapons. They constantly visited the district, and not infrequently bought from the very field labourers whom we had trained to distinguish these specimens, by overbidding us, and so running up the prices.

The combined energies of these gentlemen would, I believe, obtain from the same area quite three times the large number of stone, flint, and bronze tools and weapons that have been collected by my brother and myself, now exhibited in the Museum at Driffield. If this be the case it should be asked, What has become of so great a number? In attempting to answer the question I will briefly refer to each collector's labours.

(1.) The late Edward Tindall, of Bridlington, not only commenced to collect more than 50 years ago, but he held the almost unrivalled access to a field rich from both a geological and archæological point of view. Besides he was personally a diligent collector, so much so as to call forth at times uncomplimentary remarks from superficial observers. On one occasion whilst gathering specimens in a field near the sea at Bridlington two ladies were passing, and he overheard one remark to the other, " Look here, that poor old [meaning demented] man we saw last year is here again, picking up stones and throwing them down again." These "stones," of course, were the rejected specimens. I myself when similarly engaged have been accused of mushroom gathering.

Mr. Tindall obtained a great number of specimens, but he was always ready to dispose of them whenever any collector, no matter where from, wished to buy. Nevertheless he generally had on hand a considerable number of both geological and archæological specimens, and amongst them were often some choice ones. After his death in 1877, at the age of 63, the collection he had then on hand was sold. Part of this was obtained by Mr. Robert Gatenby, of Bridlington; but what became of the remainder I do not know.

(2.) The late Mr. George Pycock, of Malton, made a collection, which he sold many years ago to the late Dr. Rooke, of Scarborough, and it is now in Dr. Rooke's son's private museum at Scarborough.

(3.) The Rev. Canon Greenwell, of Durham, amassed a large number of valuable specimens (independently of those he obtained from his excavations of the barrows), the greater number of

which were gathered from the surface of the Wold hills and the
immediate neighbourhood. These the Canon sold in July, 1895,
to Dr. Sturge, of Nice, and they are now in the south of
France, to the great loss of East Yorkshire.

(4.) The late Frederick Porter, of Yedingham, had gathered
together several hundreds of stone and flint tools, among which
were some good specimens. He disposed of a portion of them,
I believe, to the Rev. Canon Greenwell, but I do not remember
what became of the others when he removed from Yedingham
to Jersey in 1868.

(5.) The late Mr. Charles Monkman, of Malton, was for
a considerable time an energetic collector. Some of his best
specimens fell into the hands of the Rev. Canon Greenwell, and
a few were obtained for the York Museum.

After his death, on April 13th, 1875, the remainder were
quickly disposed of by his wife ; but I am ignorant of their
present whereabouts.

(6.) The late Charley Hartley, of Malton, obtained many
specimens of both flint and stone implements from the same
collecting ground. These, about the year 1875, he sold to the
late Mr. William Robinson, of Houghton-le-Spring, near Durham,
who had a large collection gathered from all parts.

Mr. Hartley afterwards made a second collection which, after
his death on September 7th, 1883, was disposed of, and the best
of these specimens were also bought by Mr. Robinson.

(7.) The late Rev. James Robertson, Curate-in-charge of
Barton-le-Street, also for several years periodically visited the
part of the Wolds from which I obtained my collection, and he
procured a large number of pre-historic relics. Most of these,
I have been told, he disposed of in 1876 to Mr. John Evans
(now Sir John Evans), Nash Mills, Hemel Hampstead, and in
1877 my brother, the late Robert Mortimer, purchased the remaining portion of Mr. Robertson's collection for £25.

(8.) Mr. George Edson, also late of Malton, was a very
careful and industrious collector of all kinds of specimens of

archæological interest, both as an agent to Canon Greenwell and on his own account. When leaving Malton he sold his collection by auction on May 8th, 1891, and many choice specimens were disposed of to various purchasers, a few going to the York and Malton Museums.

(9.) The late Thomas Allerson, of Norton, near Malton, was, like Mr. Edson, constantly being brought into contact with the farm servants and other field labourers when on his business journeys in this neighbourhood, most of whom had then become well skilled in distinguishing the value of different specimens. They were also quite ready to take advantage of the extra prices to be obtained from the rival purchasers then in the market. So keen was this competition at one time that to retain our hold of the market we distributed handbills, offering rewards, consisting of money and a free pass to the Leeds Exhibition in 1868, to those who would supply us with the greatest number of articles of various kinds. In 1873 Mr. Allerson had obtained a considerable number of pre-historic relics, which he then wished to sell, and he offered them to me. These I purchased of him, and they are now in the Museum at Driffield.

(10.) My friend Mr. Thomas Boynton, of Bridlington Quay, has a large and choice collection of stone, flint, and bronze weapons, of local origin; as well as a few from the south of England and other districts. He also possesses many very beautiful flint and chert arrow-heads and various instruments from America and other countries. He is frequently adding to his very fine collection, which should certainly be purchased to remain in East Yorkshire.

(11.) The late Rev. Thomas J. Monson, of Kirby Underdale, was merely one of Canon Greenwell's collecting friends in this neighbourhood, and after he had purchased a few specimens picked up by the farm servants he forwarded them to the Canon without having any further interest in the matter, or knowing more about them. The probability is that the district was not very much impoverished by his labours.

(12.) Mr. Robert Gatenby, of Bridlington, has obtained several hundreds of flint, stone, and bronze specimens, a few of which are very fine ones. He is still adding to the number.

(13.) Mr. Samuel Chadwick, late of Malton, who emigrated to New Zealand in 1895, was a very energetic collector of both fossils and implements. His business occupation brought him frequently among the farm labourers and quarrymen in the rural districts. This gave him exceptional opportunities for obtaining a considerable quantity of specimens, and for a considerable time he was my most active rival. That Mr. Chadwick made good use of these facilities the contents of the Malton Museum give ample proof. This fine and large collection, gathered almost entirely from the neighbourhood, also is sufficient evidence of his energy and knowledge as a diligent collector. Besides those placed in the Malton Museum he supplied many specimens to the York Museum, and a few to other places.

There were also a few minor collectors whose united labours have assisted to impoverish this neighbourhood.

For the last few years almost the only local collectors I have had to compete with are Mr. Thomas Boynton, Bridlington Quay; Mr. Robert Gatenby, Old Bridlington ; and I may add Sir Tatton Sykes, Bart., of Sledmere.

COLLECTIONS FROM THE BARROWS.

Hitherto I have only referred to the collections of specimens which have been obtained from the surface of the land, or otherwise accidentally found.

In addition to these, four valuable collections of Ancient British and Anglo-Saxon relics have been obtained by excavating the barrows of this district.

(1.) The late Lord Londesborough explored a great number of barrows in this neighbourhood during a period of ten years, ranging from 1842 to 1852, and the principal of the articles he then discovered were placed in his museum at Grimston. After his lordship's death, when the house and estate at Grimston

were sold (about 1872), the contents of the museum were dispersed. Afterwards (in 1888), a portion of the relics were sold
by auction by Messrs. Christie, Manson & Wood, at their rooms,
King Street, St. James', London.

Of the present whereabouts of this large collection (excepting a small portion, including some rare specimens from a barrow
at Kelleythorpe, near Driffield, which at the above sale found
its way to the British Museum) I know nothing. I fear, however, that the whole of it is lost to East Yorkshire.

(2.) The late James Silbourn, of Pocklington, during the
years 1851-2 opened several of the barrows in the neighbourhood
of Huggate and Warter. Since then I have reopened nearly
the whole of these particular barrows, as I could not distinguish
before excavating which of them had been opened by Mr. Silbourn.
I found that he had placed a strip of lead on which his name
was stamped in several of the barrows he had opened.

In the spring of 1852 Mr. Silbourn, during an exploration
in stormy weather, took cold, which brought on inflammation,
and so caused his death. After this regrettable circumstance
the pottery and many other relics he had obtained from the
barrows were sold by his relations, and, like the previously-named
collection, their fate is unknown to me, excepting—as in the
previous instance—a very small portion, which is now in the
British Museum.

(3.) The Rev. Canon Greenwell, of Durham, during a period
of 30 years (1864-1894) excavated upwards of 300 barrows on
the chalk wolds, immediately adjoining my field of research. An
account of the greater number of these he published in his work
on "British Barrows" (1877). The illustrations and descriptions
in this very valuable book clearly indicate what a large treasure
of relics was then obtained. That all these have been placed
in the British Museum, and are now entirely lost to East Yorkshire, their legitimate home, is, I think, much to be regretted.

(4.) And lastly, I have myself explored nearly the whole
of a series of the Wold barrows on an area of about 80 square

miles, between 1864 and the present time. That I have safely
preserved the relics discovered during these researches the contents of the museum at Driffield will testify. I also possess
about 1,000 drawings * which my daughter has made for me, of
all the objects of interest which I have discovered; and I have
in addition a full type-written description of the results of all
my excavations.

And I may say that the procuring and arranging of this
collection has been one of the greatest pleasures of my life.

That this collection should belong to and remain in the
district has been and is my great and constant desire. Unfortunately, however, I cannot afford to offer it as a free gift; but,
to prove my great anxiety for its remaining in the neighbourhood, I have offered it to the East Riding County Council at
half its value.

Probably such a transaction by a County Council might
seem to be a little in advance of the times; nevertheless, a time
will come when such a thing will be done, and if the East
Riding County Council accept this offer they will never regret
being one of the pioneers in such an advanced and enlightened
step.

From the memoranda I have just given it is sad to observe
that of all the collectors I have referred to, including myself,
only six are now living. It is also to be lamented that of the
fourteen collections named only four remain in the neighbourhood, these being in the Driffield and the Malton Museums
respectively, and those belonging to Mr. Thomas Boynton and
Mr. Robert Gatenby. Of the other ten, nine are mainly absorbed
by public and private museums in distant parts of the country,
or have otherwise disappeared; whilst a great portion of one (the
most important of the ten) has been removed so far as the South
of France.

It is still more to be regretted that three of the most
valuable collections of the four named explorers of the barrows

---

* Some of these were exhibited when the paper was read.

(viz., that of the late Lord Londesborough, the Rev. Canon Greenwell, and the late James Silbourn) have been dispersed and are lost to their native East Yorkshire.

Such, unfortunately, must be the fate of all private collections, if not permanently fixed during the life of their original owner, as it far too frequently happens that that which one generation gathers the next generation scatters.

I have said "more to be regretted" because it is possible that some future collector might obtain a small collection of specimens from the surface of the land, but to make another collection from the barrows of this district would be quite an impossibility, as they are practically exhausted.

From these lamentable facts it is evident that the neighbourhood has been deprived of a great number of its precious relics, which were a valuable legacy left by our ancient forefathers, and by right should have remained and belonged to the present and all future occupants of the district.

These valuable remains are almost the only reliable records of the customs and mode of living of our remote ancestors. They are the fossil history of the district, and they must always be of the greatest interest to the neighbourhood in which they have been found. It is, therefore, our bounden duty to provide, as far as possible, for their safe keeping in the district. Nevertheless, I have shown that, unfortunately, during the last thirty-five years this district has been immensely impoverished of its archæological treasures. And it is much to be regretted that even at the present time the tendency is to favour the removal to distant collections any relics which are found in this neighbourhood, rather than assist to retain them in the district to which they belong by inheritance. Such instances have recently come under my notice.

At present, only three of all the eighteen collections I have referred to, viz., fourteen consisting of specimens obtained from the surface of the land, and four from the excavations of the barrows, remain in East Yorkshire.

Surely the East Riding possesses some governing body that, before it is too late, will see the wisdom of permanently possessing these, and handing them down to future governing bodies as a source of education and a treasure of permanent value to the district. When this is accomplished, and it is known that this collection belongs for ever to the district, it will be a centre of donations of relics found in and belonging to the neighbourhood (rather than the specimens be sent to distant collections, where they can only be of minor value), and in time it ought to and will become a large and very valuable possession.

In Memoriam.

RICHARD REYNOLDS, F.C.S.

Mr. Richard Reynolds, who was elected a member of this
:ty in the year 1864, served as a member of its Council
: 1870, and was elected a Vice-President at the Annual
:ing at Bradford in 1894, died at his house, Cliff Road,
e Park, Leeds, on April 5th, 1900. He was known and
∍d by all students of science in Leeds, and one might almost
in Yorkshire. A short account of his useful but unostenta-
ı career will be of interest to those who worked with him
so many years.

Mr. Reynolds came of an old Quaker stock, being descended
ı John Gurney, the "prisoner of Norwich," who was shut
luring three years in gaol for refusing to take a prescribed
. He was born at Banbury in 1829, being the eldest son
n apothecary, who died when the boy was only four years
At fourteen Richard Reynolds left school, and was appren-
ı to James Deane, a chemist on Clapham Common. Had
training been prolonged he would have made an excellent
lar, for his aptitude for science was remarkable, and his
vledge of books was in after years that of a cultivated man.
spite of a scanty education he was able to take the first
:s in both botany and chemistry at the very first examina-
held by the Pharmaceutical Society. This early distinction,
some relationship between the Deanes and the Harveys, may
· brought Reynolds to Leeds, where he soon became partner
he late Thomas Harvey. The firm of Harvey & Reynolds
me very prosperous, but it is remarkable not only for its
nercial success, but for the public services of members of
firm. William West, F.R.S., Thomas Harvey, and Richard
nolds kept up for at least eighty years a succession of
vated and public-spirited citizens—all good friends of science
education. Both West and Reynolds became in succession
ırers on chemistry at the Leeds Medical School, and Hon.
etaries to the Leeds Philosophical and Literary Society.

The great service which Reynolds was privileged to render to his own generation and to his adopted county was connected with the foundation of the Yorkshire College. He was Hon. Secretary to the College during its first critical years, and but for his diligence, sagacity, and knowledge of men, the College could hardly have surmounted its early difficulties. For ten years he was its mainspring, and no sacrifice of time and labour seemed too great, if only he could thereby carry the great project one step nearer to complete realisation. The Yorkshire College is, to those who know its inner history, a lasting monument to his indefatigable, though, of course, not unaided exertions.

Mr. Reynolds was active in other directions also. In 1881 he was called upon to preside over the Pharmaceutical Conference during its meeting at York, and wherever he saw a prospect of public usefulness, he was ready with his counsel and support. In private life he was the same amiable and modest man that we knew so well in public—energetic without fuss, well-informed without parade. Leeds and Yorkshire want a succession of such men, but we are not so sanguine as to anticipate that the want will be regularly met.   L. C. MIALL.

SECRETARY'S REPORT, 1899.

The Society continues in a prosperous condition, and the Report of the year's work is of exceptional value and interest.

The roll now consists of 6 honorary members, 53 life members, and 115 subscribing members, a net increase of 9 on last year's record.

It is with deep regret that we have to report the decease of four members, Messrs. Ernest Haworth and J. F. Ianson, of Wakefield; Mr. Henry Nelson, of Leeds, who had been a member for 24 years and died on May 20th at the advanced age of 85; and the Earl of Wharncliffe, who for 60 years had taken a deep interest in the welfare of our Society (having been elected a member in 1839), and had served as a Vice-President during that period.

The first General Meeting and Field Excursion for 1899 was held at Todmorden on June 7th and 8th. The members were met at Todmorden on June 7th by Mr. Robert Law, F.G.S., who acted as leader in a very efficient manner, and proceeded by wagonette to Summit Inn. A detour was made to the adjoining moorland to trace one of the feeders of the Calder to its source. Interesting slickensided grit rocks were examined by the way. After lunch at Summit Inn the party examined a section in the Third Grits overlaid by drift full of foreign boulders in the adjoining brickyard. Thence a visit was paid to some interesting landslips on the eastern side of Snoddle Hill Reservoir, forming five or six parallel ridges covered with blocks of gritstone. The leader then took the party to view good sections in the Third Grit beds in the Light Hazel, Long Lees, and Warland Quarries He pointed out two thin coal seams between which there is a band of calliard rock full of stigmarian rootlets, and which, he said, was persistent over a wide area. In each quarry drift beds, enclosing Lake District boulders, were found overlying the grit rocks. At the Summit Brickyard the nodules and black shales were successfully searched, yielding many fossils. Most important

among these were nine specimens of fish, on which Mr. E
Wellburn, F.G.S., reports that three were *Cœlacanthus* and
different palæoniscal fishes which appear to be new to Sci
and certainly are new to the Millstone Grits. The other fi
which two good specimens were obtained, proved to be *Elonic
Aitkeni* (Traquair), which is an interesting find, having only
before been found in the Yorkshire Grits, and that spec
having been lost makes the present find of considerable va

After dinner the General Meeting was held at the V
Hart Hotel, Todmorden, under the presidency of Mr. Wi
Cash, F.G.S., Treasurer of the Society. Eleven new mer
were elected. An address was delivered by Mr. Percy F. Ker
F.G.S., on "The Physical History of the Calder," followed l
outline of "The General Geology of Upper Calderdale,
Mr. Robert Law, F.G.S. After the papers had been read
was a vigorous discussion.

On June 8th the party went by train to Portsmouth St
and ascended Green's Clough. A fine section of the whole c
Millstone Grit beds above the Upper Kinderscout is expos
this Clough, dipping south-west. As the moor was crossec
old workings for the coal seams which cropped out on the hi
were seen. At Sharney Ford are extensive tip mounds, d
the working of the flagstones, which are the equivalents o
Elland Flags of adjoining districts. In descending Duk
visits were paid to workings in the Forty Yards'
and Ganister coal-seams. The leader pointed out the plac
which several faults crossed the valley, and a foot-thick
seam in the middle of the Rough Rock was examined,
photographed by Mr. Godfrey Bingley. Baum-pots and
balls were examined for fossils, in some of which small ho
were found filled with petroleum. The weather was most fa
able and the attendance good, and before separating a very
vote of thanks was accorded to Mr. Robert Law, F.G.S., fo
able leadership.

The summer General Meeting was held at Stokesle
Friday, August 4th, and was associated with an interesting

extended Field Excursion for the examination of the northern slopes of the Cleveland Hills. On Friday, August 4th, the party were met at Sexhow Station, and the way was taken through the picturesque village of Carlton to Carlton Bank, which was ascended. The spoil-heaps of the old jet workings were passed on the way, and a good exposure of the fossiliferous Middle Lias was examined. An inspection was also made of the old workings of the Carlton Alum Works. Raisdale was crossed and several foreign boulders were picked up *en route*. The party then clambered round the steep escarpment of Cringley Moor, and an ascent made to a fine exposure of the Dogger on the flank of Cold Moor. This deposit is of great interest, being full of well-rounded pebbles of a white limestone which has not been identified with any known rock. The matrix was full of fossils, and one of the largest pebbles was a fine specimen of Thamnastræa. Crossing the moorland Bilsdale was reached and the ascent made to the Wainstones, a picturesque, weathered escarpment of the Lower Estuarine Beds.

Hasty Bank was then descended to Greenhow Park, and the party returned by wagonette to Stokesley. After dinner at the Bay Horse Hotel, the General Meeting was held under the presidency of Mr. Robert Bell Turton, of Kildale Hall. A paper "On the Roman Roads in the East Riding of Yorkshire" was read by the Rev. E. Maule Cole, M.A., F.G.S., and one "On a Peat Deposit at Stokesley" by the Rev. John Hawell, M.A., F.G.S. Mr. P. F. Kendall, F.G.S., gave a description of some recent and interesting evidence on the condition of Cleveland in the glacial period, which tended to prove that the mouth of the Tees was choked by the North Sea ice when the Teesdale glacier pushed its way into Cleveland. The highest level Drift deposits along the slopes of the Cleveland Hills consist almost entirely of erratics of a northern type, which would hardly have been the case had the Teesdale glacier obtained an earlier access to the sea. Subsequently by the retreat of the North Sea glacier the western stream was able to push out seawards and force its way along the shoreline to the south. During this period

Mr. Fox-Strangways had shown that the Vale of Pickering was blocked by the ice and became a great lake. Similar reasoning, said Mr. Kendall, would require an extra morainic lake in Eskdale, in Kildale, and at many other points along the edge of the ice. Evidences of these lakes were found all round the Cleveland Hills. The overflows were over soft rocks and the lake-levels consequently fell too rapidly for the formation of marked beaches, but the presence of numerous overflow valleys revealed the history of the glacial lakes and their varying drainage in a very interesting manner. These papers were followed by a discussion, in which Messrs. F. F. Walton, J. W. Stather, E. Hawkesworth, and W. L. Carter took part, the readers of the papers responding. A vote of thanks was passed to the Chairman, the Leader, and the Readers of Papers, and Mr. Turton briefly replied.

The second day's Field Excursion was taken by wagonette to Great Ayton, where a visit was paid, under the Rev. John Hawell's guidance, to the workings in the Cleveland dyke. These are now carried some distance beneath the surface by mining operations giving an interesting series of exposures. After searching the old ironstone spoil heaps for fossils, and examining a gravel pit, in which shell fragments were found, the ascent of Roseberry Topping was made and the extensive view admired. Thence a detour was made over Ayton Moor, on which some interesting barrows and entrenchments were seen, to Lonsdale, and thence through the wood to Kildale. Near Kildale Railway Station a deposit of shell-marl of post-glacial age, full of shells, was examined. In the overlying peat deposit antlers of the reindeer have been found. The path through Kildale Wood was then taken to Dundale Beck, and the return journey made by wagonette to Stokesley.

In connection with the Yorkshire Naturalists' Union Excursion on Bank Holiday Monday, an extension of this geological route was arranged. The party, led by the Rev. John Hawell, ascended Hasty Bank, and examined a good exposure of Middle Lias. The watershed was then crossed to view a pretty little overflow valley into Bilsdale, and then began a long and tiring moorland tramp round the Ingleby Greenhow embayment. Botton Head tumulus, the highest point in Cleveland, was ascended; an exposure of erect Calamites was examined, and some of the party had a peep into the beauties of Basedale.

The meeting of the Yorkshire Naturalists Union was held in the schoolroom at Ingleby Greenhow, presided over by the President of the Union, Mr. Wm. West, F.L.S., of Bradford, and was attended by several of our members.

A smaller party arranged an extra excursion on Tuesday, August 8th, under the guidance of Mr. P. F. Kendall, F.G.S., to the moorlands north of Eskdale, to examine the evidences of glacial action. From Commondale Station the route by the brick works was taken to High Moor, where exposures of gravel containing boulders of the northern type were examined. In the highest of these, about 810 feet above O.D., a small boulder of rhomb-porphyry was found by Mr. J. W. Stather.[*] Thence a traverse was made across Moorsholm Moor, and several interesting notch-like valleys were inspected cutting across the usual drainage. These valleys were explained as lines of drainage for the water along the ice-front at various stages in its advance and retreat. The return to Danby Station was made by a wide overflow valley, which must have carried a vast volume of water at the time of its excavation, but which is now peat-logged and hardly carries any running water at all. At the lower end of this valley of glacial overflow there is a great delta deposit reaching down towards Danby. After visiting a deposit of finely laminated mud at the Danby Brick Works, which had been deposited by the Eskdale extra-morainic lake, the party separated with hearty thanks to the genial and indefatigable leaders for the splendid series of excursions which they had arranged.

---

[*] This boulder is figured in Dr. A. R. Wallace's "Studies, Scientific and Social," Vol. I., p. 86. No other Scandinavian boulder has been found at so high an altitude in England.—P. F. K.

At the April Council Meeting a letter was read
Mr. J. H. Howarth, F.G.S., embodying a suggestion for a
complete examination of the underground waters of Malham
Clapham by means of delicate chemical tests, and convey
generous offer from Mr. George Bray, of Headingley, to
the necessary costs of such an investigation. This sugg
was warmly adopted by the Council, and a Committee, c
ing of the members of the Council, together with Messr
Bray (Leeds), F. Swann, B.Sc. (Ilkley), S. W. Cuttriss (L
and Walter Morrison, M.P. (Malham Tarn), with power t
to their number, was appointed to conduct the investig
Subsequently Messrs. W. Ackroyd, F.I.C. (Halifax), B. A. B
F.I.C. (Leeds), J. W. Broughton (Skipton), J. A. Bean ('
field), and Professors Smithells and Procter and Dr. J. (
of the Yorkshire College, were added to the Committee.
meetings of the Committee were held in Leeds, and a
meeting was called at Malham, for June 21st and 22
carry out the arrangements agreed upon. Sub-committee
appointed to carry out special gauging, chemical, and geo
investigations. The whole of the work was carried out
immense pains and care, and has resulted in a distinct
to the knowledge of the underground waters of Malham and
movements. A small sub-committee, consisting of Messrs.
Fennell, F.G.S., J. A. Bean, F. W. Branson, F.I.C., W. Ac
F.I.C., P. F. Kendall, F.G.S., J. H. Howarth, F.G.S., and
Carter, F.G.S., was appointed to draw up an abstract and a
for its presentation to the British Association at their
meeting. By desire of the Sub-committee the abstract was
by Mr. J. H. Howarth, approved at a subsequent meetin
presented to the Geological Section of the British Asso
by Mr. P. F. Kendall, who also was authorised to apply for
stantial grant in aid of the continuation of these investi
The results arrived at were embodied in the following set
clusions, which were unanimously adopted by the Sub-commi

  1. The observations upon the gauges showed that a t
water let in at the Tarn Water Sinks during a period of 2

caused an increased flow of water at Aire Head of much smaller
volume, but longer duration, indicating that there must have
been a "backing-up" of water in the intervening limestone.

2. The unexpected delay in the appearance of the chemicals
at the outlets has been due to a low and quiescent water-flow,
and their appearance in every case (except Scalegill Spring) has
succeeded a comparatively heavy rainfall, whence it appears that
the chemical solutions which had been detained underground
were all subsequently flushed out at the same time.

3. Under conditions of summer flow, such as prevailed on
June 22nd—

> (a) The water descending at the Smelt Mill Sink emerges
> at Malham Cove and not at Aire Head, Gordale Beck, or
> Scalegill Spring.
>
> (b) The sinks below Malham Tarn are connected with
> Aire Head and not with Gordale Beck; and under certain
> conditions, as detailed in the report, ammonium sulphate
> put in at Tarn Sinks emerges at the Cove.

4. The Tranlands Beck Water Sink is connected with Scalegill Spring and not with Aire Head.

5. The geological observations show that the underground
waters follow master-joints; that master-joints north of the
Middle Fault run north-west and south-east; that the Smelt Mill
water follows these joints directly, and that probably the Tarn
water follows these joints to about the Middle Fault near Grey
Gill.

South of the Middle Fault the master-joints change to nearly
north-east and south-west, and it is inferred that the Tarn water
on crossing the fault adopts these joints, dips with the Pendleside
Limestone under the Shales, and appears with the limestone at
Aire Head.

The Gordale stream is absorbed, and carried by master-joints
to the south-east, and it is probable that it supplies the great
springs on the east side of Gordale.

The thanks of the Council are heartily given to all the
gentlemen who have so freely contributed of their professional

experience, time, and money to attain these eminently satisfactory results. To Messrs. W. Ackroyd, F. W. Branson, B. A. Burrell, and F. Swann, for the patience and care with which the extended series of chemical tests were carried out; to Messrs. Fennell and Bean for much labour in connection with the fixing of the gauges; to Mr. Fennell for preparing a large map and drawing a gauge diagram and index-map for the British Association; and to Messrs. Kendall, Howarth, and Simpson for careful surveys of the limestone, our special thanks are due. Your Council would also gratefully acknowledge the generous contributions of material for the testing of the underground waters by Mr. George Bray, without which it would have been impracticable to carry out these interesting investigations, and the advice and help in many ways of Mr. Walter Morrison, M.P., and of Messrs. Winskill and Townend, of Malham. Mr. Godfrey Bingley also has put the Council under a debt of gratitude by taking a series of valuable photographs of the investigations.

We have to announce with regret the resignation of Mr. J. Stubbins, F.G.S., a member of the Council for some years.

The position of Hon. Auditor left vacant at the Annual Meeting was filled up by the Council, who appointed Mr. J. H. Howarth, F.G.S., of Bradford.

The Rev. W. L. Carter, M.A., F.G.S., was elected Representative Governor of the Yorkshire College, and Mr. William Gregson, F.G.S., was appointed our Representative on the Corresponding Society's Committee of the British Association at their Dover meeting.

In considering the arrangements for 1900, the Council recommend that General Meetings and Excursions should be held either in the neighbourhood of Keighley or Clitheroe, and that an Excursion should be taken to the Cheviots in July, to examine the rocks *in situ* from which so many of our Yorkshire erratics have been derived.

The Proceedings, Vol. XIII., Part 4, which concludes the volume, was issued to the members in August. The thanks of

e Council are tendered to the gentlemen who have written
pers, drawn plans, lent negatives, or in any other way con-
iced to the value of the part, and to Mr. W. H. Crofts, of
ull, for kindly redrawing to an altered scale the sections
ustrating the late Mr. J. Spencer's paper on the geology of
ilderdale.

Our Proceedings as usual have been forwarded to leading
cientific Societies in various parts of the world, and publica-
ons in exchange have been received from the following
cieties :—

British Association.
Royal Dublin Society.
Royal Geographical Society.
Royal Society of Edinburgh.
Royal Physical Society of Edinburgh.
Royal Society of New South Wales.
Department of Mines, Sydney, N.S.W.
Nova Scotian Institute of Science.
Royal Institution of Cornwall, Truro.
Bristol Naturalists' Society.
Cambridge Philosophical Society.
Essex Naturalists' Field Club.
Edinburgh Geological Society.
Geological Association, London.
Geological Society of London.
Leeds Philosophical and Literary Society.
Liverpool Geological Society.
Liverpool Geological Association.
Hampshire Field Club.
Hull Geological Society.
Herefordshire Natural History Society.
Manchester Geological Society.
Manchester Geographical Society.
Manchester Literary and Philosophical Society.
University Library, Cambridge.
Yorkshire Naturalists' Union.
Yorkshire Philosophical Society, York.
American Philosophical Society, Philadelphia, U.S.A.
American Museum of Natural History, New York, U.S.A.
Academy of Natural Sciences, Philadelphia, U.S.A.

Boston Society of Natural History, Boston, U.S.A.

Kansas University, Lawrence, Kansas.

Wisconsin Geological and Natural History Survey, Madison, Wis, U.S.A.

Geological Survey of Minnesota, Minneapolis, Minn., U.S.A.

Chicago Academy of Sciences.

Museum of Comparative Zoology at Harvard College, Cambridge, Mass.

New York Academy of Sciences, New York.

United States Geological Survey, Washington, D.C.

Elisha Mitchell Scientific Society, University of N. Carolina, Chapel Hill, U.S.A.

New York State Library, Albany, U.S.A.

Wisconsin Academy of Sciences, Arts, and Letters.

Smithsonian Institution, Washington, D.C.

L'Academie Royale Suedoise des Sciences, Stockholm.

Société Imperiale Mineralogique de St. Petersburg.

Société Imperiale des Naturalistes, Moscow.

Comité Geologique de la Russie, St. Petersburg.

Instituto Geologico de Mexico.

Sociedad Cientifica "Antonio Alzate," Mexico City.

Australian Museum, Sydney.

Australian Association for the Advancement of Science, Sydney.

Natural History Society of New Brunswick.

L'Academie Royale des Sciences et des Lettres de Danemark, Copenhague.

Kaiserliche Leopold-Carol. Deutsche Akademie der Naturforscher, Halle-a-Saale.

Geological Institution, Royal University Library, Upsala.

Imperial University of Tokyo, Japan.

REVENUE ACCOUNT.

Receipts.	£	s.	d.	Expenditure.	£	s.	d.
1899.				1899.			
To Subscriptions	57	4	0	By Balance due to Treasurer 1st Nov., 1898	0	15	11
" Sale of Proceedings	0	17	9	" Chorley & Pickersgill, Printing Proceedings	64	5	2
" Transfer from Capital Account	18	5	0	" Chorley & Pickersgill, Postage of Proceedings	3	5	4
" Halifax Corporation Interest	10	3	0	" Mintern Bros., Lithographing Plates	3	10	0
" Balance due to Treasurer Nov. 1st	18	18	2	" G. West & Sons, Drawing and Lithographing Plates	6	19	3
				" J. Green, Drawing Plates	6	0	0
				" Expenses of Meetings	7	15	11
				" Postages and Petty Cash	8	3	7
				" F. Carter, Stationery and Circulars ..	4	12	9
	£105	7	11		£105	7	11

CAPITAL ACCOUNT.

	£	s.	d.		£	s.	d.
To Life Members' Subscriptions	.. £18	5	0	By Transfer to General Account	... £18	5	0
To Halifax Corporation Bond .	£350	0	0				

Audited and found correct, 15th October, 1899.

J. H. HOWARTH.

RECORDS OF MEETINGS.

Council Meeting, Philosophical Hall, Leeds, April 20th, 1899.

Chairman :—Mr. J. T. Atkinson, F.G.S.

Present :—Messrs. H. Crowther, C. W. Fennell, F. W. Branson, J. E. Bedford, E. D. Wellburn, R. Reynolds, W. Gregson, G. Bingley, P. F. Kendall, and W. L. Carter (Hon. Sec.).

The minutes of the previous Council Meeting were read and confirmed.

Letters of regret for non-attendance were read from Messrs. S. Jury, W. Simpson, and G. H. Parke.

Meetings and Excursions.—A Meeting in Upper Calderdale was decided upon for the examination of the district round Todmorden, including the sources of the Yorkshire Calder, about June 7th.

It was decided to hold the Summer Meeting and the Field Excursion in the Cleveland district, from August 5th to 8th, the headquarters to be at Stokesley.

It was resolved to hold the next Annual General Meeting at Halifax.

Resignation and Elections. The Hon. Secretary reported that Mr. J. Stubbins, F.G.S., had resigned his membership of the Society and his seat on the Council. Resolved that his resignation be accepted with regret, and that the filling up of the vacancy be postponed.

Mr. J. H. Howarth, F.G.S., was appointed Hon. Auditor in the place of Mr. Geo. Patchett, resigned.

The Rev. W. Lower Carter, M.A., F.G.S., was elected Representative Governor of the Yorkshire College.

Mr. Wm. Cash, F.G.S., was elected Delegate to the Corresponding Societies' Committee of the British Association.

Accounts.—The following accounts were passed for payment :—

	£	s.	d.
Mintern Bros.—Plates	3	10	0
G. West & Sons—Plates	6	19	3
Jas. Green—Drawing Fish Plates ...	6	0	0
F. Carter—Circulars and Stationery ...	4	12	9
	£21	2	0

Grassington Finds.—The Hon. Secretary reported that the
ondition of the case and its contents was still very unsatis-
actory, and that there was no expectation of a satisfactory home
eing found for them in Grassington.

Accordingly it was resolved that as the Finds were in great
anger of being destroyed that, if they could not be suitably
oused in Grassington, they be offered to the Council of the
eeds Philosophical and Literary Society, for safe deposit in the
eeds Museum.

Underground Waters of Craven.—A letter was read from
Ir. J. H. Howarth, F.G.S., embodying a suggestion for a more
omplete examination of the underground waters of Malham and
Clapham by means of delicate chemical tests.   Mr. Howarth's
etter conveyed a generous offer from Mr. George Bray, of Leeds,
o provide materials for such an investigation.

Resolved.—That the Council approves of the suggestion to
rrange for a careful investigation of the underground waters of
Malham and Clapham as suggested in Mr. Howarth's letter, and
ppoints the members of the Council, together with Messrs.
}. Bray (Leeds), F. Swann, B.Sc. (Ilkley), S. W. Cuttriss (Leeds),
nd W. Morrison, M.P. (Malham), to form a Committee to
onduct these investigations, with power to add to their number.

———

Meeting of the Committee for the Investigation of the
Underground Waters of Craven, Philosophical Hall, Leeds, May
}th, 1899.

Chairman :—Mr. F. W. Branson, F.I.C.

Present :—Messrs. J. E. Wilson, W. Ackroyd, F. Swann, G. Bray, C. W. Fennell, J. E. Bedford, S. W. Cuttriss. G. Bingley, P. F. Kendall, J. W. Stather, E. D. Wellburn, J. J. Wilkinson, J. W. Broughton, J. H. Howarth, and W. L. Carter (Hon. Sec.).

The Hon. Secretary read the minute of the Council constituting the Committee.

Letters of regret for non-attendance were read from Messrs. W. Gregson, J. T. Atkinson, R. Reynolds, and W. Morrison, M.P.

The Hon. Secretary proposed that Messrs. B. A. Burrell (Leeds), W. Ackroyd (Halifax), and J. W. Broughton (Skipton) be added to the Committee. Carried unanimously.

Resolved.—That this Committee shall undertake the further investigation of the underground waters of Malham and Clapham.

After a lengthened conversation as to the tests to be used and the method of their application, the following suggestions were agreed upon :—

(1.) That the streams leading to the various water-sinks be gauged at the time when the tests are made.

(2.) That common salt be the chief chemical test to be used: and that a lithium test be associated with it if necessary.

Resolved.—That Messrs. F. W. Branson, W. Ackroyd, F. Swann, G. Bray, and W. Morrison, M.P., be a Sub-Committee to make a preliminary survey of the water-sinks and outlets at Malham, and to report on the best arrangements for making the necessary chemical tests.

Resolved.—That Messrs. C. W. Fennell and S. W. Cuttriss be a Sub-Committee to make a preliminary survey of the streams which feed the water-sinks and of the outlets at Malham, and to report as to the best method of gauging the flow of water at the time of the tests.

Meeting of the Committee for the Investigation of the underground Waters of Craven, Philosophical Hall, Leeds, May 9th, 1899.

Chairman :—Mr. Godfrey Bingley.

Present :—Messrs. C. W. Fennell, F. W. Branson, G. Bray, W. Cuttriss, H. Crowther, W. Ackroyd, and W. L. Carter (Hon. Sec.).

The minutes of the previous meeting of the Committee were read and confirmed.

Letters of regret for non-attendance were read from Messrs. T. Morrison, M.P., J. W. Broughton, W. Gregson, and Professor Smithells.

The Hon. Secretary proposed that the following names be added to the Committee :—Professors Smithells and Procter, and Dr. J. Cohen, of the Yorkshire College. Carried unanimously.

Reports of the Sub-Committees.—Mr. F. W. Branson reported that all the members of the Chemical Sub-Committee had met at Malham on May 27th, and had been met by Mr. Winskill, Mr. Morrison's agent, who had afforded them every courtesy and assistance.

Experiments had been made by the introduction of fluorescein at the Tarn water-sinks, but though the outlets at the Cove and Aire Head had been carefully watched, no coloration of the water had been observed.

Mr. Branson, however, reported that by passing carbonic acid gas through a solution of fluorescein the colour almost disappeared, but returned on the addition of an alkali to the solution.

The Sub-Committee had carefully estimated the percentage of chlorine in each of the streams, and the results of their calculations showed that the salt test would work very satisfactorily.

Mr. C. W. Fennell reported that the engineering Sub-committee had not yet been able to visit Malham on account of the unfavourable weather.

114   RECORDS OF MEETINGS.

It was then arranged that tests should be put in at the Smelt Mill and Tarn water-sinks, and that the outlets at the Cove and Aire Head should be specially examined.

A meeting was arranged at Malham on June 21st and 22nd, to carry out these tests.

General Meeting and Field Excursion, Todmorden, June 7th and 8th, 1899.

June 7th.—The members were met at Todmorden by Mr. Robert Law, F.G.S., and proceeded to Summit Inn by wagonette.

A visit was paid to the adjoining moorland to trace one of the feeders of the Calder to its source. After lunch at Summit Inn visits were paid to the Brick Works, the landslip at Snoddle Hill Reservoir, and to Light Hazel, Long Lees, and Warland Quarries in the Third Grits.

After dinner the General Meeting was held at the White Hart Hotel, Todmorden, under the presidency of Mr. William Cash, F.G.S.

The following new members were elected :—

A. E. Dalzell, Halifax.
Hugh Oliver, Brighouse.
Ernest George Annis, L.R.C.P., Huddersfield.
Abbey & Hanson, Huddersfield.
Robert B. Turton (Life Member), London.
Chas. H. Bould, Leeds.
Alderman A. Crossley, Todmorden.
A. C. Slater, B.Sc., Pudsey.
William Omerod, J.P. (Life Member), Todmorden.
Thos. Walshaw, Wakefield.
J. A. Bean, Wakefield.

The Chairman delivered an address.

An address was given by Mr. Percy F. Kendall, F.G. on "The Physical History of the Calder."

A paper was read by Mr. Robert Law, F.G.S., on "1 General Geology of Calderdale."

The papers were followed by a discussion.

June 8th.—The party went by train to Portsmouth Station ıd examined the section of Millstone Grit beds in Green's ough. The moor was crossed to Sharney Ford, and in scending Dulesgate some workings in the Forty Yards Mine d Ganister coal seams were visited. A foot-thick coal seam the middle of the Rough Rock was also examined and otographed.

After dinner at the White Hart Hotel, a very hearty vote thanks was passed to Mr. Robert Law, F.G.S., for his able dership.

———

Special Meeting and Field Excursion at Malham, June 21st 1 22nd, 1899, under the direction of the Underground Waters' mmittee.

After dinner at the Buck Hotel, Malham, on June 21st, '- Walter Morrison, M.P., in the chair, it was decided to put nmon salt into the Smelt Mills sink, and ammonium sulphate the upper Tarn sink. The flow of water from the Tarn was be regulated to a normal amount, but at one o'clock p.m. ı increased volume of water was to be sent down. Arrange-ents were made for the putting in of the chemicals, for the atching of the outlets at the Cove and Aire Head, and for ıe working of a central testing station at the Buck Hotel.

On Thursday, June 22nd, these arrangements were carried ut, but up to the morning of June 23rd, when the Committee eft Malham, there was no trace of chlorine beyond the normal, nd no variation in the ammonia was detected, either at the love or Aire Head.

One pound of fluorescein was also put into Tranlands Beck, ıe lower part of which was dry. No change was found either ı the main stream or at Aire Head springs, but the next day ıe fluorescein was found to issue at a spring at Scalegill Mill, bout half a mile to the south.

Before leaving the Committee arranged for more salt to be ıtroduced into the Smelt Mill sink, and for a series of samples

ı 2

to be taken from the Cove, Aire Head, and Gordale streams, and to be forwarded to Leeds for examination.

Meeting of the Committee for the Investigation of the Underground Waters of Craven, Philosophical Hall, Leeds, July 4th, 1899.

Chairman : Mr. Godfrey Bingley.

Present : —Messrs. F. W. Branson, F. Swann, B. A. Burrell, P. F. Kendall, J. H. Howarth, C. W. Fennell, S. W. Cuttriss, H. Crowther, W. Simpson, Professor Smithells, and W. L. Carter (Hon. Sec.).

The minutes of the previous meeting of the Committee were read and confirmed.

Letters of regret for non-attendance were read from Messrs. W. Morrison, M.P., and W. Ackroyd.

Reports of the work done at Malham were given by Messrs. F. W. Branson and C. W. Fennell.

After some discussion it was resolved :—That the tests be continued at the Smelt Mill and Tarn sinks with a prolonged flow of water from the Tarn, and that samples be collected at intervals for two or three miles down the stream.

The following Sub-Committee was appointed to continue the tests :—

Messrs. F. W. Branson, F. Swann, B. A. Burrell, and W. Ackroyd to form a Special Tests Committee, and in addition, Messrs. G. Bingley, S. W. Cuttriss, C. W. Fennell, J. A. Bean, P. F. Kendall, J. H. Howarth, G. Bray, and W. L. Carter (Hon. Sec.).

It was resolved that the name of Mr. J. A. Bean, of Wakefield, be added to the General Committee.

General Meeting and Field Excursion, Stokesley, August 4th and 5th, 1899.

August 4th.—The Rev. John Hawell, M.A., F.G.S., met the members at Sexhow station. The party lunched at Carlton, and ascended Carlton Bank, examining an exposure of Middle Lias and the old Alum Workings. The western branch of Raisdale was crossed and the escarpment of Cringley Moor examined. A fine exposure of the Dogger on the side of Cold Moor was visited, and a descent made by Hasty Bank and Greenhow Park to the wagonettes, which took the party to Stokesley.

After dinner the General Meeting was held at the Bay Horse Hotel, Stokesley, under the presidency of Mr. Robert Bell Turton, of Kildale Hall.

The following new members were elected :—

William H. Uttley, Sowerby Bridge.

W. N. King, P.A.S.I., Wakefield.

A. E. Greaves, Wakefield.

The Chairman delivered an address.

A paper was read by the Rev. E. Maule Cole, M.A., F.G.S., on "The Roman Roads in the East Riding of Yorkshire."

An address was given by Mr. P. F. Kendall, F.G.S., on "The Glacial Features of Cleveland."

A paper was read by the Rev. John Hawell, M.A., F.G.S., on "A Peat Deposit at Stokesley."

A discussion followed.

A vote of thanks was passed to the Chairman, the Leader, and the Readers of the papers. Mr. Turton briefly responded.

August 5th.—The party went by wagonette to Great Ayton, where workings in the Cleveland Dyke were examined, and the old spoil mounds from the ironstone workings were searched for fossils. An ascent of Roseberry Topping was made, and Ayton Moor crossed to Lonsdale, and the path through the wood taken to Kildale. Mr. Hawell pointed out the shell-marl near Kildale station, and a walk was taken through Kildale Wood to Dundale Beck, where the carriages for Stokesley were in waiting.

By invitation of the Yorkshire Naturalists' Union the members of our Society took part in their excursion and meeting on Bank Holiday, August 7th.

Meeting of the Malham Sub-Committee, Leeds Philosophical Hall, August 18th, 1899.

Chairman :— Mr. P. F. Kendall, F.G.S.

Present :— Messrs. C. W. Fennell, F. W. Branson, G. Bingley, J. H. Howarth, S. W. Cuttriss, J. Winskill, B. A. Burrell, and W. L. Carter (Hon. Sec.).

Letters of regret for non-attendance were received from Messrs. W. Morrison, M.P., F. Swann, W. Ackroyd, and G. Bray.

The report of the Chemical Sub-Committee was presented by Mr. F. W. Branson.

The report of the Engineering Sub-Committee was received from Messrs. J. A. Bean and C. W. Fennell.

Mr. J. H. Howarth read a preliminary geological report.

This report was supplemented by Mr. P. F. Kendall by explanations of the geological structure of Malham, and its relation to the flow of underground water, especially with reference to Grey Gill and the "Burst."

Mr. Cuttriss reported experiments in connection with Messrs. Burrell and Townend at Grey Gill. A flush of water had been sent down from the Tarn, but no rush of water had been heard at Grey Gill. Mr. Winskill reported a similar experiment with a larger flow of water, but with similar negative results.

Messrs. Kendall and Cuttriss gave interesting information about the structure of Grey Gill Cave, and the presence of water in it during wet seasons.

After discussion it was decided to make excavations in the screes below Grey Gill Cave and at the uppermost "Burst," in order that chemical tests might be introduced at those points; and that the Tarn water-sinks should be cleared of loose blocks so that the underlying rock might be examined.

It was also resolved that a large map of the Malham area and diagrams of the chlorine, rainfall, and waterflow curves obtained during the investigations should be made for exhibition at the British Association (Dover) Meeting.

Meeting of the Committee for the Investigation of the ıderground Waters of Craven, Philosophical Hall, Leeds, August th, 1899.

Chairman:—Mr. J. T. Atkinson, F.G.S.

Present:—Messrs. G. Bingley, W. Ackroyd, P. F. Kendall, H. Howarth, S. W. Cuttriss, W. Simpson, H. Crowther, . D. Wellburn, and W. L. Carter (Hon. Sec.).

The minutes of the previous Committee Meeting were read ıd confirmed.

Letters of regret for non-attendance were read from Messrs. '. Morrison, M.P., C. W. Fennell, W. Gregson, H. R. Procter, W. Branson, and G. Bray.

The reports of the Engineering, Chemical, and Geological ıb-Committees were presented.

The following resolutions were carried:—

1. That Mr. J. H. Howarth be requested to draw up an abstract of the report for the British Association (Dover) Meeting.
2. That Messrs. F. W. Branson, W. Ackroyd, C. W. Fennell, J. A. Bean, J. H. Howarth, P. F. Kendall, and W. L. Carter be a Sub-Committee to complete the full report and to arrange for the abstract for the B.A. Meeting.
3. That Mr. P. F. Kendall be requested to present the report to the Geological Section at Dover.
4. That Mr. Kendall be authorised to apply for a grant of £70 from the British Association for the continuance of the investigations.
5. That the question of investigations in the Ingleborough area be referred to the Sub-Committee appointed in Resolution 2.

Meeting of the Malham Sub-Committee, Philosophical Hall, ›eds, September 6th, 1899.

Chairman:—Mr. F. W. Branson, F.I.C.

Present :---Messrs. W. Ackroyd, C. W. Fennell, P. F. Kendall, J. H. Howarth, and W. L. Carter (Hon. Sec.).

Mr. J. H. Howarth read the abstract of the reports which he had prepared. After full discussion and some alterations it was adopted and ordered to be printed.

It was resolved that 150 copies of the abstract with water-flow and chlorine curve diagrams and an index map be printed and forwarded to Mr. Kendall at Dover.

Council Meeting, Philosophical Hall, Leeds, Oct. 10th, 1892.
Chairman :-- Mr. J. E. Bedford, F.G.S.
Present :--Messrs. P. F. Kendall, F. W. Branson, G. Bingley, E. D. Wellburn, W. Simpson, J. J. Wilkinson, and W. L. Carter (Hon. Sec.).

Letters of regret for non-attendance were read from Messrs. J. H. Howarth, J. W. Stather, R. Reynolds, H. Crowther, and W. Cash.

The minutes of the previous Council Meeting were read and confirmed.

The following accounts were passed for payment :—

	£	s.	d.
F. Carter (Circulars and Stationery) ...	3	11	0
Chorley & Pickersgill (Proceedings) ...	41	7	2
	£44	18	2

Annual Meeting.-- A letter was read from the Marquis of Ripon regretting his inability to attend the Annual Meeting.

Resolved. That His Worship the Mayor of Halifax be invited to preside at the meeting and dinner.

Resolved. That the meeting be held at Halifax on November 2nd, and the dinner at the Swan Hotel.

Resolved.—That an excursion be arranged to examine the Mytholmroyd drift deposits, and that Messrs. P. F. Kendall, R. Law, W. Simpson, and E. D. Wellburn be a Sub-Committee to arrange for the opening of suitable sections for examination.

Officers and Council.—Mr. R. Law, F.G.S., was nominated the seat on the Council rendered vacant by the resignation Mr. J. Stubbins, F.G.S. With this alteration the previous r's list was adopted for nomination at the Annual Meeting.

The Secretary read an abstract of the Annual Report, which approved.

Meetings and Excursions in 1900.—The place selected for first General Meeting and Field Excursion was either Sheroe or Keighley, the second excursion to be to the viots, to examine *in situ* the rocks from which it is believed many of the East Riding erratics have been derived.

Grassington Finds.—Letters were read from Mr. J. Ray Eldy imating that the Case and Finds were now housed in the wn Hall, Grassington, were in good condition, and were suit-y looked after.

It was resolved that a communication be sent to the assington Parish Council offering them the Finds if they uld have them properly taken care of and would give the orkshire Geological and Polytechnic Society access to them at ty time.

The report of the Underground Waters' Committee was resented and passed.

—————

Annual General Meeting, Town Hall, Halifax, November 2nd, 399, the Mayor of Halifax (Alderman J. T. Simpson) in the air.

Letters regretting absence were read from Lord Ripon, the wn Clerk, Borough Surveyor, and Waterworks Engineer of alifax, and Messrs. J. W. Stather, H. Waterworth, R. Reynolds, d M. B. Slater.

The Annual Report was read by the Hon. Secretary.

The Financial Statement was presented by the Treasurer.

Resolved.—That the Annual Report and Financial State-ent as presented be adopted, and that the best thanks of the xciety be given to the Officers and Council for their conduct

of the affairs of the Society during the past year: prope
Rev. C. T. Pratt, M.A., seconded by Mr. B. A. Burrell.

The following new members were elected : —

W. H. Stewart, Wakefield.
George Bray, Leeds.
J. Young Short, Thirsk.
W. B. Crump, M.A., Halifax.
Edward Collinson, Halifax.
Richard Edgar Horsfall, Halifax.
Raymond Berry, Hipperholme.
J. R. Appleyard, Halifax.

Officers and Council.—Messrs. E. Hawkesworth and
Dwerryhouse were appointed scrutineers for the ballot
Council.

Resolved. - That the Marquis of Ripon, K.G., be
President : proposed by Mr. Jas. Booth, J.P., seconded
Walter Rowley, F.G.S.

Resolved. - That the Vice-Presidents, Treasurer, Hor
tary, Auditor, and Local Secretaries as nominated be re
proposed by Mr. J. T. Atkinson, F.G.S., seconded by Mi
Sutcliffe.

Vice-Presidents :

Earl Fitzwilliam, K.G.
Earl of Wharncliffe.
Earl of Crewe.
Viscount Halifax.
H. Clifton Sorby, LL.D., F.R.S.
Walter Morrison, M.P.
W. T. W. S. Stanhope, J.P.
James Booth, J.P., F.G.S.
F. H. Bowman, D.Sc., F.R.S.E.
W. H. Hudleston, F.R.S.
Richard Reynolds, F.C.S.
J. Ray Eddy, F.G.S.
David Forsyth, D.Sc., M.A.

Treasurer:
William Cash, F.G.S.
Hon. Secretary:
William Lower Carter, M.A., F.G.S.
Auditor:
J. H. Howarth, F.G.S.
Local Secretaries:
Barnsley—T. W. H. Mitchell.
Bradford —J. E. Wilson.
Driffield—Rev. E. M. Cole, M.A., F.G.S.
Halifax—W. Simpson, F.G.S.
Harrogate—Robert Peach.
Huddersfield—Samuel Jury.
Hull—John W. Stather, F.G.S.
Leeds—H. Crowther, F.R.M.S.
Middlesbrough—Rev. J. Hawell, M.A., F.G.S.
Skipton—J. J. Wilkinson.
Thirsk—W. Gregson, F.G.S.
Wakefield —C. W. Fennell, F.G.S.
Wensleydale —W. Horne, F.G.S.

Thirteen names having been nominated for the twelve seats on the Council, a ballot was taken and the following were declared elected:—

Council:

W. Ackroyd, F.I.C.
J. F. Atkinson, F.G.S.
J. E. Bedford, F.G.S.
Godfrey Bingley.
F. W. Branson, F.I.C.
J. H. Howarth, F.G.S.
P. F. Kendall, F.G.S.
R. Law, F.G.S.
G. H. Parke, F.L.S., F.G.S.
Walter Rowley, F.G.S.
F. F. Walton, F.G.S.
E. D. Wellburn, F.G.S.

An address was delivered by the Chairman.

The Reports of the Committee for the Investigation of the Underground Waters of Malham were read:—

1. The Engineering Report, by Mr. C. W. Fennell, F.G.S.
2. The Chemical Report, by Mr. F. W. Branson, F.I.C.
3. The Geological Report, by Mr. J. H. Howarth, F.G.S.

A paper on "The Composition of some Malham V
was read by Mr. B. A. Burrell, F.C.S.

A paper on "Megalichthys" was read by Mr. E. D. W
F.G.S.

A paper on "The Glacial Geology of Bradford, i
evidence obtained from recent excavations of a limestor
on the south side of the valley," was read by Mr.
Monckman, D.Sc.

A discussion took place after each paper.

A paper on "A Contribution to the History of the
Corals of Yorkshire" was communicated by Mr. R. F
F.G.S.

A paper on "The Geology of Clapham and Distr
communicated by Professor T. McK. Hughes, F.R.S.

A series of lantern slides descriptive of the Fiel
sions to Clapham and Todmorden was exhibited by Mr
Bingley.

Resolved. That the best thanks of the Society
to His Worship the Mayor of Halifax for presiding
Annual Meeting, and for his kindness in granting th
his reception room ; also to the readers of the paper
Mr. Godfrey Bingley for his excellent exhibition o.
pictures ; proposed by Mr. R. M. Kerr, seconded by
Ackroyd, F.I.C.

His Worship the Mayor briefly responded.

———

The members dined together at the Swan Hotel, i
presidency of Alderman J. T. Simpson (Mayor of Halif

[IV.]   [PART II.

PROCEEDINGS

OF THE

YORKSHIRE

)LOGICAL AND POLYTECHNIC SOCIETY.

EDITED BY W. LOWER CARTER, M.A., F.G.S.,
AND WILLIAM CASH, F.G.S.

1901.

INGLEBOROUGH.

PART I.   PHYSICAL GEOGRAPHY.

MCKENNY HUGHES, M.A., F.R.S., F.G.S., WOODWARDIAN PROFESSOR
OF GEOLOGY AT THE UNIVERSITY OF CAMBRIDGE.

ι most studies there are two simple ways of giving a student
a of the methods and leading facts.   One is by explaining
rinciples and stating the results of observation in some
e order, generally with a view to establishing positions from
)f which the advance to the next is most easily effected.
;her method is to take some limited portion of the subject,
·oncrete example, some complex object, and describe it fully,
g such explanations of each difficulty as may be possible
ιt much previous knowledge.   This latter method is some-
employed in teaching language by attempting first the in-
·ation of selected passages instead of beginning with the
·nts of grammar, or in Science by the description of some
·ntative form.   It is always usefully employed in the case
se who have some preliminary knowledge of the elements
 subject.

In Geology, however, the method does not appear
systematically tried, though perhaps there is no oth
which this method can be so well applied or be suit
a large proportion of its students.

In the following Memoir I have endeavoured to
some notes on one well-defined area in such a manner
to me will be most useful for students who wish t
selves of this concrete method of teaching the principl

I had no trouble about the selection of a distri
long arrived at the conclusion that, of all the di
world that it has fallen to my lot to visit, there
compare to Ingleborough and its surroundings for
and variety of its problems, or the clearness and a
the evidence upon which we must depend for their

There are few subjects, moreover, with regard
term student has so wide a significance as in Geol
too young to collect, and to learn to collect intellige
too old to follow the progress of this ever-expandin
carry on the arrangement of and, with experience
speculate upon the significance of facts observed a
amassed in the earlier more vigorous years of life.

Having regard, then, to the requirements of
getting up the subject as part of their early educati
of those who wish to investigate geological phenom
selves collectively as Field Clubs or Scientific Soci
who rush off alone to take a short holiday in the
intellectual character in the open air, I offer this s
tion, in the hope that it may forward their wishes.

I have adopted the stratigraphical rather than th
arrangement, because I think it far more useful to
details of any district in that manner, and becau
those who can pay only one visit to each locality :
trouble to get up some of the details beforehand.

The position of Ingleborough is known to m
the North of England. It is the grand terraced n
whose base you run by rail all the way from Sett

Plate XXII). It is the bold bluff that travellers from Lancaster northwards see on the north and east standing out like a huge citadel in front of the fells of the West Riding. It is the great brown flat-topped mass along the eastern flanks of which the Settle and Carlisle Railway climbs, giving the traveller a final view of its northern slopes just before he plunges into the great tunnel near the source of the Ribble.

Ribblesdale, Chapel-le-dale, and the valley of the Wenning, almost enclose the mass that may be referred to Ingleborough. Its base spreads over an area of about 30 square miles. It rises

Fig. 1.

THE PLAIN OF THE HOWGILL FELLS SEEN FROM THE WEST SLOPE OF
INGLEBOROUGH.

Showing the sea-plain at about 2,00) feet above sea-level.

2,373 feet above the sea—which is seen from its summit opening out on the south-west in Morecambe Bay, between the Claughton Fells and the lower hills of Arnside and Grange.

Turning the other way, we see that it is one of many similar masses which close up to form the Great Plateau of the West Riding, Whernside and Penyghent being isolated and forming mountains more or less resembling Ingleborough, while Widdle and Dod are less completely hewn out. Only one summit dominates Ingleborough, namely, the hog-backed Whernside, which rises 41 feet higher.

The hummocky mass of the Howgill Fells rises to about the same elevation (Fig. 1), and, carrying our eye along the sky-line further west, we see range after range reaching the

A paper on "The Composition of some Malham Waters" was read by Mr. B. A. Burrell, F.C.S.

A paper on "Megalichthys" was read by Mr. E. D. Wellburn, F.G.S.

A paper on "The Glacial Geology of Bradford, and the evidence obtained from recent excavations of a limestone track on the south side of the valley," was read by Mr. James Monckman, D.Sc.

A discussion took place after each paper.

A paper on "A Contribution to the History of the Mesozoic Corals of Yorkshire" was communicated by Mr. R. F. Tomes, F.G.S.

A paper on "The Geology of Clapham and District" was communicated by Professor T. McK. Hughes, F.R.S.

A series of lantern slides descriptive of the Field Excursions to Clapham and Todmorden was exhibited by Mr. Godfrey Bingley.

Resolved. That the best thanks of the Society be given to His Worship the Mayor of Halifax for presiding over the Annual Meeting, and for his kindness in granting the use of his reception room; also to the readers of the papers and to Mr. Godfrey Bingley for his excellent exhibition of lantern pictures: proposed by Mr. R. M. Kerr, seconded by Mr. W. Ackroyd, F.I.C.

His Worship the Mayor briefly responded.

––––––

The members dined together at the Swan Hotel, under the presidency of Alderman J. T. Simpson (Mayor of Halifax).

PROCEEDINGS

of the

YORKSHIRE

GEOLOGICAL AND POLYTECHNIC SOCIETY.

Edited by W. LOWER CARTER, M.A., F.G.S.,
and WILLIAM CASH, F.G.S.

1901.

INGLEBOROUGH.

PART I.   PHYSICAL GEOGRAPHY.

T. MCKENNY HUGHES, M.A., F.R.S., F.G.S., WOODWARDIAN PROFESSOR
OF GEOLOGY AT THE UNIVERSITY OF CAMBRIDGE.

In most studies there are two simple ways of giving a student
idea of the methods and leading facts.   One is by explaining
e principles and stating the results of observation in some
finite order, generally with a view to establishing positions from
:h of which the advance to the next is most easily effected.
e other method is to take some limited portion of the subject,
ne concrete example, some complex object, and describe it fully,
ering such explanations of each difficulty as may be possible
thout much previous knowledge.   This latter method is some-
nes employed in teaching language by attempting first the in-
·pretation of selected passages instead of beginning with the
diments of grammar, or in Science by the description of some
presentative form.   It is always usefully employed in the case
those who have some preliminary knowledge of the elements
the subject.

In Geology, however, the method does not appear to have been systematically tried, though perhaps there is no other subject in which this method can be so well applied or be suitable for such a large proportion of its students.

In the following Memoir I have endeavoured to put together some notes on one well-defined area in such a manner as it appears to me will be most useful for students who wish to avail them-selves of this concrete method of teaching the principles of Geology.

I had no trouble about the selection of a district, for I have long arrived at the conclusion that, of all the districts in the world that it has fallen to my lot to visit, there is not one to compare to Ingleborough and its surroundings for the grandeur and variety of its problems, or the clearness and accessibility of the evidence upon which we must depend for their solution.

There are few subjects, moreover, with regard to which this term student has so wide a significance as in Geology. Few are too young to collect, and to learn to collect intelligently; and few too old to follow the progress of this ever-expanding study or to carry on the arrangement of and, with experience as a check, to speculate upon the significance of facts observed and collections amassed in the earlier more vigorous years of life.

Having regard, then, to the requirements of those who are getting up the subject as part of their early education, as well as of those who wish to investigate geological phenomena for them-selves collectively as Field Clubs or Scientific Societies, or those who rush off alone to take a short holiday in the pursuits of an intellectual character in the open air, I offer this small contribu-tion, in the hope that it may forward their wishes.

I have adopted the stratigraphical rather than the geographical arrangement, because I think it far more useful to work out the details of any district in that manner, and because I feel that those who can pay only one visit to each locality must take the trouble to get up some of the details beforehand.

The position of Ingleborough is known to most visitors to the North of England. It is the grand terraced mountain along whose base you run by rail all the way from Settle to Ingleton

(Plate XXII). It is the bold bluff that travellers from Lancaster northwards see on the north and east standing out like a huge citadel in front of the fells of the West Riding. It is the great brown flat-topped mass along the eastern flanks of which the Settle and Carlisle Railway climbs, giving the traveller a final view of its northern slopes just before he plunges into the great tunnel near the source of the Ribble.

Ribblesdale, Chapel-le-dale, and the valley of the Wenning, almost enclose the mass that may be referred to Ingleborough. Its base spreads over an area of about 30 square miles. It rises

Fig. 1.

THE PLAIN OF THE HOWGILL FELLS SEEN FROM THE WEST SLOPE OF INGLEBOROUGH.

Showing the sea-plain at about 2,000 feet above sea-level.

2,373 feet above the sea—which is seen from its summit opening out on the south-west in Morecambe Bay, between the Claughton Fells and the lower hills of Arnside and Grange.

Turning the other way, we see that it is one of many similar masses which close up to form the Great Plateau of the West Riding, Whernside and Penyghent being isolated and forming mountains more or less resembling Ingleborough, while Widdle and Dod are less completely hewn out. Only one summit dominates Ingleborough, namely, the hog-backed Whernside, which rises 41 feet higher.

The hummocky mass of the Howgill Fells rises to about the same elevation (Fig. 1), and, carrying our eye along the sky-line further west, we see range after range reaching the

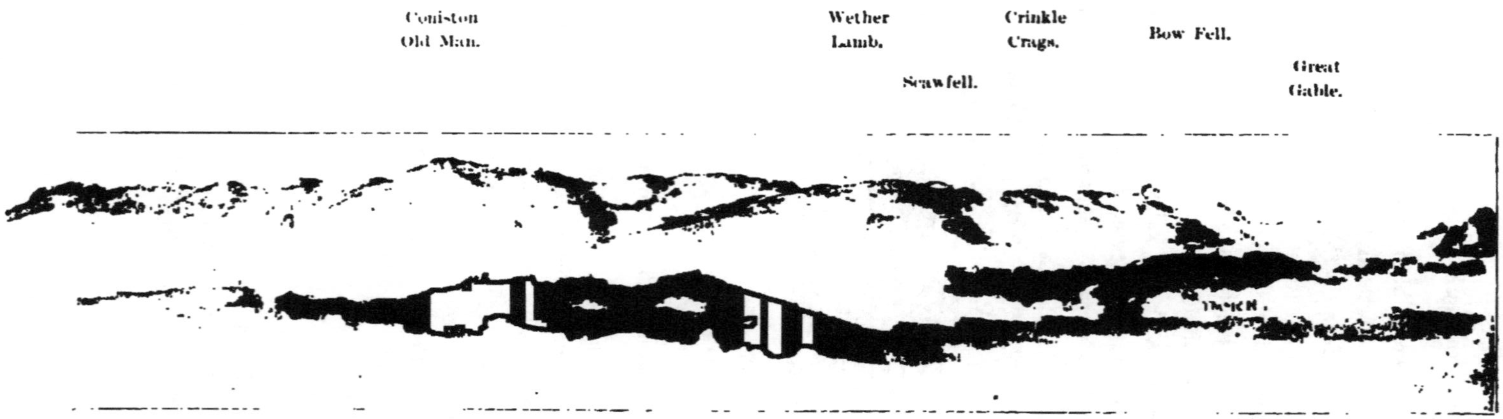

Coniston
Old Man.

Wether
Lamb.

Scawfell.

Crinkle
Crags.

Bow Fell.

Great
Gable.

Fig. 2.

VIEW OF THE SOUTHERN GROUP OF LAKE MOUNTAINS FROM ESDMOOR, SEVEN MILES SOUTH OF KENDAL.

Showing the sea-plain at a little over 3,000 feet above sea level.

ie general level up to the base of the Lake Mountains. These
their turn are obviously fragments of a higher plateau, the
rage elevation of which is some 3,000 feet above sea level;
le that on which we stand and which runs up to the base of
higher Lake Mountains is a little over 2,000 (Fig. 2).

What is the origin of this very marked feature in the land-
>e? We know from the geology of the district that none of
se mountains to which I have called attention, either of those
ch touch the level of the Lake District Plateau, or of those
ch belong to the West Yorkshire Plateau, owe their present
line to original deposition. Nor is it due to any hard bed
rn to which denudation, sub-aerial or mixed, had reduced the
eral surface level. Not only have the valleys which separate
> mountain from another been scooped out by various agents
denudation, but the tops have been planed off by denudation
some kind, and we do not in any case see the original highest
ds.

In regarding these great plateaux, we are clearly face to
ce with some phenomenon connected with the greater operations
: Nature — something upon which depended the modelling of
ir highest mountain groups, and the interpretation of which
ight to give us the key to the great succession of events of
hich Geology treats.

There are, however, certain complex operations that come
der our observation at the present day which will fully explain
e existence of relics of wide-spread plateaux of this character.
ong the shore we see the waves twice a day in accordance
th wind and tide and local conditions rolling along the *débris*
at falls from the cliffs, or is carried down by streams. It uses
> boulders and pebbles and sand as ammunition with which
batter down the rocks. It carries on this waste only to
depth of some 60 or 100 feet below sea level, for deep ocean
rrents do not contribute much to this sort of work.

The sea is always at it, and, if the relative level of land and
ter remained steady, all the dry land would in time be carried
wn and spread out below the waters of the sea. There is

plenty of room for it all there, for it would take about 36 times all the land that stands above the sea to fill up its bed.

The waves plane all down for some 60 or 100 feet more, and there the surface is protected from further waste.

The plain of the Yorkshire Fells is one of these sea-plains now lifted up some 2,000 feet more or less above sea level. If we want evidence that it is not merely a stage in sub-aerial waste, determined by the same hard and widespread bed which has arrested the action of the frost and ice and rain and streams, we have only to examine the sequence of rocks more closely to obtain the proof. The bed that forms the hill tops is not always the same. Even from Ingleborough to Penyghent we creep on to higher beds, and if we trace the beds further north we shall still find different members of the series capping the Fells. The evidence is clear enough, even in the nearly horizontal strata of the Carboniferous rocks of the district north of Clapham. But we have further proof, and clearer, if we just cross to the north-west from Ingleborough on to the Silurian Fells near Sedbergh (Fig. 1). There the rocks are no longer nearly horizontal, but roll up and down in faulted folds, yet the tops of the hills are all planed off to the level of the sea-plain, which is touched by the Carboniferous Fells north of Clapham. From that it may be traced always at about 2,000 feet above sea level to the base of the Lake Mountains, which are an island, itself the last remnant in that part of England of a now higher sea-plain, undulating at about 3,000 feet above sea level. This is not the only example of these two sea-plains. In Wales the lower or 2,000 foot plain touches the tops of all the higher mountains of South Wales, and, leaving Plinlimmon as an island in Central Wales, laps round the higher or 3,000 foot plain of Snowdonia, just as our 2,000 foot plain runs up to the 3,000 foot plain of the Lake District.

Surely we have here a grand subject for further research. What is the age of these two plains ? What basement bed derived its pebbles from the shore of the sea that arrested denudation at the 2,000 foot plain of Yorkshire ? and what formation was laid down in the sea that crept across the 3,000 foot plain of Cumberland? Which way did that sea advance, from north, or south, or east, or west?

Geologists, who regarded these phenomena chiefly from the point of view which is forced upon us as we stand upon the coast line, and watch the tremendous power of the waves as they batter the cliffs and lash the shore, called the level surface so produced a *Plain of marine Denudation.*

But there is another set of agents at work reducing all the protuberances of the earth's surface. The air, and rain and rivers, and glaciers, and changes of temperature and moisture are breaking, dissolving, transporting everything down to the sea. There it can do no more; the sea arrests all further sub-aerial waste and reserves for itself the work of removing the last 60 or 100 feet. Geologists who have regarded the great work of degradation chiefly from this inland point of view, have called the level surface down to which the whole land is thus reduced by sub-aerial agencies the *Base-level of Erosion.* Another name suggested by American Geologists for it is *Peneplain;* a word which they would define to mean a region of faint relief, the penultimate result of long-continued action of denudation on a once larger land-mass, whose ultimate result is a base-levelled plain.

Of course it is to both of the agencies above mentioned, acting simultaneously throughout long ages, that we must refer the tremendous results that we have forced upon our attention as we look around from the top of Ingleborough. We will refer to these great plateaux by the shorter term *Sea-plain;* to distinguish them from the *River-plains* or *Bed-plains,* of both which also we have examples round Ingleborough.

It is possible that there might be traces of the action which formed these sea-plains. Fissures filled by *débris* of the Poikilitic and Jurassic sea were found by Charles Moore in the Carboniferous Limestone, near Bristol.

Why should we not find in cracks and fissures on the top of Ingleborough, or of some other parts of our ancient sea-plain, the *débris* washed in by the sea that reduced them to this level?

The sea that planed off the top of Ingleborough lashed the rocky base of the Lake Mountains, which then, however, did not rise more than a thousand feet above its level. But it is to be feared that the denudation which has been going on ever since that time has completely swept away whatever traces were left upon that rocky shore. However, there we see shores which were washed by the sea that planed off the top of Ingleborough.

We have, as I have already pointed out, another fragment of that ancient land in North Wales, where the Snowdonian group represents it, attaining about the same height, viz., a little over 3,000 feet, while all round it there stretches the great 2,000 foot sea-plain, corresponding to that of which the top of Ingleborough forms part.

Just consider for a moment what this means. To reconstruct the upper sea-plain so as to unite Snowdonia with Lakeland you must put back 1,000 feet of rock over all the north-west of England and the whole of Wales, as well as over the intervening sea, in which the mountains of the Isle of Man are the only relic of the former extension of either sea-plain over this area.

Now, denudation implies a corresponding deposition. Where is the great formation built up of the material carried down to the sea when the Ingleborough sea-plain was formed? It must be later than Carboniferous, because Carboniferous rocks were being planed off. Was it Jurassic with its Poikilitic basement bed, or was it Cretaceous, or does it belong to that age of volcanic activity and vast denudation, the Miocene? Or must we refer it to the time of great erosion which immediately preceded glacial conditions here, and so make it correspond to the Osarkian or to the Champlain of America?

As we look out south over the range of hills that trends away to the east from Lancaster and to the flat-topped isolated mass of Pendle, we ask, have we here other outlying fragments of our great West Riding Plateau? But we soon find that their summits do not attain to anything like the level of Ingleborough. This difference of elevation is too great to allow us to consider them now as part of the West Riding Plateau, but may we peculate upon their having originally formed part of the same lateau? And, if so, we have to admit that since the planing ff of this great sea-plain all the land south of the Craven faults as dropped many hundred feet. Mr. Tiddeman is of opinion 1at this downthrow was going on in Carboniferous times. f we could prove that the Claughton Fell level belongs to the Vest Riding sea-plain, then we should have to admit that the ownward movement on the south, whether by fault or by radual southward slope, or by both, still went on long after he formation of the newest Carboniferous rocks.

The Ingleborough sea-plain is newer than the great faults hat run from the Eden Valley down Ravenstonedale to Lunes-ale, for the sea planed across the faults that throw the Car-oniferous rocks of Mallerstang and the rest of the Yorkshire 1oorlands on the east against the Silurian of the Howgill Fells n the west (see Fig. 1), leaving them both as parts of the ngleborough sea-plain.

If we could make a guess as to the approximate age of the ngleborough sea-plain, to what age can we assign the much more ncient sea-plain of Snowdonia and Lakeland? It is a joy to lie n a clear day on the top of Ingleborough and think these uestions out.

If the origin of these sea-plains is such as I have described, hey must be part of the most constant and continuous operations f Nature. They must be always in process of formation, and nust always have been formed. We ought, therefore, to find races of them in the rocks.

Here on Ingleborough, from which we have the clearest view f two wide sea-plains which form part of the existing surface of he ground, we have also the most stupendous exhibition of a imilar plain belonging to a far more remote Geologic past, and, s we were able to trace evidence of the newer plains far afield, nd even to find in Wales representatives of both our higher and ur lower sea-plain, so also we have satisfied ourselves by an xamination of the crags round Ingleborough that there is a imilar sea-plain buried under the mass of Carboniferous rocks of which it is built up; and we are able to follow the sea bottom of which it formed part to closely adjoining districts, where valleys were filled up by *débris* from the ancient plain, and further out still to where vast deposits were being accumulated beneath the sea in an area of depression long before that sea had swept across the bare rock on which later the deposits were heaped up out of which Ingleborough was carved out.

We step down from the top of Ingleborough on to the great shelves and ledges of the mountain limestone (Plate XXII.), the explanation of which we will consider later on. We then cross the whole of the Mountain Limestone down to its base, and there we find this other sea-plain of far more ancient date than that on which we stood on the top of Ingleborough.

Here we see the Carboniferous rocks resting on the up-turned edges of all the older rocks that make up the country between us and Helvellyn. Here we can study the character of the surface of that old sea bottom. Generally speaking the rocks were evenly planed off, but the tougher rocks, such as the gritty sandstones of Austwick, or those that presented the bed faces to the waves so that they could not be undermined, resisted the various denuding agents more than the slaty or differently inclined beds, and remain in long ridges. We can follow the base of the Carboniferous rocks along the sides of Dale Beck and Moughton and Ribblesdale, and often see that these ridges run through with the strike of the rocks from one of those valleys to the other. It was generally a clean wave-swept surface, with few troughs in which the *débris* from the land could be caught. But there are a few hollows, and those very suggestive. In the first place we notice that finer material is preserved in the deeper depressions only, but sometimes very large boulders remain on the flat, rocky sea bottom, as if the last current had been strong enough to carry away the finer material, but not to remove the large blocks, or the gravel sheltered by them. Curiously enough these are seen in the base of the Mountain Limestone under Norber Brow, on the top of which isolated boulders of Glacial age are perched and challenge comparison.

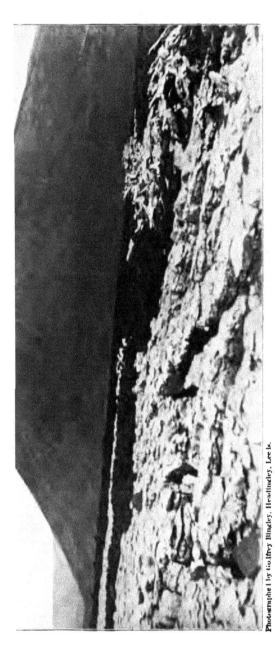

Photographed by Godfrey Bingley, Headingley, Leeds.

THE SUMMIT OF INGLEBOROUGH FROM RAVENSCAR.

When we come to examine the basement deposits of the
ntain Limestone in detail we find many other curious facts
require explanation.   For instance, we find that where the
ghs or valleys in the old sea bottom are considerable, the
sit that fills them is generally red, but the conglomerate or
stones that occur above the level of the depression or extends
where over the general surface of the older rocks is never red.
This is first of all a question for the chemist.   Where we
find beds containing carbonate of lime in red rocks, as in
cornstones of the old red sandstone, or evidence that there
been carbonate of lime present though now dissolved away,
ı the lenticular beds full of casts of fossils in the rusty red
iary beds of Kent, these subordinate beds are green.
Therefore we should not expect to find red beds in the main
of the Mountain Limestone, except the red earthy residuum
to late action of surface waters.
It seems not improbable that in the case before us the
ized superficial deposits of the adjoining pre-Carboniferous
when they were swept down and preserved at once in a
y or trough, some of which were perhaps sub-aerial, retained
red colour, but that when the material had long been
ed in the surf of the encroaching sea, all the little pellicles
əd oxide which coated the grains of sand were removed, and
whole mass lost its colouring matter.
Most of the carbonate of lime was probably derived from
nisms which grew on the spot, but the other sediment seems
ave come from far and to have been well rolled and washed,
ıay be inferred from the constant recurrence of bands of
tz pebbles in the lower part of the Mountain Limestone, as
seen near Thornton Force, for instance.   They had travelled
ır that all the softer sedimentary or other rock in which the
tz veins occurred had been ground down to sand and mud,
only the quartz, rolled into small, perfectly smooth pebbles,
ived the long journey.   The various forms of life so abundant
he overlying limestone had not yet migrated into this area ;
efore fossils are rare in the red *débris* swept into the hollows.

There was no muddy bottom yet for shells to live on, and corals were scoured off the exposed bare rock surface. Only a few fish swam over and left their remains entombed. In one of these hollows I found *Lophodus levissimus* L.Ag. and *Copodus cornutus* L.Ag.

It is interesting to note that these are closely akin to well-known Devonian forms, if not identical species.

We ask what was going on over the surrounding district when this sea-plain was being formed here, and how far does the sea-plain extend? We soon find some evidence bearing on this point as we follow it to the north and west. Instead of resting on a smoothly undulating surface of bare rock, the Mountain Limestone has at its base enormous banks of gravel and sand now deeply stained red.

Along the valley of the Lune, by Kirkby Lonsdale, Barbon, Sedbergh, and Tebay, it lies in a manner suggestive of a long valley with tributaries coinciding in direction with the present drainage system, but it does not resemble the gravel of river terraces. At the mouth of Ullswater high hills are wholly composed of it, and it plunges under the Carboniferous rocks on the east. Here it seems probable, from the form, the composition, and arrangement of the material, that the gravel carried down the steep valleys out of the heart of the Lake Mountains was, some of it perhaps, distributed along the shore, but was mostly swept into the seaward depths of the submerged valleys. The material seems always to be derived from the neighbouring rocks. In these thick masses of sediment no fossils have been found, except derivative fossils in the fragments of older rocks.

Follow the base of the Carboniferous across the sea to the Isle of Man and to North Wales, and we find evidence of similar conditions having prevailed there.

But if we travel on into South Wales we get beyond the sea-plain and its marginal valleys, and there find widespread sands, always with white quartz pebbles, underlying the Mountain Limestone and passing unconformably across the Old Red of Herefordshire and the Silurian and Bala of Carmarthenshire.

Why fossils are so scarce, so obscure, and of such small use
urposes of correlation in these South Wales beds I cannot
but in that respect also they resemble our pockets of con-
:rate, sandstone, and shale under Ingleborough, or the great
onglomerate of the borders of the Lake District.

Cross the Bristol Channel, and there you find the equivalents
ie "Brown stones" of Carmarthenshire split up by shales
limestones in which there are plenty of fossils, but in the
sandstones of the Devonian they are more scarce.

Now, we have got a suggested correlation.

The Old Red of Hereford is not Devonian, but the Devonian
bergavenny creeps across the Old Red unconformably. There
a mountainous district in the North of England and North
s, while South Wales and Devonshire sunk beneath the
nian Sea. By-and-by that sea cut down the mountains of
northern part of our island and determined the level of the
lain on which Ingleborough rests while the Lake District
stood above it.

If this be the true story, this sea-plain at the base of Ingle-
igh is of Devonian age, and the pockets of sediment which
i it are the last bits of sediment which can be referred to
Devonian before the widespread changes which then took
ushered in the conditions which allowed of the deposition
ie Lower Carboniferous Series.

Many are the questions in Physical Geography which may
udied in this wonderful district.

One to which I have had to refer, and which seems at first
: connected with the sea-plains which we have been consider-
is the origin of the ledge of Mountain Limestone on which
ipper half of the mountain rests as on a table (Plate XXII.).
l the mass of Yoredale shales, sandstones, &c., and overlying
stone Grit were removed the resulting feature would be much
what we have now, namely, a capping of hard rock which
it or might not have a corresponding flat top seen on the
ining hills. What we have to look into is this: Is this
ice of limestone the margin of a sea-plain which washed the

base of the steep slopes of Ingleborough, Whernside, and Penyghent, just as we have inferred that the sea which planed off the top of Ingleborough lashed the base of the Lake Mountains! We have here the question of the distinction between a sea cliff and an escarpment. Fortunately, the example before is easily studied. The rock is one that records the evidence in the most satisfactory way.

The Mountain Limestone everywhere yields along the bedding planes so as to give rise to a bare jointed surface locally known as Clints or Helks, but this is especially the case at the top of the formation where the overlying Yoredale shales are swept off the hard limestone platform on which they rest. If we examine the limestone where newly exposed we notice that the joints are closed and the surface smooth, but, as we leave the margin of the covering deposits, the joints are more and more opened out until the top bed is represented only by a series of long, bolster-like masses, the crevices between which commonly extend down through bed after bed to a depth of from 5 to 15 feet (Fig. 3). In the deep shadows of these fissures, into the bottom of which the heat of the sun never strikes, many a rare fern and flower grows, and every here and there we find a line of funnel-shaped holes opening out into channels in the mass of the rock below. These swallow-holes or pot-holes are apt to lie in rows, each set at a corresponding distance from the margin of the impervious shale or clay that rests upon the limestone. Elsewhere we see how they are formed. The water that falls on the fissured limestone runs into the joints, where it falls and never can be gathered into runlets. That which falls on the impervious beds above forms streams and rivulets, and where these reach the cavernous rock they open out the fissures, and soon make a way for themselves by chemical and mechanical action, and rush out through caves of their own making to join the rivers in the valley below.

Sometimes the accidents of the mode of distribution of the drift and other superficial deposits, or the occurrence of a belt of broken rock have caused the stream to seek the same inlet long

ər the impervious shale has been cut back far from where it
ended when the hole originated.   But, as a rule, the first set
holes is deserted and left dry when, by the cutting back of

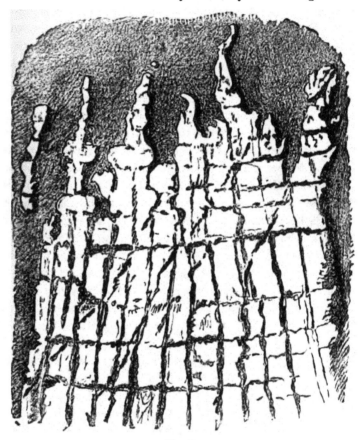

Fig. 3.

GROUND-PLAN SHOWING MODE OF WEATHERING OF MOUNTAIN LIMESTONE.
(1 inch = 2 feet 8 inches.)

[*By permission from the Quart. Journ. Geol. Soc., Nov. 1886.*]

impervious beds, the streamlets gain access to an inner circle.
us row after row of such holes remain witness to the constant,
ugh intermittent, recession of the impervious beds on which
water collected that formed them.

This is a sub-aerial action.   It is not the way the sea acts
on the limestone on the shore where the water cannot seek lower
levels through the saturated rock.   This, then, is evidence that
the steep slope above the terrace of limestone is an escarpment
and not a sea cliff.

There is other evidence also.   If we go to the Howgill Fells,
for instance, on the top of which we found the extension of
the West Riding Plateau so clearly marked (see Fig. 1), and
where, if the limestone shelves of Ingleborough had been due to
the sea, we might expect to find also some traces of the action
of shore waves, we cannot detect any cliff corresponding to that
which follows the top of the Mountain Limestone of Ingleborough.
Moreover, the base of the cliff always corresponds with the top
of the Mountain Limestone, even when the movements of the
strata have thrown that sometimes much higher, sometimes much
lower, while local differences in thickness, texture, and composition affect it irrespective of level.   So that for all these reasons,
viz., that the surface of the bared limestone shows evidence of
the gradual sub-aerial cutting back of the overlying shales; that
there are no ancient sea cliffs at the corresponding level in the
neighbouring Silurian mountains; and, further, because in this
region the level follows the rise and fall of the base of the Yoredale Rocks, and does not appear to have cut horizontally across
whatever bed was there, as should be the case were it a sea cliff,
we must infer that the steep slope of the Yoredale Rocks above
the terrace of Mountain Limestone on Ingleborough is a sub-aerial
escarpment and not a sea cliff.

What part in its formation was played by ice action we
must reserve till we come to the consideration of the glacial
phenomena.

In speaking of these great expanses of level rock, we have
had so far no occasion to refer to river-plains.   Yet we are not without the most striking examples of river denudation round the base
of Ingleborough.   The transverse strath, drained by the Wennin
that bounds it on the south, affords much matter for inquiry or
speculation.   But before we speculate upon its origin, let us look

down from the crags round Ingleborough upon the long, straight valleys of Kingsdale, Chapel-le-dale, and Ribblesdale, which are easier of explanation. These three valleys represent three stages in the history of the cutting back of valleys into a mountain mass. The Carboniferous rocks dip gently in a northerly direction, so that the rim of Mountain Limestone is higher and higher the further south we trace it. If, then, anything should sweep the surface of the Mountain Limestone, or of any one and the same bed in it, quite bare, the water would accumulate on it, deepening to the north, until a gorge was cut back from the rim to tap it.

Thus we find in Kingsdale an alluvial flat on one of the lower horizons of the limestone, and the water cutting a little gorge through the rim at Thornton Force (Plate XXIII.), and so eating its way back to tap the valley above. In the case of this great jointed limestone the water does not all wait till it reaches the fall, but, working down into the cracks and opening them out by chemical and mechanical action, often carries all the water away through the crevices so formed, while the water tumbles over the top of the rock only when, after heavy rain, there is more than the subterranean channels can carry. At the north end of Chapel-le-dale, where the valley changes its character and the limestone is much covered by impervious drift, this action is very striking. The greater part of the water of the stream is generally lost in a grand chasm known as Weathercote Cave, and only in very heavy floods fills this to the brim, and overflowing runs on through the surface channel. The water that disappears in the gravel at the bottom of Weathercote Cave boils out below in Jingling Pot and Hurtle Pot, and supplies the stream that runs down Chapel-le-dale.

On the floor of Kingsdale there is Mountain Limestone. In Chapel-le-dale, however, denudation has removed all the limestone, and the valley lies on Silurian and Bala, and perhaps some older rocks. The basement bed of the Carboniferous is seen some way up the hill on either side. But as the rim of the Mountain Limestone arrested denudation and let the stream wind about

B

from side to side, alluvial flats were produced here
where there were in old times small tarns, and where
rain flooded meadows may now be seen.

The same conditions must once have prevailed
dale. There, however, the rim of limestone has
cut back, but the occurrence of beds of greater
in the older rocks has kept up the barrier, and
from the first falls below Dale House, by
cascade, to join the Greta at Ingleton.

On the other side of Ingleborough there is
valley, now covered by the pretty artificial lake
constructing a high dam just above the village of

Further east we find the small valley of Cram
(Plate XXIII.), which is of the same type as Cl
This valley lies chiefly in Silurian and older rocks, be
much easier to see the reason for each interruption in th
of the features and for the barrier which arrested
back of the stream at its lower end. Hard bands of
so as to present themselves to the denuding agent in
that made them least accessible to its action, are se
the valley near White Stone Lane and barring the
of Southwaite.

One little tarn has been filled up with shell mar
and as these are both useful, the one for dressing th
other for fuel, excavations have been made which
whole story of their origin and infilling.

Further east still is Ribblesdale, with its barrier
Moor and Great Stainforth, and rapids and water
This valley has been cut down further through
horizontal limestone rock. Part of it was certainly on
by a tarn, in which the trees and nuts brought down
drifted chiefly to the south-west corner, where they are
found in the peat. Much of this valley is covered by
alluvial mud, till we follow it up to the great mass
matter about Horton which the river has not yet h
remove or level.

THORNTON FORCE.

CRUMMACK DALE.

These dales running north into the base of Ingleborough
·nish clear examples of almost every type of river denudation.
ere are the rapids and waterfalls cutting back to alluvial
iins over which streams wandered about, widening their valleys,
t doing little to cut their beds deeper ; there are the lakes
)wever caused) being filled with alluvium and peat.

Although they are on a small scale, all the most important
·enomena are represented here, and thus we have as part of
gleborough examples of the sea-plain, of the bed-plain, and of
e river-plain all clearly defined and accessible.

I have not touched upon the effect of glacial action on
iese features which are primarily due to other causes, but we
ust remember that the ice of the glacial epoch gathered on the
ights of the old sea-plain, crushed its way over the Mountain
imestone ledges round the flanks of Ingleborough, and, later on
ill, pushed long fingers of ice down the deeper valleys, leaving, as
receded, the moraines which still determine or modify denudation.

I cannot name any district in which the ordinary details
denudation and many of its more exceptional operations can
so well studied as within the area which we have included
ler Ingleborough or in its immediate neighbourhood.

The condensation of the moisture of the winds on the cold
:s working where no rain can reach ; the action of water
e or less charged with acids on the limestones ; the fantastic
i s which are thus produced ; the effect of this action on a
>r scale in the formation of pot-holes, and of underground
i nels and of valleys by the falling-in of caves ; all claim
attention of the geological student. The breaking up of
t masses of jointed rock under the influence of frost and
masses that in the thaw are carried over the frozen snow
fills the place where the talus should rest, and form
entic masses lying some way in front of the cliff ; the cut-
back of gorges by the removal of block after block, first
iched by complex denudation, then lifted out of their bed
the hydrostatic paradox and hurled over the edges of the
; all these, too, can be studied here.

The pre-Carboniferous sea-plain was of course neve
surface, but the harder rocks stood out like the enam
tooth of an elephant while the softer dentine was wo
We cannot, however, assume that the ridges of Silu
Bala rocks which we observe to-day exactly measure the
inequalities of the sea bottom, but they must appro
represent the relative heights. In Chapel-le-dale they
little short of 900 feet, and only fall to between 700
feet. In Crummack Dale, where the tough Silurian se
cross the valley, they rise to only a few feet short of 1,
but where the Bala shales come out from below the
south of Norber, the ancient surface on which the Carb
rocks were deposited falls to 700 feet above sea le
Fig. 4).

Similarly in Ribblesdale the tough sandstones are
throw the base of the Carboniferous rocks up, while tl
beds form troughs into which the earliest Carboniferous
was swept. Thus the grits and sandstones of Great S
rise to near 1,250 feet above sea level.

From these observations we should infer that the tr
valleys, such as that on the north side of and parallel to
Lane between Clapham and Austwick, or that along wl
road from Austwick to Stainforth runs on the south
Moughton Scar, really represent pre-Carboniferous E.S
W.N.W. valleys in the softer beds of the Silurian an
The great height of the base of the Carboniferous in M
Scar, above the general level of the Silurian and Bala
the low ground between Wharfe and Swarth Moor (see
is not, therefore, a proof of great denudation along tha
verse valley since those beds were exposed, but the
surface probably represents very nearly the original pre-
iferous sea bottom.

The difference of level corresponds almost exactly wi
seen in section along the west side of Crummack Dale, wl
base of the Carboniferous rocks falls from nearly 1,200 fee
west of Crummack to 700 feet south of Norber (see Figs. 4

Scar ½ mile SW of Crummack

N

Norber

Fig. 4.

Wharfe   700   800   900   1000 1100   1200   Houghton

N

Fig. 5.

The north and south valleys, Kingsdale, Chapel-le-dale, Crummack, and Ribblesdale, have apparently suffered considerable glacial
and post-glacial erosion.

FAULTS.—The existence and effect of faults has often been
mentioned in describing the relation of the various formations to
one another, but the phenomena connected with faults call for
special treatment.

Nowhere, even in this district, can the behaviour of the rocks
along a great fault be so well studied as in the gorge of the Twis
or Greet above Ingleton, at the sharp elbow made by the stream
where it is caught in the crushed and fissured rock and carried
out of its southerly course for a quarter of a mile or so to the

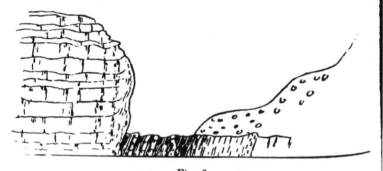

Fig. 6.

SECTION ACROSS THE TWIS OR GREET ABOVE INGLETON.

south-east. The Bala limestones and shales are exposed along the
bed of the stream, and pass under a great mass of drift on the left
bank, while on the right bank, which is the downthrow side of
the fault, the Mountain Limestone stands in a wall some 200 feet
high, above which rises a precipitous broken slope for about
200 feet more before we reach the level of the broad limestone
terrace through which the gorge has been cut (see Fig. 6).

The upper part of the face of the limestone cliff overhangs
its base in agreement with the inclination or hade of the fault
to the downthrow side, and sweeps down stream in bold curves
which represent the original winding course of the fault. What
is most striking in this section is the swelling irregular surface

the exposed side of the fault, showing how much the hade and rection of a great fault may vary within short distances. The ck is not often much shattered, but from its condition where runs into the hill near the elbow of the stream, and from the ndency to flake which is seen on many parts of the face, it ems probable that there were some outside crushed layers having rough cleavage parallel to the plane of the fault, but that these ave been removed by denudation as the gorge was being cut own.

The exterior portion of the limestone near the fault-face sumes a brownish-yellow colour, and is in places honeycombed r weathered into irregularly rounded cavities such as might be lled by geodes. A chemical examination of the changes in the fountain Limestone here as it approaches the actual fault would robably yield some interesting results.

The Bala Beds, on the other hand, being composed chiefly f shale instead of massive rock, are crushed and twisted in all lirections, and the harder bands are thrust through the softer. Several dykes traverse the series, and from the manner in which he soft shales are moulded round them it is clear that they also, being of a more unyielding nature and unable to accommodate hemselves to the general kneading up of the mass as readily as he shales, were broken and thrust in among them.

This proves that they were intruded at an earlier date than he movements which crushed up the Bala Beds, that is, they nust be earlier than the fault. As the cleavage of the Green Slates and Coniston Limestone series was contemporary with the olding by which they were upturned, the dykes, being somewhat guided in direction by the cleavage plains, would therefore appear o belong to that enormous interval during which the folded Green Slates and Coniston Limestone series were being reduced o the "peneplain" on which the Carboniferous Rocks were leposited.

The crushing that the dykes themselves have undergone is shown in the veins now filled with carbonate of lime which traverse them. This is especially noticeable in the tough grey

felspathic rock with small flakes of black mica, which forms one
of the three principal dykes seen here.

We must not imagine that the faults in a district like this
took place suddenly.   These faults are merely easements during
the folding of the rocks, and therefore, as the folding was a slow
process, commensurate with the great denudations that planed of
the land as it was raised, and with the sedimentation which was
the necessary accompaniment of that denudation, so the faults
must have been going on continuously or spasmodically while
deposits were being laid down.   They, however, may indicate
periods of more rapid movement, in which the rocks which had

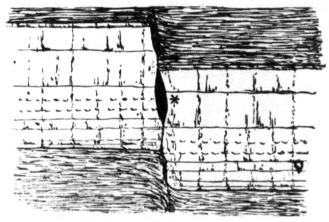

Fig. 7.

not time to bend must break, and they also point to greater
inequalities of surface and generally suggest periods of locally
changing conditions.

Perhaps there may be some reason for suspecting that the
Dam House Bridge fault near Austwick was going on during
the accumulation of the basement beds of the Silurian from the
difference in the character and thickness of the conglomerate on
the north or upthrow side of the fault as compared with that on
the south side, and from the great paleontological change which
marks the incoming of the Silurian and points to altered geographical conditions.

The fault which crosses the top of Ingleborough has a very small displacement. Indeed, it may be only a shake and crack over a pre-Carboniferous fault below. But it is well worth careful examination, as it has obviously affected denudation, and yet runs over the highest ground in the district. Crossing the south end of the Millstone Grit, it cracks the Main Limestone and the Yoredale Grit, and can be traced as a long peat-covered hollow across the Mountain Limestone below.

A point which is specially deserving of careful attention is forced upon our notice by an examination of the great cliff, which represents one wall of the fault in the Twis valley. If two curved

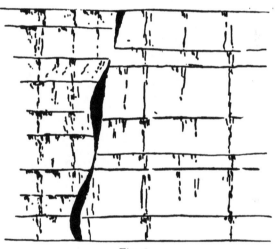

Fig. 8.

rock faces of that kind are relatively shifted, it is clear that the protuberant parts must often hold the walls of the fault apart and spaces be left which, if not filled with crushed material, will offer suitable conditions for the precipitation and crystallisation of mineral matter (see Fig. 7). As a matter of fact, we do find veins and lodes commonly occurring in lenticular cavities which appear to be formed in just that way. Sometimes when the fissure cuts across strata of various degrees of compressibility the more yielding beds are squeezed out, filling the crack completely and even to some extent creeping up between the harder strata; while new minerals are formed between the two walls of the fault only in the lenticular open spaces where the more solid strata hold the opposing sides apart, as shown in Fig. 8. Hence in such a case we have galena, for instance, occurring along the lode in such a place as that indicated by the * in Fig. 7, whereas in the intermediate stages the fault is entirely closed and no ore is found.

THE GLACIAL GEOLOGY OF BRADFORD,
ṬD THE EVIDENCE OBTAINED FROM RECENT EXCAVATIONS OF A
LIMESTONE TRACK ON THE SOUTH SIDE OF THE VALLEY.

BY JAS. MONCKMAN, D.Sc.

During the past two or three years the building trade in
ḷford has been very brisk, and consequently digging for
dations and drains has been extensively carried on, with the
ḷt that large quantities of glacial material have been exposed,
ɘ of it in quite unexpected positions.

The largest and most important of these masses is at the
of the west side of the hill in Great Horton. It extends
ɩ Grange Road on the one side to Great Horton Station on
other, and from the Westbrook to the Escarpment of Horton
ḷ.

On the Six-inch Geological Map this is marked as a sand-
ɘ outcrop, and in fact it had all the appearance of such, but
ɴ the builder proceeded to remove the soil and fill up the
ɔy denuded by the brook, there was no stratified rock, but
ɘad a great thickness of boulder clay. At the north end
ʳ removed 12 feet of this, and then dug 10 feet lower to
ɩ a drain, but did not get to the bottom. Further south
excavation was not so deep. On the opposite side of the
ḷk there is not much clay, and this rapidly thins out, making
ɔpear that a pre-Glacial valley has been filled up, and after-
ḷs partly worn out again by the present brook.

The workmen, in removing the clay, threw the boulders into
s and ridges, so that I had a large collection of material
ɘrately conveniently arranged for examination.

The boulders were chiefly sandstones, with abundance of
of very coarse texture; there were also numbers of red
ɩstones (fine) and grits (coarse), dark-coloured limestone was
mon, light-coloured rarer, but still in considerable quantity,
stones and shales abundant. I obtained about a dozen speci-
ɩs of Silurian grit, two specimens of banded limestone, and
of chert with shale.

The upper part of this boulder clay was yellowish, and the lower blue. There was no appearance of stratification except at one place, where two different kinds of material were laid together in a rough sort of stratified deposit.

At Lidget Green specimens of limestone were found in a blue clay at and near the corner formed by Legrams Lane and Beckside Lane, where they dug for the foundations of the new premises of the Co-operative Society, also in the excavation a little further along the road towards Bradford (21 Note), while to the north the clay ran out, and to the south sandstones only have been found. (See 3, 4, 22.)

Limestone is recorded by the Geological Survey on the Six-inch Map at a point about one mile above Leventhorpe Mill (S.E. of the Hall).

Blue clay with limestone was found in digging for the foundation of the houses in Burnett Avenue, Manchester Road, (23) ; it is also found exposed at the sides and above the end of the tunnel Bradford to Low Moor (25).

I have not been able to find limestone south of this line, and Mr. R. T. Dawson, whose work as a contractor has given him opportunities of judging perhaps greater than most men in Great Horton, and whose knowledge of geology and interest in this subject has caused him to make and record observation for a number of years, informs me that there is abundance of grit boulders but no limestone anywhere on the hills near Horton (See 1, 2, 5, 8, and A.)

Mr. Olliver reports that he found limestones, upon blue clay with local pebbles, and overlaid by yellow clay, at Lady Roy Thornton Road (33). Lower down the road there appears to be no limestone in the drift (28, 9, 10, 11, 18, 19, B) until we get to Brewery Street (28) and the Town Hall (30), and these specimens probably came up the valley from Shipley, as did also that in East Bowling.

When I found the limestone I at first considered it to be a lateral moraine of about 600 feet elevation. Additional weight was given to this notion by the presence at Leventhorpe, in

ay extending from the Hall to the Mill, of a large quantity
pebbles and sand evidently deposited in water.   Mr. J. E.
son explains these and other similar beds by supposing that
ice, by blocking up the outlet in the lower part of the valley,
ed back the water until it rose high enough to pass over the
st part of the ridge at Wibsey Bank, which is about 600 feet
re sea level.   In this way a lake was formed at Leventhorpe,
sand and pebbles were deposited by the streams flowing down
rnton and Bell Dean valleys.

There is abundant evidence that ice came through Chellow
n, but so far I have not been able to find limestones in the
. Mr. Olliver, however, found them at Lady Royd, which
the line joining Chellow with Lidget Green.   All these things
ar to show that the ice came through Chellow Dean and crossed
by Lidget Green to Grange Road, and so on to Bowling.

There are, however, points that should be taken into con-
ation :—

1. Limestone boulders are reported by the Geological
   Survey at a point about level with Leventhorpe Hall
   (29), or south of the lake deposits mentioned above.

2. Clay containing sandstone boulders, and pronounced to
   be true boulder clay by Mr. R. F. Dawson (8), is
   found on Wibsey Slack at 800 feet above sea level.

3. The hills above Leventhorpe, in the Thornton Valley,
   have the form of glaciated hills, although their
   structure would lead one to expect a steep escarpment
   of sandstone at the top, and a gentle slope for the
   softer rocks underneath; the outline (as seen from
   Daisy Hill when looking up the valley) is rounded like
   a roche moutonnée.

4. At Clayton, when the workmen were digging a mill
   dam behind Benn's Mill, they cut into a clay deposit
   of great thickness, which I regard as of glacial origin.

These facts appear to show that the ice was at one time
er than it was at the time that the Leventhorpe Lake was
ed, and that there were changes of level in the ice (34 and 24),

as indicated by the sand and pebble beds in the clay at Tyersal and at Woodroyd.

The Leventhorpe beds themselves show the same, the upper plane at Leventhorpe Hall being about level with Wibsey Bank and the lower at the mill with the gap at Laisterdyke.

It appears therefore most probable that the lake was formed when the ice was retreating.

If you refer to the map (Plate XXIV.), you will find that the places where limestone has been found lie on a fairly straight line from a point above Leventhorpe Hall, through Lidget Green, Grange Road, Manchester Road, to Bowling Tunnel and Woodroyd, and this appears to be the end of a track from some place higher up Airedale.

As there is no light-coloured limestone on the south side of the Aire Valley and no Silurian rock, the specimens found at Grange Road must have come either from the north side at Malham or from Ribblesdale.

It is difficult to explain how they could cross the Aire Valley, hence we are driven to the conclusion that the Ribblesdale glacier was forced over the low water-parting at Hellifield and so down the Aire Valley. The western moraine in Ribblesdale would then become the southern one in the latter valley, and the rocks that would fall upon the ice from the hills on the west side as it passed down by the side of Ingleborough, and those that would be added by the Crummack Dale ice, would be of the same nature as those found by me at Grange Road excavations.

I am informed by Mr. Howarth that there is evidence near Hellifield that the ice has passed over the dividing ridge.

Since writing the above, I have got some additional indications that the line is continued in the direction suggested in this paper.

I examined the workings at Many Well Springs, and found, with abundance of angular sandstones and grits that were evidently foreign, one piece of encrinital limestone in the clay. Mr. Tatham, who has charge of the farm, showed me a considerable number of pieces of weathered limestone in the walls of the fields. We examined them, and concluded that they could not have been

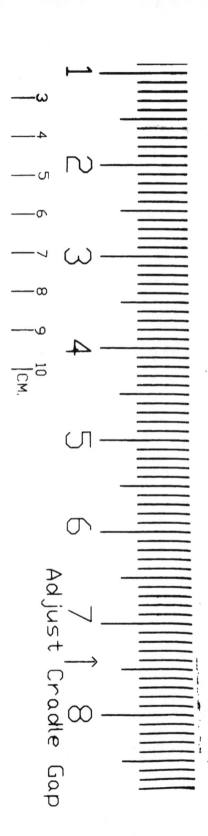

Adjust Cradle Gap

carted on to the land for farming purposes, but were probably from the clay.   Many other specimens were found higher up the hill side, but I could not be sure that they were from the clay.   This is not so satisfactory as one could wish it to be.

Since that time, Mr. W. E. Holloway led the Bradford Scientific Association over Cowling District, and pointed out a deposit of clay with limestone at the head of Lumb Clough (1,000 feet above sea level), and on the same day the Rev. J. N. Lee showed us an immense deposit in the valley below the village.

More recently, Mr. H. B. Muff has published his researches on the Upper Aire Valley, and he traces a lateral moraine down the south side until it arrives at Denholme, then turns north along the watershed and through the gap at Chellow Dean. (See paper at the British Association, Bradford.)

Later still, Mr. E. E. Gregory and I, in examining the lake deposits at Leventhorpe, have found evidence to show that prior to the formation of the lake the whole district was covered with ice, and that it came down Bell Dean.

Underlying the gravels there is a bed of very stiff blue boulder clay exposed in the bed of Pitty Beck, about 100 yards below the Thornton Road bridge; also a deposit of clay with boulders on the side of the hill (600 feet above sea level) and 200 yards below Pitty Bridge.   Here we found one specimen polished and scratched, the striæ running E. and W. or in a line with the valley.   On both sides of the footbridge, over which the path from School Green to Clayton crosses the Thornton Beck, we found several limestones (both dark and light), a considerable quantity of gannister, and some very rough grits, some of which contained large quartz pebbles; and in the gravel pit near Thornton Road limestone and chert (rounded and angular) evidently derived from pre-existing glacial drift higher up Bell Dean.

More information is required before a full explanation of the Bradford deposits can be attempted, and the Sub-Committee formed for the purpose will be glad of any aid that can be given to them.

	LOCALITIES.	MATERIAL.	HEIGHT ABOVE SEA IN FEET.	AUTHORITY.
1	In a drain in Cecil Avenue, Great Horton	Sandstone	600	Mr. R. F. Dawson.
2	In a drain in Park Side Road, Bowling	Rough Rock	600	"
3	In a drain between Lidget Green and Paradise Green	Sandy Boulder Clay	550	"
4	At Princeville	" "	460	"
5	Opposite Elm Tree Inn, Manchester Road	" "	...	
6	At First Avenue, Killinghall Road	10 ft. Sandy Boulder Clay, 1 ft. tough Blue Clay under it (not bottomed)	...	
7	Metcalf's Wheel Pit	Clay with Limestone	...	"
8	Wibsey Slack	Sandstone, Boulder Clay	800	"
A	NOTE.—Never found or heard of limestone anywhere in Great Horton or higher up the hill.		...	
9	Behind Rhodes's Foundry, Thornton Road	Sandstones	350	Mr. W. Parker.
10	In foundations of building on the opposite side of Thornton Road			
11	Drains in Thornton Road and in various parts of Heaton	"	350	"
B	NOTE.—Not found any limestone in any part of the valley leading to Thornton, i.e., Heaton, Girlington, &c.	"	320	"
12	On the north side of the East Brook at Crebbin's Foundry	Blue Clay	...	Mr. Drake.
13	Speight's Mill, Thornbury	"	...	
14	Windhill Cragg, in a side street off Cragg Lane, near upper Lock on Canal	Striated Rough Rock. Direction of Striæ N.W. to S.E.	250	Mr. E. E. Gregory and J. Monckman.
15	Rhumbolds Moor, above keeper's house, on the footpath from Baildon to Ilkley	Grit striated from N.W. to N.E.	1,100	

	LOCALITIES.	MATERIAL.	HEIGHT ABOVE SEA IN FEET.	AUTHORITY.
16	Hanson Board School...	Sandstone blocks and flags, 3 ft. × 2 ft. × 1 in. to 2 in. on end, with much sand and one much weathered coral	500 ?	Messrs. Foulds and Forrest.
17	Frizinghall, north of Railway Station	Sand, Sandstone	300	Mr. E. E. Gregory.
18	Empire Theatre, Horton Road	Sandstones and Grits	380	Dr. J. Monckman.
19	New Street (drainage) parallel to Shear Bridge Road	Sandstones (8 to 10 ft., not bottomed)	420	,,
20	Behind Grange Road, on east side of the beck as far as the Station, Great Horton	Stiff Clay, yellow on top, blue below, contains Sandstones, Grits, Red Sandstones, Blue Limestone, Light-coloured Limestones, Banded Limestones, Chert, Silurian Grits	450 to 500	,,
21	Lidget Green, in the digging for foundations for corner shops (Co-operative Society's new place)	Sandstones and some Limestones	550	,,
22	Lidget Green, further N.	Sandstones only	550	,,
23	Burnett Avenue, Manchester Road	Blue Clay with Limestones	550	,,
24	Woodroyd Brick Works	Clay with Sandstones and a few Limestones, with a sand and pebble bed interposed	600	Dr. J. Monckman and Mr. C. Smith.
25	Top of the entrance to the tunnel to Low Moor, L. & Y. R.	Sandstones and Limestones...	525	Dr. J. Monckman.

c

	LOCALITIES.	MATERIAL.	HEIGHT ABOVE SEA IN FEET.	AUTHORITY.
26	Foundations for house at Chellow Dean	Clay with Sandstones	700	Dr. J. Monckman.
27	Excavations in Clay used for Puddle in the New Reservoirs	,, ,,	700	Mr. W. Cudworth.
28	Drain in Brewery Street	Clay with some Limestones	...	Geological Survey.
29	One mile above Leventhorpe Mill, S.E. of the Hall	Limestone Boulder	500	Mr. Webster.
30	Town Hall, excavations for the hoist	,, ,,	330	
31	Hewenden reservoir	Limestone. One specimen (encrinital) from clay, considerable number of old and 1 mus from walls		
32	Oxenhope	Limestone	677	Dr. J. Monckman.
33	Lady Royd, Thornton Road	Yellow Clay, under whic was drift with lime, and below that Blue Clay with local Sandstones	...	Mr. H. B. Muff.
34	Tyersal, near the Board School	Blue Clay with lime, with 1 ft. of gravel underneath	...	Mr. R. M. Olliver.
35	Head of Lumb Clough	Blue Clay with Limestones	1,000	Messrs. Monckman, Foulds, and Forrest.
36	Near Cowling	Clay with Limestones, &c.	500	Mr. W. R. Holloway.
37	Birkshall, Bowling, on the site of the Corporation Gasworks	Cannister beds, striated and polished. Direction S.W. to N.E.	550	Rev. J. N. Lee. / Mr. J. Dunn.

ON THE FISH FAUNA OF THE YORKSHIRE COAL MEASURES.

BY EDGAR D. WELLBURN, L.R.C.P., F.G.S., F.R.I.P.H., ETC.

INTRODUCTION.

In Yorkshire only the Lower and Middle Coal Measures are present, the Upper being absent.* On the north and west the coalfield is bounded by the Millstone Grits; on the east they sink below the Permian Limestones, and on the south—through the Barnsley beds—they become continuous with the Derbyshire coalfields.

The Lower Coal Measures extend from the Rough Rock (Millstone Grits) at the base to the Silkstone or Blocking Coal at the top, and are composed of thick-bedded, often coarse, grit rocks, with thick intermediate beds of shales, with occasional seams of coal. That some of these coal seams were laid down under marine conditions (estuarine) is certain, and especially is this the case of the Halifax Hard Bed Coal, where fish remains are associated with a distinct marine fauna consisting of *Goniatites*, *Orthoceras*, *Aviculopecten*, *Posidonomya*, &c. Higher in the series fresh water conditions appear to have prevailed, the measures being of the Lagoon type, the shales not only yielding fish remains, but also a good assortment of land plants and fresh water mollusca, and occasionally remains of *Labyrinthodonts* are found.

The Middle Coal Measures consist of rapidly alternating shales and sandstones, with frequent recurring coals. These measures were in all probability laid down in a series of lagoons or lake basins, and appear to have been of fresh water origin. Several of the seams have yielded fish remains, and one—the Cannel at Tingley, near Leeds—contains a most remarkable series of fish, many of the specimens being in a nearly perfect condition, whereas in the other coal shales—both in the Middle and Lower Measures—the fish remains are only found in a very fragmentary condition.

---

* Prof. Green gives some measures in the Conisborough district as probably belonging to the Upper Coal Measures (see Geology of the Yorkshire Coal Fields).

With the exception of the Cannel Coal at Tingley—where the fish remains are found, not only in the shale or "Hubb," but also in the coal itself—the remains are found in the shale immediately above the coal seams, being most plentiful in the shale lying directly on the coal.

HABITS OF LIFE.—When we come to consider under what conditions the fish lived, and their habits during the deposition of the Coal Measures, we are faced with a rather difficult problem, as we find Elasmobranchs, Teleostians, and in some districts even Dipnoian fishes mingled together in such a manner as to point to the fact that they must have been associated during life. Representatives of the Elasmobranchs are found not only above the Halifax Hard Bed Coal, which is undoubtedly of marine origin, but also in most of the beds higher in the measures, where the coal seams appear to have been formed under fresh water conditions. The *Chondrostian Teleostians*—represented by the families Palæoniscidæ and Platysomidæ—are also found in beds of both fresh water and marine origin, this going to prove that they—like their living representative, the Sturgeon—were able to exist under both these conditions; but when we come to the Dipnoi —fishes which are represented in the Yorkshire Coal Measures by the family Ctenodontidæ—their remains have hitherto only occurred in beds of undoubted fresh water origin. Again, the *Crossopterygidian Teleostian* genera, *Megalichthys* and *Cœlacanthus* —whose living representative *Polypterus* is at present found living in the rivers of Africa—are found in all the fish-bearing coal shales in both the Middle and Lower Measures—i.e., in beds of marine and fresh water origin. These facts may, I think, be explained by the supposition that the lakes or lagoons, in which the coals of the Middle and the greater majority of the seams of the Lower Measures were laid down, were at one time in direct communication with the sea, but that subsequently, owing to some elevation of the land or some other cause, they became shut off from the sea, this producing salt water lakes containing fish of a marine type. Then it seems reasonable to suppose that the water, being fed by rivers and streams from the land, would become

dually less and less salt, until no trace of its marine origin
nained, and that the sharks and any other marine fauna present
uld adapt themselves (as they have been proved to do at the
sent day) to their gradually changing surroundings, and that
a conditions now being favourable, fresh water fish would emi-
ate from the surrounding rivers and streams, so that we should
ntually have a fresh water lake with a fish fauna of both fresh
ter and marine types.

In studying the life history of the Yorkshire Coal Fishes,
at of *Cœlacanthus* stands out as of special interest.   These fish
a present in great numbers in the Cannel Coal at Tingley, which
al appears to have been formed in a shallow lake of no great
tent, the coal rapidly thinning out in all directions ; and it seems
ghly probable that at certain seasons semi-stagnant and even
ied-up conditions prevailed, and it was probably to meet these
tter conditions that the fish *Cœlacanthus* were provided with
hat may be termed their greatest " physiological peculiarity,"
eir swim bladder, as it was probably by this means that they,
e the lung fishes of the present day, were enabled to live
iring the dry seasons, their swim bladder acting—for the time
ing—as a lung.   That the swim bladder did play such a part
rendered highly probable by their peculiar anatomical structure.

REMARKS ON THE FISH REMAINS.

SUB-CLASS : ELASMOBRANCHII.

Order : ICHTHYOTOMI.

Family : Pleuracanthidæ.

Genus : *Pleuracanthus* Agassiz, 1837.

The Yorkshire coalfields have yielded the spines of several
ecies of this genus, many being in a fine state of preservation,
it others show all stages of erosion, some being destitute of
nticles and blunt-pointed, and in these latter specimens, the
perficial smooth layer having been removed, the spines appear
a fibrous texture.   To the author *Compsacanthus triangularis*
avis; *Compsacanthus major* Davis; *Phricacanthus biserialis* Davis;

and the following species of *Pleuracanthus*, viz., *erectus* Davis;
planus Agassiz; *Wardi* Davis; *pulchellus* Davis; appear to have
been founded on spines of *Pleuracanthus* in various stages of
erosion.

The following appear to be good species :—

> *P. lævissimus* Agassiz, 1837.
> *P. robustus* Davis, 1880.
> *P. cylindricus* Agassiz, 1843.
> *P. alatus* Davis, 1880.
> *P. alternidentatus* Davis, 1880.
> *P. tenuis* Davis, 1880.
> *P. denticulatus* Davis, 1880.
> *P. horidus* Traquair, 1881.*
> *P. Wardi* Davis ?, 1880.

Form. and Loc.: See Table of Distribution.

> Genus: *Diplodus* Agassiz, 1843.

It certainly appears to the author that this genus should be
merged with that of *Pleuracanthus*. Teeth of *Diplodus gibbosus*
Agassiz are found in the Staffordshire Coal Measures in such
a close relation to *P. cylindricus* as to leave little doubt that they
belong to one and the same fish; again, in the Yorkshire Measures,
where *P. cylindricus* Ag. is somewhat rare, teeth of the type
D. gibbosus Ag. are less commonly met with than those of the
type *D. tenuis* A. Smith Woodward. These facts appear to
point strongly to the conclusion that the teeth *D. gibbosus* pertain
to *P. cylindricus*, and those of the type *D. tenuis* to *P. lævissimus*
and others.

Sp. *D. gibbosus* Agassiz, 1843.

Not very widely distributed or common.

Sp. *D. tenuis* A. S. Woodward, 1889.

Widely distributed and common.

> Family : Cladodontidæ.
> Genus: *Cladodus* Agassiz, 1843.

---

* There is a spine in the Natural History Museum, Cromwell Road, labelled
P. horidus Traq. (from the Better Bed Coal), Clifton, near Halifax.

Sp. *Cladodus* species ?

Teeth rare, distribution very limited.

Genus: *Phœbodus* ? St. John and Worthen, 1875.

One tooth from the Better Bed Coal Shale, Low Moor, appears to show the characters of this genus, the principal lateral cones of the tooth being as large as the median cone.

Order: SELACHII.

Sub-order: Tectospondyli.

Family: Petalodontidæ.

Genus: *Janassa* Munster, 1832.

Sp. *J. linguæformis* Atthey, 1870.

Teeth rare and distribution very limited.

Sp. *J. sulcatus* sp. nov.

Type: Teeth, author's collection.   After having carefully compared these teeth with the specimens of *Janassa* in the British Museum (Natural History), Cromwell Road, and with the figures and descriptions of Munster* and others, I am of the opinion that these teeth belong to a new species, for which I propose the specific name "*sulcatus*," as their chief peculiarity is a deep, well-marked sulcus on their anterior surface, the crown is thin and petal-shaped, the base broad and crossed by several well-marked transverse ridges.   Teeth rare, distribution limited.

Genus: *Petalodus* Owen, 1840.

Sp. *P. hastingsiæ* McCoy, 1855.

Teeth not common, distribution very limited.

Sp. *P. ornatus* sp. nov.

Type: Teeth, author's collection.

The teeth in general form resemble those of the last species, but differ in the fact that the anterior surface is ornamented with a series of sharply-cut grooves arranged in festoons, one series being median and two lateral.   I have compared the teeth with the Petalodont Teeth in the British Museum (Natural History), Cromwell Road, and as I have not seen any teeth showing the above characters, I venture to consider the species as new.

Form. and Loc.: Better Bed Coal, Low Moor.

---

* Beitr. Petrefalct i., p. 67, pl. iv., Figs. 1, 2.

Genus : *Ctenoptychius* Agassiz, 1838.

Sp. *C. apicalis* Agassiz, 1838.

Teeth not very common, but fairly well distributed.

Genus : *Callopristodus* Traquair, 1888.

Sp. *C. pectinatus* Agassiz, 1838.

Teeth rare, distribution very limited.

Sub-order : Atterospondyli.

Family : Cochliodontidæ.

Genus : *Helodus* Agassiz, 1838.

Sp. *H. simplex* Agassiz, 1838.

Teeth not rare, distribution not extensive.

Genus : *Pleuroplax* A. Smith Woodward, 1889.

Sp. *P. Rankinei* Handcock and Atthey, 1872.

Teeth moderately common above the Better Bed Coal, but not common elsewhere.   Distribution somewhat limited.

Sp. *P. Attheyi* (W. J. Barkas, 1874).

Teeth rare, distribution limited.

Family : Cestraciontidæ.

Genus : *Sphenacanthus* Agassiz, 1837.

Sp. *S. hyboides* (Egerton, 1853).

Spines and Teeth : Spines common and fairly well distributed ; teeth not common.

Sp. *S. æquistriatus* Davis, 1879.

Spines not common, distribution limited.

Sp. *S. minor* Davis, 1879. .

Spines not common, distribution limited.

Sp. *S.* sp. nov.

Rare, distribution very limited.

Order : ACANTHODI.

Family : Acanthodidæ.

Genus : *Acanthodes* Agassiz, 1833.

Sp. *A. Wardi* Egerton, 1866.

Fragmentary remains moderately common and well distributed.

Sp. *A. major* Davis, 1894.

To the author it appears to be certain that there are two species of Acanthodes in the Yorkshire Coal Measures, one being a very much larger species than *A. Wardi*. I have seen spines of this species fully seven inches in length, and after having examined some hundreds of these spines of Acanthodes, I have not been able to trace any intermediate forms which would suggest that *A. Wardi* was a young form of the larger species, and this being so it—to me—seems justifiable for the present to retain *A. major* Davis for the larger forms.

Fairly common, but distribution limited.

Genus: *Acanthodopsis* Handcock and Atthey, 1868.

Sp. *A. Wardi* Handcock and Atthey, 1868.

Part of a jaw with teeth.

Rare, distribution very limited.

ICHTHYODORULITES.

Genus: *Homacanthus* Agassiz, 1845.

Sp. *H. microdus* McCoy, 1848.

Two spines in the author's collection show well the characters of this species.

Form. and Loc.: Shale above the Crow Coal, Leeds, and Better Bed Coal, Low Moor.

Genus: *Hoplonchus*, 1876.

Sp. *H. elegans* Davis, 1876.

There are two distinct forms of the spines of this species, one being straighter, longer, and more robust than the other, which is much more arched and slender. They are probably the anterior and posterior dorsal fin spines of the fish. To the author this genus appears to be distinct from the last.

Not common, distribution limited.

Genus: *Ostracanthus* Davis, 1879.

Sp. *O. dilatus* Davis, 1879.

Only the type specimen is known, and to the author it appears to be probable that the Ichthyolite may be the basal portion of a spine of some species of *Pleuracanthus*.

Form. and Loc.: Cannel Coal, Tingley.

Genus : *Lepracanthus*, 1869.

Sp. *L. Colei* Owen, 1869.

Not common, distribution limited.

Sp. *L. rectus* Wellburn, 1899.

Type : Author's collection.

Rare, distribution very limited.

Genus : *Euctenius unilateralis* (W. J. Barkas, 1874).

Rare, distribution very limited.

SUB-CLASS : DIPNOI.

Order : SIRENOIDEI.

Family : Ctenodontidæ.

Genus : *Ctenodus* Agassiz, 1838.

Sp. *C. cristatus*, 1838.

Not common ; teeth, ribs, scales, &c. ; distribution limited.

Genus : *Sagenodus* Owen, 1867.

Sp. *S. inæqualis* Owen, 1867.

Teeth rare, and distribution very limited.

SUB-CLASS : TELEOSTOMI.

Order : CROSSOPTERYGII.

Sub-order : Rhapidistia.

Family : Rhizodontidæ.

Genus : *Rhizodopsis* Young, 1866.

Sp. *R. sauroides* (Williamson, 1837).

In the author's collection are good specimens of this fish from the Cannel Coal, Tingley. In a fragmentary condition, the fish is widely distributed, but is nowhere very common.

Genus : *Strepsodus* Young, 1866.

Sp. *S. sauroides* (Binney, 1841).

Only found in a fragmentary condition, but in the author's collection are fine specimens of the Clavicle, Infraclavicle, scales, teeth, &c. Distribution wide.

Sp. *S. sulcidens* Handcock and Atthey.

In the Brighouse Museum there is part of a fine *Mandibular ramus**—seen from the inner side—showing three fine laniary

---

* See Wellburn, Proc. Yorks. Geol. and Polytec. Soc., Vol. XIV.,
Part I., p. 86.

teeth. This specimen was found in the shale above the Better Bed Coal, Low Moor. Very rare, distribution very limited.

Family : Osteolepidæ.

Genus : *Megalichthys* Agassiz, 1844.

Sp. *M. Hibberti* Agassiz, 1844.

The Yorkshire Coal Measures have yielded some of the finest known specimens of this fish; Agassiz's Type and the beautiful fish in the Leeds Museum having been found in these measures. Common and widely distributed.

Sp. *M. pygmœus* Traquair, 1879.

In the author's collection are several fine *Mandibular rami* and other fragmentary remains of this fish. Rare, and not widely distributed.

Sp. *M. intermedius* A. S. Woodward, 1891.

Present in a fragmentary condition. Rare, not widely distributed.

Sp. *M. coccolepis* Young, 1870.

Present, but rare; distribution very limited.

Sub-order : Actinistia.

Family : Cœlacanthidæ.

Genus : *Cœlacanthus* Agassiz, 1844.

Sp. *C. tingleyensis* Davis, 1884.

The Cannel Coal at Tingley has yielded many beautiful specimens of this fish, but the fish is not peculiar to that district, being found in other districts.

Not common except at Tingley, distribution not wide.

Sp. *C. elegans (lepturus* Ag.) Newberry, 1856.

This species is one of the most common, and has the widest distribution of any fish in the Yorkshire Coal Measures, but is generally found in a very fragmentary condition.

Sp. *C. Phillipsi* Agassiz, 1844.

The type of this species was found in a " Baum Pot " above the Halifax Hard Bed Coal; it is the only specimen known.

Sp. *C. elongatus* Huxley, 1866.

In the author's collection are specimens from Tingley, showing well the characters of this fish. Rare; only found in above locality.

Sp. *C. robustus* Newberry, 1856.

Fragmentary remains from the **Better Bed Coal, Low Moor,** show the characters of this fish.   Rare.

Sp. *C. granulostriatus* sp. nov.

Sp. *C. distans* sp. nov.

Sp. *C. Woodwardi* sp. nov.

Sp. *C. tuberculatus* sp. nov.

Sp. *C. spinatus* sp. nov.

Sp. *C. corrugatus* sp. nov.

> Order : ACTINOPTERYGII.
>
> Sub-order : Chondrostei.
>
> Family : Palæoniscidæ.
>
> Genus : *Gonatodus* Traquair, 1877.

Sp. *G.* (*Molyneuxi* Traquair ? 1888).

> Rare, only found in **Better Bed Coal Shale.**
>
> Genus : *Cycloptychius* Young, 1865.

Sp. *C. carbonarius* Young, 1866.

> Fragmentary remains.
>
> Rare, distribution very limited.
>
> Genus : *Rhadinichthys* Traquair, 1877.

Sp. *R. monensis* Egerton, 1850.

A very beautiful specimen of this fish*, in the author's collection, was found in the shale above the **Barnsley Thick Bed Coal.** The fish in a fragmentary condition is widely distributed, and is not uncommon.

Sp. *R. Planti* Traquair, 1888.

> Fragmentary remains.
>
> Rare, distribution very limited.

Sp. *R. macrodon* Traquair, 1886.

> Fragmentary remains.
>
> Rare, distribution very limited.

Sp. *R. Handcocki* Woodward and Serborn, 1890.

> Fragmentary remains.
>
> Rare, distribution very limited.

---

* See Wellburn, Geo. Mag., New Series, Dec. IV., Vol. V., No. IX.

Genus: *Elonichthys* Giebel, 1848.

١. *E. Aitkeni* Traquair, 1886.

١. *E. Egertoni* (Egerton), 1850.

١. *E. semistriatus* Traquair 1877.

١. *E. caudalis* Traquair ? 1877.

١. *E. oblongus* Traquair ? 1877.

١. *E. Traquairi* sp. nov.

All the above species of *Elonichthys* are—with the exception of a single specimen of *E. Aitkeni* Tr., which shows the lower half of the fish for about two-thirds its length—found in a fragmentary condition. None of them are common, and their distribution is limited.

Genus: *Acrolepis* Agassiz, 1833.

١. *A. Hopkinsi* McCoy, 1844.

Present mostly in a fragmentary condition, but one fine specimen—which is in the British Museum (Natural History), Cromwell Road — was found in a "Baum Pot" in the shale above the Halifax Hard Bed Coal.

Rare, distribution limited.

Family: Platysomatidæ.

Genus: *Mesolepis* Young, 1866.

١. *M. Wardi* Young, 1866.

In the author's collection there is a specimen of *Mesolepis* from Tingley; the posterior two-thirds of the body is shown, also the dorsal and caudal fins, but the ventral is absent. From the general form of the fish I place it in this species. Fragmentary remains are rarely found in Better Bed Coal shale, Low Moor.

Rare, distribution limited.

١. *M. scalaris* Young, 1866.

There are two specimens from Tingley in the author's collection, which, from the deep form of the body, appear to belong to this species.

Rare, distribution very limited.

Genus : *Cheirodus* McCoy, 1848.

Sp. *C. granulosus* (Young, 1866).

Fragmentary remains.

Very rare, distribution very limited.

Sp. *C. striatus ?* (Handcock and Atthey, 1872).

Some scales from the Better Bed Coal shale show the characters of this fish. They are ornamented with vertical striæ, and the articular peg is strong and broad, extending nearly the whole width of the scale.

Very rare, distribution very limited.

Genus : *Platysomus* Agassiz, 1835.

Sp. *P. Fosteri* Handcock and Atthey, 1872.

In the author's collection are two very fine specimens of this fish from Tingley. One shows the head and the body—with the exception of a small portion of the ventral surface—to the commencement of the caudal fin.

Not common, distribution limited.

Sp. *P. parvulus* Williamson, 1849.

Fragmentary remains.

Rare, distribution limited.

Sp. *P. tenuistriatus* Traquair ? 1879.

A group of scales from the Barnsley Thick Bed Coal shale appears to show the characters of this species ; they are ornamented with *very* fine vertical striæ, and the scales are high and narrow.

Rare, distribution very limited.

Sp. *P.* sp. nov ?

The scales are ornamented with widely-spaced, irregular, branching vertical striæ. Rare, distribution very limited.

NOTE.—It is the intention of the author to describe the new species in detail later.

I cannot conclude without expressing my warmest thanks to the following gentlemen for their great and kindly help, viz. :— Dr. R. H. Traquair, F.R.S., Dr. Smith Woodward, F.L.S., John Ward, Esq., F.G.S., and Messrs. J. W. Bond, of Leeds, and W. Hemingway, of Barnsley.

This table lists the distribution of the fish fauna across the Lower Coal Measures of Yorkshire. The species (columns) are arranged under the sub-class **Elasmobranchii**; the coal seams form the rows.

SUB-CLASS: ELASMOBRANCHII.

Genus: *Pleuracanthus.*
- Sp. *P. laevissimus* Ag.
- *P. robustus* Davis
- *P. cylindricus* Ag.
- *P. alatus* Davis
- *P. alternidentatus* Davis
- *P. tenuis* Davis
- *P. denticulatus* Davis
- *P. Wardi* Davis?
- *P. horridus* Traq.

Genus: *Diplodus.*
- Sp. *D. gibbosus* Ag.
- *D. tenuis* A. S. Woodward

Genus: *Cladodus.*
- Sp. *C. mirabilis* Ag.

Genus: *Phœbodus.*
- Sp. ?

Genus: *Janassa.*
- Sp. *J. linguiformis* Atthey
- *J. clavata* McCoy
- *J. sulcatus* sp. nov.

Genus: *Petalodus.*
- Sp. *P. hastingsiœ* McCoy
- *P. ornatus* sp. nov.

LOWER COAL MEASURES — coal seams (rows):

Coal Seam	Species occurrences
Stanley Scale Coal, Wakefield.	*P. laevissimus*; *D. gibbosus*
Yard Coal.	*P. laevissimus*
Kent Thick Coal, Barnsley.	*P. laevissimus*
Cannel above Barnsley Thick Coal.	*P. laevissimus*; *D. gibbosus*
Barnsley Thick Coal, Barnsley.	*P. laevissimus*; *P. robustus*; *P. cylindricus*; *D. gibbosus*; *D. tenuis*
Haigh Moor, Castleford.	*P. laevissimus*
Cannel or Adwalton stone Coal, Tingley.	*P. laevissimus*; *P. robustus*; *P. cylindricus*; *P. alatus*; *P. alternidentatus*; *D. gibbosus*; *D. tenuis*
Cannel above Middleton Main, Toug, Bradford.	*P. laevissimus*
Middleton Main Coal.	*P. laevissimus*; *P. robustus*; *D. gibbosus*; *P. ornatus*
Cannel above Silkstone.	*P. laevissimus*
Silkstone or Blocking Coal.	—
Black Shale Third above Beeston Coal, Leeds.	*P. laevissimus*
Shale above a Thin Coal 34 feet above Crow Coal, Leeds.	*P. laevissimus*
Cannel Coal above Black Bed, near Low Moor.	*P. laevissimus*; *P. robustus*; *D. gibbosus*; *D. tenuis*
Black Bed Coal, &c., Low Moor.	*P. laevissimus*; *D. gibbosus*; *D. tenuis*; *P. ornatus*
Better Bed Coal, Low Moor.	*P. laevissimus*; *P. alatus*; *P. alternidentatus*; *P. tenuis*; *P. denticulatus*; *P. Wardi*; *P. horridus*; *D. gibbosus*; *C. mirabilis*; *J. linguiformis*; *J. clavata*; *J. sulcatus*; *P. hastingsiœ*; *P. ornatus*
38 Yards Hard Coal, Halifax.	*P. laevissimus*
Hard Bed Coal, Halifax, &c.	*J. clavata*
Soft Bed Coal, Halifax.	*P. laevissimus*

	MIDDLE COAL MEASURES											LOWER COAL MEASURES							
	Stanley Royd Coal, Wakefield.	Yard Coal.	Kent Thick Coal, Barnsley.	Cannel above Barnsley Thick Coal.	Barnsley Thick Coal, Barnsley.	Haigh Moor, Castleford.	Cannel (Coal, Tinsley), Stone (Coal, Adwalton).	Cannel above Middleton Main, Tong, Bradford.	Middleton Main (Coal).	Cannel above Silkstone.	Silkstone or Blackstone Coal.	Black Shale Thill above Beeston Coal, Leeds.	Shale above a Thill Coal, 34 feet above Crow Coal, Leeds.	Cannel Coal above Black Bed, near Low Moor.	Black Bed Coal, Low Moor, &c.	Better Bed Coal, Low Moor.	36 Yards Hard Coal, Halifax.	Hard Bed Coal, Halifax, &c.	Soft Bed Coal, Halifax.
(Genus: *Ctenoptychius.*)																			
Sp. *C. apicalis* Ag.	×				×								×		×	×			×
(Genus: *Callopristodus.*)																			
Sp. *C. pectinatus* Ag.	×			×	×		×								×	×			×
(Genus: *Helodus.*)																			
Sp. *H. simplex* Ag.					×		×					×				×			
H. sp.?					×											×			
(Genus: *Pleuroplax.*)																			
Sp. *P. Rankinei* H. & Atthey					×		×	×							×	×			
P. Attheyi W. J. Barkas																			
(Genus: *Sphenacanthus.*)																			
Sp. *S. hybodoides* Eg.																			
S. aequistriatus Davis					×		×	×							×	×		×	×
S. minor Davis				×			(×)					(×)				(×)		×	
S. sp. new																×			
(Genus: *Acanthodes.*)																			
Sp. *A. Wardi* Eg.																			
A. major Davis.																			
(Genus: *Acanthodopsis.*)																			
Sp. *A. Wardi* H. & Atthey																			
ICHTHYODORULITES.																			
Homacanthus microdus McCoy																			
Hoplonchus elegans Davis																			
Oaracanthus dilatus Davis																			

SUB-CLASS DIPNOI.
  Genus Ctenodus,
Sp. C. cristatus Agassiz ...
  Genus : Sagenodus.
Sp. S. inaequalis Owen ...
SUB-CLASS : TELEOSTOMI.
  Gns : Rhizodopsis.
Sp. R. sauroides Williamson ...
  Genus : Strepsodus.
Sp. S. ... les Bin.
Sp. S. sulcidens H.&A.
  Genus : Megalichthys.
Sp. M. Hibberti Ag.
  M. intermedius A. S. Woodward...
  M. pygmaeus Traq. ...
  M. coccolepis Young ...
  Genus Coelacanthus.
Sp. C. tingleyensis Davis
  C. elegans Newb.
  C. Phillipsi Ag.
  C. elongatus Hux.
  C. robustus Newb. ? ...
  C. grandostriatus sp. nov.
  C. distans sp. nov. ...
  C. tuberculatus sp. nov.
  C. spinatus sp. nov.
  C. corrugatus sp. nov
  C. Woodicardi sp. nov.
  Genus : Acrolepis.
Sp. A. Hopkinsi McCoy
  Gns : Elonichthys.
Sp. E. Aitkeni Traq. ...
  E. Egertoni Eg. ...
  E. semistriatus Traq.

	MIDDLE COAL MEASURES											LOWER COAL MEASURES							
	Stanley Rnle Coal, Wakefield.	Yard Coal.	Kent Thick Coal, Barnsley.	Cannel above Barnsley Thick Coal.	Barnsley Thick Coal, Barnsley.	Haigh Moor, Castleford.	Cannel or Adwalton Stone Coal, Thnrley, Tong, Bradford.	Cannel above Middleton Main, Tong, Bradford.	Middleton Main Coal.	Cannel above Milk-stone.	Silkstone or Blocking Coal.	Black Shale Thin, above Beeston Coal, Leeds.	Shale above a Thin Coal 34 feet Above (Trow Coal, Leeds.)	Cannel Coal above Black Bed, near Low Moor.	Black Bed Coal, &c Low Moor.	Better Bed Coal, Low Moor.	36 Yards Band Coal, Halifax.	Hard Bed Coal, Halifax, &c.	Soft Bed Coal, Halifax.
Sp. E. caudalis Traq.	+	+	+	+		+		+		+		+	+	+		(×)	+		+
E. oblongus? Traq.				×	×	×	×				×		×	×	×	(?)		×	×
E. Traquara sp. nov.					×									×		×			
Genus Rhadinichthys.													×	(×)		× ×			
Sp. R. monensis Eg.					×	×	×								×	×			
R. Planti Traq.?																(×)			
R. Handcocki W. & Shehorn							× ×							× ×		× ×			
R. macrodon Traq.							•.									(?)			
Genus et Sp. Cenatodus (Molyneuxi Traq.?)				×	×		×							(×)	×	× (×)		×	×

THE OCCURRENCE OF FISH REMAINS IN THE LIMESTONE SHALES
(YOREDALE) AT CRIMSWORTH DEAN (HORSE BRIDGE CLOUGH),
NEAR HEBDEN BRIDGE, IN THE WEST RIDING
OF YORKSHIRE.

BY EDGAR D. WELLBURN, F.G.S.

INTRODUCTION.

At the above locality a valley has been cut out across the
ip of the strata—through the Kinder Scout or lower beds of
he Millstone Grits and the underlying Limestone Shales (Yore-
ales). A few years ago there was—in these latter beds—a good
xposure of "Black Limestone," which had been cut through by
1e stream. At the present time, however, there is none of this
mestone to be found, and this is unfortunate, as it was literally
rowded with fossils in a beautiful state of preservation (*Goniatites*,
autilus, *Orthoceras*, *Aviculopecten*, &c.), and in rare instances.
sh remains were also found. Most of these latter, having found
1eir way into private collections, have been—with the following
xceptions—lost sight of.

DESCRIPTION OF FISH REMAINS.

SUB-CLASS : ELASMOBRANCHII.

Order : ICHTHYOTOMI.

Family : Cladodontidæ.

Genus : *Cladodus* Agassiz, 1843.

'. *mirabilis* Agassiz, 1843.

A specimen in the collection of Dr. Wheelton Hind, F.G.S.,
10ws portions of two teeth of this species, one being seen in
apression.

The teeth are of moderate size, the height of the central
mes of the crown of the more perfect one being 10 mm. The

base is semicircular, thick, strong, and extends horizontally more
or less at right angles to the cones of the crown, and its under-
side is concave at the centre. The root is not well shown. Of
the crown of the teeth, only the central cusp is shown in one
tooth, the other showing (in impression) the central and one
lateral cone. The characters of the cones are well shown in the
former tooth, which is very well preserved; it is thick, circular,
and very strongly attached to the base, abruptly tapering, slightly
inclined backwards, and from each side of the cone — slightly
below its apex—a sharp cutting edge runs laterally on to the base.
The lateral cone is not well shown, but it appears to lean slightly
outwards, and it—as well as the central cone—shows well-marked
longitudinal striæ running from the base upwards.

Form. and Loc.: Limestone Shales (Yoredales) Crimsworth
Dean.

Order : SELACHII.

Sub-order : Asterospondyli.

Family : Cestraciontidæ.

Genus : *Orodus* Agassiz, 1838.

O. elongatus Davis, 1883 (ex Agassiz M.S.).

Another specimen in the collection of Dr. Hind shows
several more or less imperfect teeth of the above species.

The teeth are very long in proportion to their width, one
—although the whole length is not shown—being extremely so,
the length of the part shown being 22 mm, whereas the greatest
width is only 3 mm. In this specimen the central prominence
is imperfect, but the lateral ridges extending from it towards
the extremities of the tooth are well shown, especially on the
longer lateral prolongation, and from this lateral carina secondary
ridges branch out irregularly on either side, the longer ones
reaching the basal margin, the shorter ones soon disappearing, or
in some instances two converge and unite to form a V-shaped
fold, which also soon disappears. The basal margin is smooth
and slightly sulcated, and the base in this tooth appears to have
a porous structure and to be slightly wider than the crown.

Another tooth—seen in impression—shows that the central cone or prominence was moderately elevated, and that it had a well marked ridge running across it from anterior to posterior margin, and still another fragment of a tooth shows that the root was large and of an open porous structure.

Form. and Loc.: Limestone Shales (Yoredale), Crimsworth Dean.

SUB-CLASS: TELEOSTOMI.

Order: ACTINOPTERYGII.

Sub-order: Chondrostei.

Family: Palæoniscidæ.

Genus: *Elonichthys* Giebel, 1848.

E. Aitkeni Traquair, 1886.

Fragmentary remains of this species are recorded by the late Mr. James Spencer, of Halifax*.

Form. and Loc.: Limestone Shales (Yoredale), Crimsworth Dale.

I cannot conclude without expressing my best thanks to Dr. Hind for the privilege of describing the specimens in his collection.

---

* Proc. Yorks. Geol. and Polytec. Soc., Vol. XIII., Pt. IV., 1898.

NOTES ON THE GEOLOGY OF CLITHEROE AND PENDLE HILL.

BY R. H. TIDDEMAN, M.A., F.G.S., OF H.M. GEOLOGICAL SURVEY.

(Prepared for the General Meeting and Field Excursion at Clitheroe. May 25th and 26th, 1900.)

The Geological Bibliography of this district is not extensive. Parts are alluded to in Phillips's "Geology of Yorkshire." More information is given in the "Geological Survey Memoirs, Burnley Coalfield," 1875. The Glaciation of the district was treated of in *Quart. Jour. Geol. Soc.*, 1872, "On the Evidence for the Ice Sheet in Lancashire, Yorkshire, and Westmorland," by the writer of these notes. The Geology is contained in "Geological Survey, One-Inch, Sheet 92, S.W." (New Series, Sheet 68). This may be had coloured for solid rocks only, or with the overlying Drift, the former giving a much better idea of the arrangement of the rocks than the latter. A section across Pendle is shown in "Horizontal Sections, Sheet 86." The Sections of the Carboniferous Series from the Ribble at Clitheroe to the top of Pendle give, on the whole, a very good idea of the type of rocks which prevails in the area south of the Craven Faults, and indeed spreads at least as far as Derbyshire.

Clitheroe abounds in limestone hills, and is much quarried, and we get there two groups :—

(*a.*) The Black Limestones, very well bedded, dark, and bituminous, showing an exceedingly regular strike and crop; and above them

(*b.*) The White Limestones of Salt Hill, Worsa, Gerna, which form rather protuberances, swelling mounds of less distinct bedding, but crammed with well-preserved fossils. These "knolls" are regarded by the writer as owing their form to the original growth of deposition; but Mr. J. E. Marr, F.R.S., attributes not their form only, but the crystalline nature of their component rock and the excellent preservation of the fossils to earth movements. ("On Limestone Knolls in the Craven District of Yorkshire and elsewhere," *Quart. Jour. Geol. Soc.*, Vol. LV., p. 327.)

Salt Hill is hardly a typical reef-knoll, but appears to be a spit or shoal, almost entirely composed of the *debris* of crinoids. A similar deposit is to be seen near Cracoe in Craven. With this exception, the knolls between Clitheroe and Downham are fairly good specimens.

.) *The Shales-with-Limestones* occupy the rising country between the knolls and the sharper rise of Pendle. Fossils, both animal and vegetable, occur in this series, but not so abundantly. The group has a thickness in places of 2,500 feet, and is composed of clayey and sometimes sandy shales, with impure clayey limestone, in many alternations of soft and harder beds. They are fairly well seen in Worston Brook above Worston.

.) *The Pendleside Limestone* gives the first steep scarp, and shows a thick series of light-coloured, mostly brown, limestones, with many thin beds and interbedded shales. Chert beds abound in them.

.) Not far above the Pendleside Limestone we come upon a second and often a third escarpment composed of the Pendleside Grit, mostly a fine hard sandstone with shale partings. The rock is in places exceedingly compact, almost like a gannister, but without the abundant vegetable forms so characteristic of the latter. Occasionally, but very rarely, this rock is a conglomerate with quartz pebbles. It retains glacial scratches well.

:) The next feature is not a scarp but a hollow groove, evidencing soft rock. Such are the *Bowland Shales*, which, underlying the bolder Pendle Grit, give features and distinctive character to most of the principal ranges of hills in Craven south of the Craven Faults, and throughout the district of Bowland. This is very marked in the range of Pendle, which reaches from near Chorley in Lancashire to Skipton in Craven, a range which only just sufficiently deviates from a straight line to ensure beauty and variety. It forms the southern side of a compound anticlinal arch, of which the northern side is far more broken and irregular.

The Bowland Shales are about 700 feet thick. A great part of them contains fossils such as *Posidonomya Gibsoni*, *Mytilus*, *Aviculopecten*, *Pinna*, *Goniatites*, *Orthoceras*, and occasionally scales and teeth of fossil fishes. Remains of vegetable organisms may also be found.

(*g.*) The top of Pendle is formed of the Pendle Grit, the lowest of the Millstone Grit series. The Pendle Grit was formerly called the "Yoredale Grit," so named in Derbyshire on the supposition that it might eventually be traced all the way thence to Wensleydale and the river Yore. We know now that the Yoredale Series is terminated on the South by the Craven Faults, and that no true Yoredale Series exists South of these strong physical limits. The Grits answering to the Pendle Grit were first called the "Shale Grit" by Farey, in Derbyshire, and a very excellent name it was; for the grits are always giving place to shales, and *vice versâ*. Still, their general characters are fairly persistent.

When geologists have mastered the rocks on the way to the summit, and noted the section, they have not completely realised all that can be gained from the excursion. The view from the top is a magnificent display of the Physical Geology of a wide region. The long range on which you are standing, fading away in the distance, carries the eye from the western sea to the heart of Yorkshire. The Grit Fells to the northwest represent the other side of the great compound arch, of which Longridge Fell to the west is but a detail. The Ribble and Hodder Valleys show the softer beds underlying, which at this distance look like a great plain. To the south-east two or three main ridges represent the Millstone Grits, with a broad furrow representing the Sabden Shales in their midst. Beyond

again is the Burnley Coalfield and the murk of busy Lancashire,
and still further, if not thereby obscured, the Pennine Chain of
Hills running up north from Derbyshire, and meeting E.N.E.
the range on which you stand.   At N.N.E. the White Cliffs
of Malham Cove are to be seen; and at N. 14° W., if in luck,
you will see Ingleborough, Whernside, and Co., the other side of
Craven.   Then after taking all this in, if your minds are not
satiated, take a look back in time, and consider a little patch
of red Permian rock lying in the valley below you at Clitheroe,
resting on beds low down in the Carboniferous Limestone
Series, and bear in mind that a similar Permian patch rests on
the Coal Measures at the upper end of the Carboniferous Series,
just at the foot of far-off Whernside.   That will set you thinking
on the great movements of the crust, and the immense waste
of rocks and transference of material between the Carboniferous
and Permian times, and you will probably come to the conclusion that there is something in Geological Time, and that
Ribblesdale, in part at any rate, was not made yesterday.

MILLSTONE GRITS AND COAL MEASURES, WITH SUPERFICIAL DEPOSITS.

To those interested in these beds the following hints may
be useful:—Take train to Whalley, where the Old Abbey is
itself worth seeing.   Keep along the east bank of the Calder, or
along the Padiham Road above.   A striking roche moutonnée
is in the gorge, a rounded mass of Millstone Grit on which
the last Abbot of Whalley was hung, but that is across the
stream.   Below the road there used to be gravel pits, showing good
sections in glacial gravels and sands, with a few marine shells.
Portfield, a Roman camp, lies on some of this.   The Sabden
Shales, a very thick series, lie beneath the road to Sabden;
parts of them in the banks of Sabden brook contain fossils.
Nearer to the Coalfield, by going across country, you get a good
view of the Upper Millstone Grits, the Third Grits of Hull
and Green, with some beds like gannister, and answering to the
Brooksbottom Series of Binney.

Lower down the slope you come to the First Grit or Rough Rock, a soft, coarse, pebbly conglomerate, and beneath it in places a coal, and then Haslingden flags, thinly bedded, rippled flags. Other flags of better quality lie above the Rough Rock.

Then the Lower Coal Measures come on and show in places *marine* fossils. The Arley Mine is taken as the base of the Middle Coal Measures, and is one of the best seams of the Coal-field, but the coals are not within the limit of a day's excursion, and are moreover much covered with Drift.

Drifts. Those who wish to see drifts might study the long gravel mound or esker which runs from Waddington to Bashall, and takes up its course again on the far side of the Hodder, three-quarters of a mile north of Stoneyhurst.

Whitewell (nine miles) is well worth a day's excursion from Clitheroe, or a week-end stay. The scenery is beautiful, and the Geology distinctly good. The reef knolls may be studied, and fossils knocked out in abundance, and there are good sections of the beds above the limestone. The north-west side of the Hodder will be found best to work at, but a trip from Whitewell Hotel to the Trough of Bowland, on the mountain pass to Lancaster, may be strongly recommended.

NOTES ON THE VOLCANIC ROCKS OF THE CHEVIOT HILLS.

ʀ HERBERT KYNASTON, B.A., F.G.S., OF THE GEOLOGICAL SURVEY
OF SCOTLAND.

(*Prepared for the Meeting and Field Excursion, at Wooler,
July 13th to July 17th, 1900.*)

TERATURE.

Geol. Mag., 1883, pp. 100, 145, 252, 344.
Do.   1885, pp. 106-121.
Papers by Mr. J. J. H. Teall on Specimens
of Andesite, Porphyrite, Quartz-felsite, and
Granite.
Mem. Geol. Survey (Sheet 108 N.E,). The Cheviot Hills
(English side), by C. T. Clough.
Sir Arch. Geikie, Ancient Volcanoes of Great Britain,
Vol. I., pp. 337, 338.
Trans. Edin. Geol. Soc., Vol. VII., pp. 390-415. Contri-
butions to the Petrology of the Cheviot Hills, by
H. Kynaston.

The volcanic district of the Cheviot Hills consists essentially
a central core or plug of granite, occupying an area of about
square miles, surrounded by a more extensive area of ande-
c lava-flows. Both the granite and the lava-flows are traversed
numerous sills and dykes of porphyrite and quartz-felsite. The
ᴀs are of Lower Old Red Sandstone age, and rest uncon-
nably upon the Silurian, and are overlain by Carboniferous.
will describe briefly the principal varieties of igneous rocks
with in the district.

1.) *The Andesites.* The lavas, formerly termed porphyrites, are throughout of Andesitic composition and structure. As a rule they are highly vesicular and a good deal weathered. Fresh varieties are, however, occasionally met with, and we may mention the glassy andesite, seen in the river Coquet, about three miles above Alwinton. The rock has been described by Mr. Teall in the *Geological Magazine.* It is black, with the lustre of a pitchstone, and traversed by narrow reddish veins of silica. It is an enstatite-andesite, and contains numerous small porphyritic crystals of enstatite in addition to augite and plagioclase felspar, in a glassy groundmass.

The more common types of andesite may be well studied in any of the crag or burn sections in that portion of the district surrounding the inner granite mass. They are augite-and-enstatite-andesites, varying in colour from almost black to purple, according to the degree of weathering which they have undergone. In texture they appear sometimes compact, with numerous porphyritic felspars, and sometimes very vesicular, the vesicles being filled with silica, usually in the form of agate. These vesicular or amygdaloidal varieties are well seen in the southern parts of the Cheviots in the neighbourhood of Alwinton, and small agates, representing the amygdules of disintegrated rocks, are common as rounded pebbles in the bed of the Coquet. The lavas are also well seen in the lower part of the Langlee valley, near Wooler, in the Carey and College burns further west, and in many other parts of the district.

Volcanic rocks of fragmental origin do not play any considerable part among the erupted products of the Cheviots. Exposures of tuff are seen to the west of Ingram, which in parts is a reddish volcanic breccia of striking appearance. In the churchyard at Ingram a slab of this rock composes the headstone to the grave of Mrs. Allgood, the wife of the former rector.

2.) *The Granite.* The central and more mountainous portion of the district consists entirely of granite. It is well seen on the slopes of the Cheviot Hill, Hedgehope Law, Dunmoor Hill, in the Harthope burn above Langleeford, and in most of the burn sections within the granite area. The rocks forming the more central part of the mass are coarse in texture, and have a greyish or sometimes pinkish appearance. Towards the margin the rock usually becomes finer-grained, and sometimes is of a dark grey colour and resembles a fine-grained diorite. The dioritic varieties are well seen in the neighbourhood of Linhope, and in the Linhope burn, a short distance above the shepherd's house. Fine-grained varieties also occasionally occur, as for instance in the crags on the N.W. side of Cheviot, which are more acid than the normal coarse granite.

The typical Cheviot granite consists of quartz, orthoclase, plagioclase, biotite, and augite. It is one of the few British examples of an augite-granite, and shows affinities in this respect to the augite-syenites of Monzoni, in the Tyrol. Another characteristic feature of the rock is the marked tendency of the quartz and orthoclase to be intergrown so as to form micropegmatite, and this is so constant in the finer-grained varieties that they may aptly be termed granophyres. The dioritic varieties merely differ from the more acid types in containing a higher proportion of the ferromagnesian constituents. In some specimens of the normal type enstatite has been detected, in addition to augite.

Although of the same general geological age, the granite is later than the Andesites, and has been intruded into them. The surrounding lavas have in consequence undergone contact metamorphism, and a microscopic study of the rocks has shown that this extends to at least half a mile from the granite margin. Junction sections between the granite and the andesite may be seen at

various points along the margin, especially in the river
Breamish above Linhope, and in the Linhope burn, near
Linhope House. At the Tathey crags, near Threestone-
burn House, the Andesite is traversed by numerous
veins of granite, and has been considerably altered by
it.   The andesite, thus altered, has a slightly more
lustrous and less compact appearance than the lavas
further from the granite, owing to the re-crystallisation
of the groundmass and the development of secondary
biotite.

[These altered lavas have been described in a paper by
myself in the eighth volume of the "Transactions of
the Edinburgh Geological Society."]

Small crystals of tourmaline, usually only to be detected
with the microscope, are often found in the marginal
varieties of the granite, in some of the more acid dykes
and veins, and in fault-breccias in the granite area.

(3.) *The Dykes and Sills.*   These may be divided into two
classes :—(*a*) the Porphyrites, and (*b*) the Quartz-felsites.
I use the term porphyrite not in its older significance
as referring to altered (decomposed) andesite, but as
signifying an intrusive rock representing the dyke-phase
of the andesitic magma, and bearing the same relation
to the andesites as the liparites and quartz-felsites do
to the rhyolites.

The dykes and sills of both classes are found cutting both
the andesites and the granite. The porphyrites are by
far the more numerous of the two.   Petrologically they
may be described as augite-biotite-porphyrites, containing
plagioclase as the dominant felspar, while augite and
biotite in varying proportions constitute the ferro-
magnesian elements.   In the field the rocks vary in
colour from purple to brick-red, and the porphyritic
felspars, and frequently also the biotite, are generally
conspicuous in a hand specimen. They are less affected.
as a rule, by weathering agencies than the lavas, and so

often stand out in bold relief when occurring amongst them. They are common throughout the entire area of the Cheviots, and frequent examples may be seen in the river Breamish, in the Common burn, and in the Carey burn.

The quartz-felsites are not so common as the porphyrites, and usually occur in, or in the neighbourhood of, the granite. Several examples of the normal type may be seen in the river Breamish, above Linhope; while a beautiful example of a biotite-quartz-felsite, showing relationships to some of the porphyrites, may be seen in the river Coquet, a quarter of a mile above Shillmoor, and again in the Ridlees burn, east of Quickening Coat. (For description see Teall, *Geol. Mag.*, 1885, p. 107, and "British Petrography," p. 343.)

Besides quartz-felsites we also find granophyres and microgranites cutting the granite, and these rocks shade into one another through intermediate varieties. In fact, by collecting a large number of specimens from the dykes of this area, it is possible to show intermediate varieties between the quartz-felsites and the porphyrites, so that it is extremely probable that the rocks have a common origin.

To glance briefly at the sequence of events which have marked the volcanic history of the Cheviots. The first period of volcanic activity was evidently marked by the eruption of immense quantities of andesitic lava. This, which we may call the extrusive phase, was followed, as is proved by the phenomena of the granite and the dykes, by the intrusion of material partly of intermediate and partly of acid composition. This intrusive phase commenced with the intrusion of the augite-granite into the contemporaneous lavas. The intrusion of the dykes and sills followed the consolidation of the granite, and constituted the latest phase of the volcanic activity of the district of which we have any record. Petrological evidence, moreover, has shown that the andesites, the granites, and the dykes have genetic relationships to one another, and we may conclude therefore that in the Cheviot district we are dealing with three successive phases, belonging to the same geological period, of one original magma—(1) an extrusive phase, characterised by outpourings of andesitic lavas ; (2) a plutonic phase, characterised by the intrusion of the granite; and (3) the fissure or dyke phase.

It is possible that the central granite mass occupies the site of the main focus of the volcanic activity of this district, and that it has thus been intruded into the lower portion of the old vent or group of vents from which the surrounding andesitic lavas were discharged.

The volcanic cone of Lower Old Red Sandstone times, which reared itself where Cheviot and Hedgehope Law now stand, has long since passed away. But the ceaseless action of denuding forces has laid bare for us the very heart and core, so to speak, of this ancient volcanic pile, with its surrounding accompaniment of dissected lava-flows and dykes.

THE FLORA OF THE CARBONIFEROUS PERIOD.

BY ROBERT KIDSTON, F.R.S.E., F.G.S.

FIRST PAPER.

I have pleasure in complying with the request of your council to read before your Society a short account of the flora of the Carboniferous Formation, and in so doing shall, so far as possible, avoid technical language, as I address myself more specially to those who, though they have not previously given serious study to the subject, may have a wish to know more about the Fossil Plants which formed such a prominent feature in Carboniferous times, and who, one would fain hope, may be induced to give some attention to a branch of botany than which there is none that would more repay careful observation.

There has long been undoubted evidence of the occurrence of Algæ and Fungi in Carboniferous times in Britain, and recently I have met with a fossil in rocks of Calciferous Sandstone age so similar in appearance to *Fegatella*, that the Liverworts must now be added to our Carboniferous plants. I shall not, however, enter into a detailed description of these fossils, which are of rare occurrence, but pass to those groups which occupy more prominent place and of which there is more certain knowledge. A fossil which has been referred to the mosses was described from the French Coal Measures by MM. Renault and Zeiller, but hitherto no representative of this class has been met with in Britain.

In the present paper we shall therefore reserve our remarks the *Ferns*, *Equisetites*, and *Calamites*, leaving the *Lycopods*, *Sphenophylls*, *Cordaitecæ*, and *Coniferæ* for a future time.

Before proceeding further, it is necessary to point out that many fossil plant genera are quite provisional, for palæobotanists have seldom the *data* for the definition of a genus in the clear

E

and full manner which one demands in the case of genera founded on existing species. Notwithstanding the difficulties of the subject, by careful collecting and study much has been done in elucidating the structure and form of Carboniferous plants, and in some rare cases our knowledge is little less perfect than if we had been able to study the growing plant. Such results have only been attained by much study and careful observation, and are generally the result of the united labours of several workers—one laying the foundation and another building thereon. Thus the science of palæozoic botany has grown and, I doubt not, will grow.

I.—FERNS.

If we only consider the mere form of the frond and the arrangement of the veins, many fossil ferns have a considerable superficial resemblance to certain recent species; still this resemblance must not be regarded as affording any evidence on which to presume a generic relationship. The fact, however, remains that the same type of pinnule form and nervation which is found amongst Carboniferous ferns is seen amongst those existing at present, and also the same mode of circinate vernation (Plate XXVI., fig. 1).

In Carboniferous ferns the main rachis sometimes divided into two arms, as in *Calymmatotheca bifida* L. & H. sp. (Plate XXV., figs. 2, 3), and this dichotomous division even more frequently occurs in the pinnæ, which are once forked, or end in a pair of forks.

This character is rare in recent ferns in their native condition, but frequent in cultivated forms, resulting in the dichotomous or crested varieties of garden origin.

Among Carboniferous ferns the principal families are the *Sphenopteridea*, *Neuropteridea*, and *Pecopteridea*. These will be briefly described.

SPHENOPTERIDEÆ.

Considerable latitude of character is shown by the ferns included in this family. The pinnules may be more or less oval, entire or lobed, and contracted at the base into a short stalk, or cuneate, and even almost filiform. Nervation radiating from the base of the pinnule and frequently dichotomising. The chief genus is *Sphenopteris* Brongt.

SPHENOPTERIS Brongt. The general characters of one of the sections of the ferns commonly included in *Sphenopteris* is well represented by *Sphenopteris obtusiloba* Brongt. (Plate XXV., fig. 1). The pinnules are oval entire, lobed, or divided into 3-5 segments—their form varying according to their position on the pinna. The dichotomising veins radiate fan-like from the base of the pinnules (Plate XXV., fig. 1a).

Another section of *Sphenopteris* has pinnules with more or less cuneate segments, of which *Sphenopteris furcata* Brongt. may be taken as a typical representative (Plate XXVII., fig. 2). The segments of the pinnule are narrow, linear, with a veinlet running into each tooth (Plate XXVII., fig. 2a). In both of these sections there are some species with very small pinnules.

Although one must be very careful in generalising, still it seems as if the linear or cuneate pinnuled forms were more characteristic of Lower Carboniferous rocks, while those with rounded lobed pinnules were more typical of the Upper Car- boniferous. Both types, however, occur together in all the divisions of the Carboniferous Formation.

Many species originally included in *Sphenopteris* have had special genera provided for their reception. In some cases the characters are derived from the mode of division of the pinnæ —characters dependent on the vegetative system. It appears to me very doubtful if any real advantage is derived from the creation of such genera, as they cannot be regarded as other than provisional, and personally I prefer retaining the ferns placed in these genera in *Sphenopteris*. As examples of the genera to which I refer, *Palmatopteris* Potonié (of which *Sphenopteris furcata* is the type), and *Diplothmema* Stur may be mentioned.

The other class of genera which have been taken from *Sphenopteris* Brongt. hold, however, a very different position, as they are founded on characters which are derived from their fructification, but before referring to these more fully it is necessary very shortly to consider the fructification of existing ferns.

Recent ferns are divided into two great classes—the *Isosporous Ferns*, or those with one kind of spore, and the *Heterosporous Ferns*, or those with two kinds of spore—macrospores and microspores. With the latter class, however, we have nothing to do at present.

Returning to the Isosporous Ferns, these again form two great sections. First, those whose sporangia are provided with a prominent ring of cells, called an *Annulus* (Plate XXVIII, fig. 2), and those whose sporangia are destitute of this structure.

The first section contains the great majority of recent ferns, of which the common Polypody and Male fern may be mentioned as well-known representatives. The second group contains the *Marattiaceæ*, which comprises few genera and a small number of species, all of which are natives of more or less tropical areas.

In Carboniferous times both these groups are represented, though the exannulate ferns seem to have outnumbered those with annulate sporangia.

Let us now return to *Sphenopteris*. Many of the species originally included in that genus have in recent years been found showing their fructification, and for these new genera have been created. Among British Sphenopteroid forms a few are known to possess annulate sporangia, and of such are *Corynepteris* Baily and *Oligocarpia* Göppert. In the former the sporangia are placed in groups of five or six, united at the base around a common centre, and collectively form a globular mass or *sorus*; in the latter they form little circular heaps composed of a number of independent sporangia. Isolated annulate sporangia are frequent in the Yorkshire and Lancashire "Coal Balls," and also occur in the material from Pettycur, Fife, which is situated in the lowest division of the Carboniferous Formation (Calciferous Sandstone Series).

The exannulate form of fructification is illustrated by several
era, which are characterised by the form and arrangement of
sporangia. Among these may be mentioned *Renaultia* Zeiller,
are the small oval sporangia are situated on the veins towards
margin of the pinnules (Plate XXVIII., fig. 4). They open
a longitudinal cleft. The fruiting pinnules are little modified
a those of the barren frond.

In *Urnatopteris* Kidston the barren (Plate XXIX., fig. 1)
fertile (Plate XXIX., figs. 2 and 3) fronds are dissimilar,
is, only some of the fronds bear sporangia, and on these
pinnules are entirely deprived of the limb—the sporangia
ag arranged in two rows, one on each side of the rachis.
sporangia are pointed-oval, and open at the summit by a
ll round pore (Plate XXIX., fig. 6). Each sporangium is
but in their structure they have considerable resemblance
the sporangia of *Danæa* only in that genus the sporangia
united to each other to form a *synangium*. Though I only
ntion these two Sphenopteroid exannulate types, others are
wn.

UROPTERIDEÆ.

The most important genus of this family is *Neuropteris*
ngt. (Plate XXVIII., fig. 3. *Neuropteris gigantea* Sternb.).
pinnules are generally more or less oval or tongue-shaped,
articulated to the rachis, from which they are easily
ached. Each pinnule had a central vein, from which are
en off lateral divided veinlets (Plate XXIX., fig. 4). On
e species of *Neuropteris*, possibly on the majority, between
points of insertion of the lateral pinnæ or towards the base of
frond, immediately below the pinnæ, the main rachis bore
iform or orbicular pinnules (Plate XXVIII., fig. 3, a, a),
etimes of large size; these, before their true origin was known,
e supposed to belong to a distinct plant, and were named
lopteris by Brongniart. The fructification of *Neuropteris* is
erfectly known, but in the case of *Neuropteris heterophylla*
ngt. it was borne on long pedicels which terminate the pinnæ.

Linopteris Presl. (*Dictyopteris* Gutbier not Lamouroux) though rare in Britain, must not be omitted. In the form of the frond and pinnules it is similar to *Neuropteris*, and specimens not showing the nervation might easily be overlooked as belonging to that genus, but it is at once distinguished by the nervation, the veinlets of which unite among themselves to form a net-like reticulation (Plate XXX., fig. 2. *Linopteris obliqua* Bunbury sp. x 3½).

Though this genus is certainly not common in Britain, it may be more common than supposed through being passed over for *Neuropteris*.

PECOPTERIDEÆ.

This family holds an important place among palæozoic ferns. *Pecopteris* Brongt. is the chief genus and contains many large and fine species. It is chiefly represented in the Upper Coal Measures, and *Pecopteris arborescens* may be regarded as the type (Plate XXVII., fig. 3).

The pinnules in *Pecopteris* are attached to the rachis by the whole of their base. They have straight sides and rounded apices. The pinnules are sometimes united among themselves at the base and possess a strong central mid-rib, from which—according to the species—are given off simple or dichotomously divided veins which extend to the margin.

The fructification of many of these ferns consists of four or five exannulate sporangia arranged in a stellate group, from which circumstance the genus *Asterotheca* has been proposed for them, though not generally adopted (Plate XXVII., fig. 4).

In the Middle Coal Measures, *Pecopteris* (*Asterotheca*) is represented by few species, but *Pecopteris* (*Asterotheca*) *Miltoni* is fairly plentiful. Though this species also occurs in the Upper Coal Measures, it is there associated with many other *Pecopterids* which are not found below that horizon.

The fronds of *Pecopteris* were of very large size and most probably some of the tree fern stems were the trunks of *Pecopteris*.

The *Pec. plumosa* Artis sp. (= *Pec. dentata* Brongt. Plate XXVII., fig. 1, Plate XXXI., figs. 1-4), so common in the Middle and Upper Coal Measures, forms the type of the genus

Dactylotheca Zeiller. This is characterised by the ovoid-pointed sporangia, which are placed singly on the veins and open by a longitudinal cleft (Plate XXXI., fig. 3). The barren pinnules vary greatly in form, being entire, lobed, or crenate, according to the position they hold on the frond. On the main rachis, at the point of insertion of the pinnæ, are curious, much-divided outgrowths, called aphlebia (Plate XXXI., fig. 2). These were originally supposed to be a climbing fern (*Schizopteris adnascens* L. & H.) which had used the frond of *Dactylotheca* as a support. These *aphlebia* are an integral part of the frond on which they occur, and are found on other species of ferns belonging to various genera.

Another *Pecopterid* genus, *Mariopteris* Zeiller, is extremely common in the Lower and Middle Coal Measures, but very rare in the Upper Coal Measures. The fructification is unknown, but the fern is distinguished by a double bifurcation of the rachis of the primary pinnæ. The leathery texture of the pinnules, difficult to describe but easily learnt from an examination of specimens, as well as the nervation, appears to me to add a character to the genus, which I would be inclined to restrict for *Mariopteris* (*Pecopteris*) *muricata* Schl. sp. (Plate XXXII., figs. 1 and 1a) and one or two close allies, but from which I would exclude such species as *Sphenopteris latifolia* Brongt.

The double bifurcation of the primary pinnæ, which occurs in this species, does not alone seem to me to be of much systematic importance.

ALETHOPTERIDEÆ.

The *Alethopterideæ* are closely related to the *Pecopterideæ*, but the pinnules are generally obliquely placed on the rachis, the prominent mid-rib joins the rachis near the upper margin of the pinnule, and thus gives a somewhat decurrent character to the mode of their insertion on the rachis. The lateral dichotomously divided veins are very numerous and run to the margin at almost right angles with the mid-rib. The common *Alethopteris lonchitica* Schl. sp. well illustrates these characters (Plate XXXII., figs. 2 and 3).

The fructification of *Alethopteris* is imperfectly known, but what is supposed to be a fruiting specimen of *Alethopteris Serlii* Brongt. has been described by Zeiller. This most interesting example appears to show that the sporangia were globular and arranged in rows along the veins. The fronds of *Alethopteris* attained to large size.

The genus *Lonchopteris* Brongt. (Plate XXVI., fig. 2) holds the same relationship to *Alethopteris* that *Linopteris* does to *Neuropteris*, having the same form of growth and pinnule cutting as *Alethopteris*, but is easily distinguished at first sight by the net-like reticulation of the veins (Plate XXVI., figs. 2a + 3).

ODONTOPTERIDEÆ.

The only genus of this family to which reference requires to be made is *Odontopteris* Brongt., which, however, is very rare in British Carboniferous rocks, and appears to be restricted to the Middle and Upper Coal Measures (Plate XXVIII., figs. 1 and 1a— *Odontopteris alpina* Presl. sp.).

The pinnules are more or less tongue-shaped and attached to the rachis by their broad base. They have no true mid-rib—several veins passing into the pinnules direct from the rachis, where they bifurcate once or twice.

TREE FERN STEMS.

Some of the palæozoic ferns had stems like our modern Tree Ferns and must have attained to a considerable height. In Britain the two following genera of fern trunks occur:—

Caulopteris L. & H. (Plate XXXIII., fig. 1—*Caulopteris anglica* Kidston). The frond scars are arranged in vertical rows placed close to each other. They are oval and contain, a short distance within the margin, a closed oval or horse-shoe-shaped band, which corresponds to a tract of sclerenchymatous or much indurated tissue. Within this band and near its upper end is placed the vascular bundle scar. The outer surface of the stem is usually densely clothed with aerial rootlets.

Megaphyton Artis. (Plate XXXIII., fig. 2—*Megaphyton* sp. allied to *M. anomalum* Grand' Eury.). The fronds are attached to these stems in two opposite rows, the frond on one side of the stem alternating with that on the other side.

The stem, except at the part to which the fronds are attached, was densely covered with aerial rootlets.

Caulopteris in its general aspect would much resemble one of the recent Tree Ferns, but *Megaphyton*, with its two opposite rows of fronds would have a very different aspect from any of the Tree Ferns at present existing.

Before passing from this brief consideration of the more important groups of palæozoic ferns, a few remarks must be made on their internal organisation, though this subject can only be touched on very slightly here.

The stems or rhizomes of recent ferns have no exogenous growth, that is, when the vascular bundle is once fully formed no new elements are subsequently added to it. Hence Tree Fern stems when once fully developed retain the same diameter of trunk for years.

Among fossil ferns whose structure is known, a few, generally of small size, possess the same structural peculiarities, but there is another type of palæozoic fern structure where, among other characters, an exogenous increase to the vascular system takes place. In these, after the formation of the primary vascular bundles, whose size is limited as in the first type, a cambium layer appears from which an outer circle of exogenously developed vascular tissue arises. This ring of secondary xylem or wood may increase indefinitely in size by additions from the cambium zone, the ultimate size of the stem being limited only by the life of the plant.

These Fern Stems with exogenous growth present certain anatomical characters intermediate between ferns and Cycads, and are now placed in a group to which Potonié has given the name of *Cycadofilices*. There is reason to believe, though little is known of their fructification, that they may be ferns, though in their anatomy they possess certain characters not found in existing members of this group. This discovery is one of the most interesting and important advances recently made in the study of palæozoic botany, and to the *Cycadofilices* are known to belong certain *Sphenopteris, Alethopteris,* and *Neuropteris.*

As an example of how step by step our knowledge of palæozoic botany is built up, it may be mentioned that the petioles described by Williamson as *Rachiopteris aspera* were subsequently found to belong to the stem named *Lyginodendron Oldhamium* by the same author, and further it has been discovered that *Lyginodendron Oldhamium* is the stem of the well-known *Sphenopteris Hoeninghausi* Brongt. (Plate XXIX, fig. 5). Could any better example be found of the result—or reward—of patient, plodding work, or of the provisional nature of genera founded on the vegetative organs?

CALAMARIEÆ.

Equisetites Sternberg.

A few fossils have been found in Carboniferous rocks which from their great external resemblance to the recent *Equisetum* or Horsetails, have been placed in a genus called *Equisetites* by Sternberg. These fossils are extremely rare, and as far as at present known do not go further back than the Coal Measures. One of the most interesting examples of the genus is the *Equisetites Hemingwayi* (Plate XXXIV., fig. 3), which was discovered by Mr. Hemingway, Barnsley.

The cones are oval, about one inch long and rather over half an inch broad. The outer surface of the cone is covered with hexagonal scales about one-fifth inch in diameter, with a small central point, indicating probably the place of attachment of the little pedicel by which the peltate shield was united to the axis of the cone. Nothing is known of the inner structure and arrangement of the sporangia, but the external appearance of *Equisetites Hemingwayi* is so like that of the cones of recent *Equisetum* (of which a figure is given for comparison, Plate XXXIV., fig. 4, *Equisetum hyemale*), that the affinities of *Equisetites Hemingwayi* Kidston with *Equisetum* is probably very close.

A specimen in the British Museum shows that the cones of *Equisetites Hemingwayi* were apparently sessile and borne at the nodes of a very Equisetum-like stem.

MM. Renault and Zeiller have described from the Comentry Coal Field an Equisetaceous stem, with distinct sheaths, under the name of *E. Monyi*. Some other small specimens from the Carboniferous have been ascribed to *Equisetites*, but their reference is in many cases doubtful.

CALAMITES Suckow.

The *Calamites* form one of the most prominent types of vegetation in Coal Measure times. True Calamites do occur in Lower Carboniferous rocks, that is, below the Millstone Grit, but there they take a very unimportant place and are of very rare occurrence. It is only when we reach the Upper Carboniferous that they attain their importance, both in numbers and diversity of form. Calamites reached to arborescent dimensions.

When dealing with the *Calamites*, we are under the necessity of placing the stems, foliage, and fructification of the plants comprised in this group in separate genera, as in few cases can the foliage and fruit be referred to the parent plant. In fact, even in the genus *Calamites* in which the stems are placed, there are almost certainly included plants which belong to different genera. One is led to infer this from the structure of Calamitic cones which show among themselves important structural differences. This fact must not be lost sight of, and the genus *Calamites* should be regarded more in the light of a group than of a true genus, but for practical purposes some system of classification, even if provisional, must be adopted.

The late Professor Weiss divided *Calamites* into three groups :—

I.—CALAMITINA. In *Calamitina* the branches are borne in verticils, but between each verticil there is one or more nodes from which no branches are developed.

II.—EUCALAMITES. The stems placed in this group bear branches from every node.

III.—STYLOCALAMITES. Here the stems are either unbranched or, if lateral branches occur, they are developed very irregularly.

In all these divisions the ribs on the *pith-cast* alternate at the nodes.

There is a fourth division which, however, only occurs in the Lower Carboniferous:—

IV.—ASTEROCALAMITES Schimper. In these plants the ribs do not alternate at the nodes, and the branch scars are irregularly produced.

Before considering these groups more fully it is desirable to make a few general remarks on the Calamites as a whole.

The majority of the fossils referred to *Calamites* have ribbed exteriors, such as the figures of *Calamites (Stylocalamites) Suckowii* given by Brongniart and others (Plate XXXV., fig. 3). These do not represent the exterior of the plant as originally supposed, but are merely the casts of the pith cavity. This is well seen in the figure of *Calamites (Calamitina) approximatus* Brongt. given on Plate XXXV., fig. 2, where the cast of the pith cavity is seen at *a*, and the vascular portion of the stem at *b*. Plate XXXIV., fig. 2, also shows the same characters. The true outer surface of the stem of Calamites is rarely preserved, and though very young stems may show faint ribs the older stems have almost invariably smooth barks, though on rare occasions a ribbing of the outer surface seems to occur as in some examples of *Calamites (Calamitina) verticillatus* L. & H. (Plate XXXVI., fig. 4).

The stems of *Calamites* (except possibly in the very young condition) were hollow except at the nodes, where a more or less complete diaphragm of cellular tissue extended across the cavity. The pith cavity was surrounded by a zone of vascular wedges, in the inner angle of which is a carinal canal. This woody zone increased indefinitely in size by additions from a cambium ring. The vascular wedges are separated by broad medullary rays, and the whole is enclosed in a thick cortex. Modifications of this structure occur in different members of the group, but all conform in their outstanding features to this type of stem. The aerial stems of *Calamites* spring from creeping rhizomes as seen in the figure of *Calamites Suckowii*, given on Plate XXX., fig. 1, as well as from the subterranean portion of the aerial stems (Plate XXXV., fig. 3).

Let us now return to the consideration of the three groups of stems, to which reference has just been made:—

I.—CALAMITINA.

Calamites varians Sternb. may be taken as typical of this group (Plate XXXIV., fig. 1. *Calamitina varians* var. *inconstans* Weiss).

The internodes vary in length, and the nodes bear closely-placed transversely oval leaf scars. The bark is smooth but frequently shows slight longitudinal clefts or cracks, which vary in their length and distance apart. These longitudinal cracks or lines probably arise through the splitting of the bark from the increase of the stem in girth. Between each branch-bearing node several branchless nodes intervene, their number varying, not only in the same species, but even on the same specimen. The foliage of some *Calamitinæ* consisted of acicular leaves, but whether all possessed such foliage is not known (Plate XXXVI., fig. 1).

II.—EUCALAMITES.

Calamites ramosus Artis. is representative of this section (Plate XXXVII., figs. 3 and 4). Each node gives rise to two branches, one on each side of the stem. The branches are superposed, and, though these again bear lateral branches, the plant would possess the form of a triangle. The surface of the stem is smooth. Plate XXXVII., fig. 4, shows the cast of the pith cavity; fig. 3 shows the outer surface of the species with smooth bark.

The foliage of *Calamites ramosus* consisted of lanceolate leaves, arranged in whorls and united by their bases to form a very narrow ring round the stem. This foliage was named *Annularia radiata* by Brongniart before it was discovered to be the foliage of *Calamites ramosus* (Plate XXXVII., fig. 1).

The fruit of *Calamites ramosus* is in the form of small cones which terminate the branchlets. Their structure is that of *Calamostachys*, which will be presently described.

In *Calamites (Eucalamites) cruciatus* Sternb. each node bore a verticil of somewhat distant branches.

III.—STYLOCALAMITES.

The *Calamites* in this group very rarely produced branches. *Calamites Suckowii* Brongt. is a good example (Plate XXXV, fig. 3). The outer surface of the stem was smooth, and if the nodes gave rise to branches they must have done so very rarely.

In *Calamites Cistii* Brongt., another member of this group, small scars occasionally are found on the nodes, but these probably are the scars, at least in part, of short stalked cones.

IV.—ASTEROCALAMITES Schimper.

This group is of generic value, and ranks in importance with the genus *Calamites*; it not only differs from *Calamites* in the ribs not alternating at the nodes, but also in the foliage being dichotomously divided. The fructification consisted of narrow cones, fully five inches long, which are periodically divided into sections by interposed barren whorls, so that the cone appears as if composed of a number of oblong segments resting on each other, and between which is a whorl of leaves. Each segment therefore consisted of a barren whorl, which is succeeded by 10 or 12 fertile whorls. Though specimens of the fruit and foliage are very rare in Britain, fragments of the stems are not uncommon. The genus is characteristic of the Lower Carboniferous.

Owing to our inability in the majority of cases of associating the isolated foliage branches of *Calamites* with the stems to which they belong, they are placed in the two following genera—*Calamocladus* and *Annularia*.

In *Calamocladus* Schimper (*Asterophyllites* Brongt.) the leaves are arranged in whorls. They are narrow linear or setaceous, single nerved and placed closely together. One of the commonest species is *Calamocladus equisetiformis* Schl. sp. (Plate XXX, fig. 3).

Annularia Sternberg contains those forms with whorled, single-nerved lanceolate leaves, widest near the centre like *Annularia radiata* Brongt. (Plate XXXVII., fig. 2), or with spathulate leaves like *Annularia sphenophylloides* Zenker. sp. (Plate XXXVII., fig. 1). The leaves unite at the base and form a very narrow collar round the stem.

FRUCTIFICATION OF CALAMITES.

The fructification of *Calamites* consists of narrow linear cones, attaining in some species a few inches in length, though in most cases they are of smaller size. The arrangement of the sporangia in many of these cones is still unknown, but of some a very complete knowledge is possessed. For their reception several genera have been founded, as hitherto it has been generally impossible to refer them to their parent stems.

The more important of these genera may be briefly described. CALAMOSTACHYS Schimper. (Plate XXXVI., fig. 2).

The cone is composed of alternating whorls of barren leaves or bracts and sporangiferous scales. The basal portions of the bracts unite to form an almost horizontal collar which surrounds the axis, while the free parts of the bracts rise up almost at a right angle, the whole forming a saucer-like structure. Between each of these barren whorls is a fertile whorl. This consists of slender pedicels or *sporangiophores*, which spring from the axis at right angles and terminate in peltate shields, on the inner surface of which are borne four sporangia. Both homosporus and heterosporous cones occur in *Calamostachys*.

PALÆOSTACHYA Weiss (Plate XXXIV., fig. 5; Plate XXXVI., fig. 3. *Palæostachya pedunculata* Williamson).

The cones placed in *Palæostachya* differ from those of *Calamostachys* in the sporangiophores springing from the axis immediately above the axils of the bracts and forming with the axis an angle of about 45 degrees.

For cones possessing the general appearance of *Calamostachys* and *Palæostachya*, but in which the arrangement and position of the fertile whorls are unknown, the genus *Volkmannia* Sternb. may be conveniently employed.

Macrostachya Schimper is another genus of Calamitic cones These attained to considerable size, and are much larger an broader than those of the three preceding genera. The cones a composed of alternating closely placed verticils of many bract united to each other throughout the greater portion of the length : only the short upturned extremities of the bracts remai free. Each whorl of bracts thus forms a saucer-like collar whic surrounds the thick axis of the cone. The arrangement of t sporangiophores has not been clearly made out.

Other types of Calamitic cones are known, but those mention are the principal forms which occur in British Carboniferous roc

Occasionally specimens of *Calamites* are found showing remains of their rootlets. These are—in whole or in part— fossils for which Lindley and Hutton founded the genus *Pinnul* (Plate XXXV., fig. 1—*Pinnularia columnaris*). They consis roots pinnately giving off lateral roots, which in turn bear rootlets, apparently in the same plane.

That the *Calamarieæ* and *Equisetaceæ* are closely relate beyond all doubt, and there seems to be no satisfactory reason they should not be united in one family under either of t names, preferably under that first mentioned.

The genus *Calamites* seems to have entirely disappeared out leaving any modern representative, while the less impor palæozoic genus *Equisetites* is probably the ancestor of the re *Equisetum*.

EXPLANATION OF PLATES.

PLATE XXV.

1. *Sphenopteris obtusiloba* Brongt.   Grange Colliery, Kilmarnock.   Lower Coal Measures [1560].   Three-fifths
   natural size.

1a. *Sphenopteris obtusiloba* Brongt.   Portion of a pinna
    slightly enlarged to show the nervation of the pinnules.
    Specimen received from the Rev. D. Landsborough,
    D.D.

   [*Note.*—The figures enclosed in brackets give the
   registration numbers of the specimens in the Author's
   collection.]

2. *Calymmatotheca bifida* L. & H. sp.   Lewis Burn, about
   200 yards below Lewis Burn Colliery, North Tynedale,
   Northumberland.   Calciferous Sandstone Series (Lower
   Carboniferous) [728].   Three-fifths natural size.   Basal
   portion of frond showing bifurcation of rachis.

3. *Calymmatotheca bifida* L. & H. sp.   Burdiehouse, Midlothian.   Calciferous Sandstone Series [717].   Collected
   by the late Mr. C. W. Peach.   Three-fifths natural
   size.   Upper portion of frond showing bifurcation
   and pinnæ.

3a. Pinnule enlarged.

Photographed by R. Kidston.

PLATE XXVI.

*. 1. *Spiropteris.* Braysdown, near Radstock, Somerset. Upper Coal Measures [510].   Young frond, probably of *Pecopteris* in circinate vernation.   Rather less than half natural size.

*. 2. *Lonchopteris rugosa.* Brongt.   St. Eloi, Mariemont, Belgium.   Coal Measures.   Natural size [2634]. Specimen communicated by Rev. Father G. Schmitz, S.J., Louvain.

*. 2a. Portion of pinnule showing the reticulate nervation. Magnified about $2\frac{1}{2}$ times.

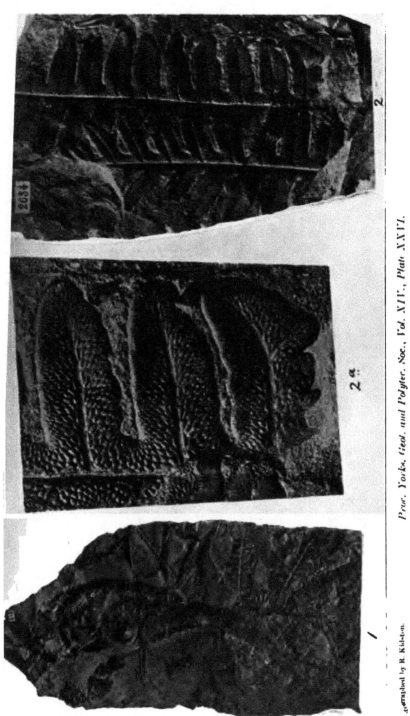

Lithographed by R. Kidston.

2

2 a.

1

PLATE XXVII.

;. 1.   *Dactylotheca plumosa* Artis. sp.   Monckton Main Colliery, near Barnsley.   Middle Coal Measures [2107].   Natural size.   Collected by Mr. W. Hemingway.

'. 1*a*.   Pinnule enlarged four times.

'. 2.   *Sphenopteris furcata* Brongt.   Cramlington, Northumberland.   Lower Coal Measures [259].   Natural size.   Collected by Mr. J. Sim.

. 2*a*.   Pinnule enlarged to show nervation.

. 3.   *Pecopteris arborescens* Schloth. sp.   Radstock, Somerset.   Natural size.   Upper Coal Measures [452].

. 3*a*.   Pinnule enlarged to show nervation.

. 4.   *Pecopteris (Asterotheca) Miltoni* Artis. sp. (after Zeiller).   Enlarged twice.

PLATE XXVIII.

Odontopteris alpina Presl sp.   Monckton Main Colliery,
near Barnsley.   Middle Coal Measures [1962].   Three-
fifths natural size.   Collected by Mr. W. Hemingway.

1. Portion of pinna enlarged to show the nervation.

. Annulate fern sporangia, in section, Pettycur, Fife [Slide
No. 550].   Magnified 50 times.   *a*. Annulus.

. *Neuropteris gigantea* Sternb.   Hill Top Colliery, Skegby,
near Hiechnael-under-Huthwaite, Notts. [206].   Three-
fifths natural size.   Middle Coal Measures.   Collected
by Mr. E. Wilson.   On the main rachis, as at *a*, are
seen the small cyclopteroid pinnules.

. *Renaultia microcarpa* Lesqx.   Blairpoint, near Dysart,
Fife.   Lower Coal Measures [773].   Collected by Mr.
James Bennie.   Pinnule showing the fructification
enlarged.

ι. Sporangium more highly enlarged to show the structure.

Plate XXIX.

g. 1. *Urnatopteris tenella* Brongt. sp.   Furnace Bank, Sauchie, near Alloa, Clackmannanshire.   Lower Coal Measures [1983].   Natural size.   Portion of barren frond.

g. 2. *Urnatopteris tenella* Brongt. sp.   Furnace Bank, Sauchie, near Alloa, Clackmannanshire.   Lower Coal Measures [1988].   Natural size.   Fruiting frond.

g. 3. *Urnatopteris tenella* Brongt. sp.   Ellismuir, Baillieston, Lanarkshire.   Lower Coal Measures [2450].   Enlarged about twice.   Collected by Mr. P. Jack.

g. 4. *Neuropteris gigantea* Sternb.   Coseley, near Dudley. Middle Coal Measures [212].   Pinnule enlarged to show nervation.

g. 5. *Sphenopteris Höninghausi* Brongt.   Tullygarth, near Clackmannan.   Lower Coal Measures [938].   Natural size.

g. 5a. Pinnule enlarged [936].

g. 6. *Urnatopteris tenella* Brongt. sp.   Furnace Bank, Sauchie, near Alloa, Clackmannanshire.   A few sporangia enlarged and showing terminal pore [1970].

3

2

4

6.

938

5

5ª.

PLATE XXX.

. *Calamites Suckowii* Brongt. Ellismuir, Ballieston, Lanarkshire. Lower Coal Measures. Four-fifths natural size.
Towards the centre of the specimen the rhizome gives
rise to three aerial stems. At the lower end of the
specimen another stem is given off. From the direction
in which the stems bend, it is apparently the under
surface of the rhizome which is exhibited. Collected
by Mr. P. Jack.

. *Linopteris obliqua* Bunbury sp. Pittston, Pa., U.S.A.
Specimen received from the late Mr. R. D. Lacoe
[1348]. Pinnule enlarged about $3\frac{1}{2}$ times to show the
nervation.

. *Calamocladus equisetiformis* Schl. sp. Cadeby Colliery,
Conisborough, Yorkshire. Middle Coal Measures [1536].
Collected by Mr. W. Hemingway.

1.

2

3

PLATE XXXI.

Dactylotheca plumosa Artis. sp.   Monckton Main Colliery,
near Barnsley.   Middle Coal Measures [2105].   Natural
size.   Collected by Mr. W. Hemingway.

Dactylotheca plumosa Artis. sp. forma *crenata* L. & H. sp.
Fruiting specimen.   Monckton Main Colliery, near
Barnsley.   Middle Coal Measures [1210].   Portion of
frond showing *Aphlebia* — the *Schizopteris adnascens*
L. & H.   Natural size.   Collected by Mr. W. Heming-
way.

.   *Dactylotheca plumosa* Artis. sp.   Monckton Main Colliery,
near Barnsley.   Middle Coal Measures [2008].   Pinnule
showing sporangia × 8.   Collected by Mr. W. Hemingway.

a. Sporangium × 25.

.   *Dactylotheca plumosa* Artis. sp. forma *dentata* sp. Brongt.
Monckton Main Colliery, near Barnsley.   Middle Coal
Measures [2112].   Pinnules enlarged.   Collected by
Mr. W. Hemingway.

Engraved by H. Kirkham

PLATE XXXII.

Mariopteris muricata Schl. sp.   Monckton Main Colliery, near Barnsley.   Middle Coal Measures [2393].   Three-fifths natural size.   Collected by Mr. W. Hemingway.

'. Portion of pinna with pinnules to show the nervation. Magnified twice.

Alethopteris lonchitica Schl. sp.   Monckton Main Colliery, near Barnsley.   Middle Coal Measures [1959].   One quarter natural size.   Collected by Mr. W. Hemingway.

. *Alethopteris lonchitica* Schl. sp.   Blairpoint, Dysart, Fife. Lower Coal Measures [2816].   Portion of a pinna enlarged twice to show the nervation.

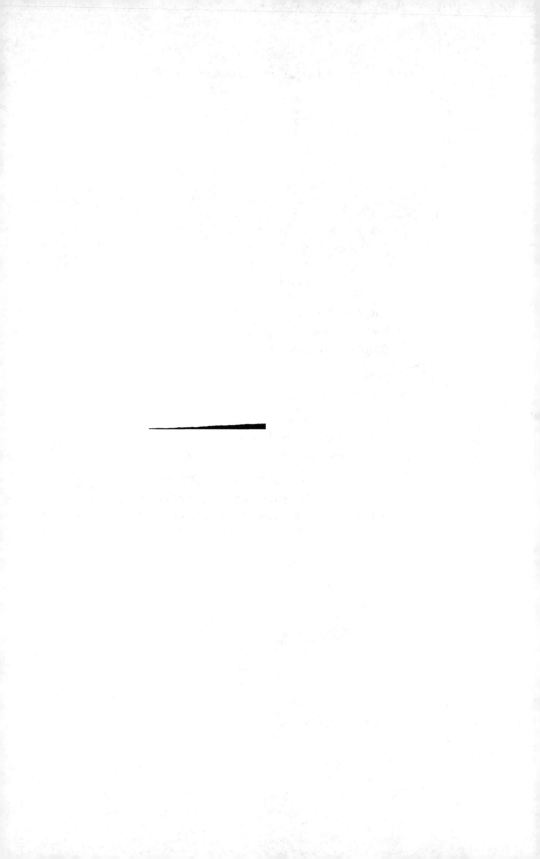

Photographed by R. Kidston.

Proc. Yorks. Geol. and Polytec. Soc., Vol. XIV., Plate XXXII.

PLATE XXXIII.

Caulopteris cyclostigma Lesqx.   Braysdown Colliery, Radstock, Somerset.   Upper Coal Measures [972].   Three-
fifths natural size. *a.* Vascular scar contained within
the sclerenchymatous band *b* of the frond scar.

Megaphyton sp. allied to *Megaphyton anomalum* Grand'
Eury.   Woolley Colliery, Darton, near Barnsley.
Middle Coal Measures [2159].   Three-fifths natural
size. Collected by Mr. W. Hemingway.   Portion of
stem showing one of the rows of frond scars.

PLATE XXXIV.

1. *Calamitina Göpperti* Ett. sp. Woolley Colliery, Darton, near Barnsley. Middle Coal Measures [1199]. Natural size. Collected by Mr. W. Hemingway. At *a* and *b* are seen two whorls of branch scars.

2. *Calamite* from "Coal Ball," Hard Bed, Halifax. Lower Coal Measures. Specimen in the collection of the late Mr. Spencer, Halifax. *a.* Cast of pith cavity. *b b.* Vascular axis with structure preserved. Natural size.

3. *Equisetites Hemingwayi* Kidston. Monckton Main Colliery, near Barnsley. Middle Coal Measures [1678]. Natural size. Collected by Mr. W. Hemingway.

4. Cone of *Equisetum hyemale*, natural size, for comparison with *Equisetites Hemingwayi*.

5. *Palæostachya pedunculata* Williamson. Blairpoint, Dysart, Fife. Lower Coal Measures [1997]. Natural size.

2

3

4

1199

1

a

b

a

PLATE XXXV.

Pinnularia columnaris Artis. sp.   Crophead Pit, Sauchie, near Alloa, Clackmannanshire.   Lower Coal Measures [2815].   Three-fifths natural size.   Probably the rootlets of a *Calamite*.

Calamitina approximata Brongt. sp.   Woodhill Quarry, Kilmaurs, Ayrshire.   Lower Coal Measures [1551]. Three-fifths natural size.   At *a* is seen the cast of the pith cavity, and at *b* the impression of the vascular cylinder.

Calamites Suckowii Brongt.   Oaks Colliery, near Barnsley. Middle Coal Measures [2218].   Three-fifths natural size. Collected by Mr. W. Hemingway.   Pith cast of stem *a*, giving off another stem *b*, also only represented by the pith cast.   At *c* are seen some rootlets.

PLATE XXXVI.

Termination of a Calamite belonging to the section *Calamitina*, showing the long narrow foliage. Three-fifths natural size. Dolly Lane, Leeds. Middle Coal Measures. Collected by Mr. J. W. Bond.

Diagrammatic representation of *Calamostachys*, showing barren and fertile whorl.

Diagrammatic representation of *Palæostachya*, showing barren and fertile whorl.

Calamitina verticillata L. & H. sp. Oaks Colliery, near Barnsley, Yorkshire. Middle Coal Measures [2148]. Three-fifths natural size. Collected by Mr. W. Hemingway. At the upper end of the specimen is seen a verticil of branch scars. The fossil shows a ribbed exterior which in this species appears to represent the outer surface of the plant.

Photographed by R. Kidston.

Proc. Yorks. Geol. and Polytec. Soc., Vol. XIV., Plate XXXVI.

PLATE XXXVII.

Annularia sphenophylloides Zenker sp.   Small branch showing leaf-whorls. Camerton, Somerset. Upper Coal Measures [2304]. Collected by Mr. G. West. Three-fifths natural size.

1. Leaf enlarged.

Annularia radiata Brongt.   This is the foliage of *Calamites ramosus* Artis.   Three-fifths natural size. Lochwood Colliery, Easterhouse, Lanarkshire.   Lower Coal Measures [2426]. Collected by Mr. P. Jack.

1. Leaf enlarged.

Calamites ramosus Artis.   Dolly Lane, Leeds.   Middle Coal Measures [2699].   Three-fifths natural size. Collected by Mr. J. W. Bond. This example shows the outer surface of the stem, which is smooth.

Calamites ramosus Artis.   Devonside, near Alloa, Clack-mannanshire.   Lower Coal Measures [2817].   Three-fifths natural size.   Collected by Mr. J. F. Lyon. This specimen is the cast of the pith cavity.

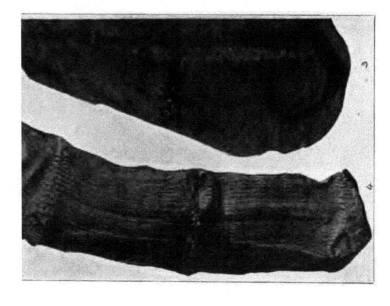

REPORT ON THE DRIFT DEPOSITS AT MYTHOLMROYD.

BY ROBERT LAW, F.G.S., AND WM. SIMPSON, F.G.S.

)n the south side of the river Calder, at Mytholmroyd, is
,teau or terrace raised some 20 to 50 feet above the
it banks of the river, from which it is separated by a
v tract of alluvial holm land. This terrace is about half
e long, by 300 to 400 yards wide, and is on the 330 to
'eet Datum level.

t is bounded on the west by Stubbs Clough and Hawks
h, and on the east it extends to within about 100 yards
ᵊ Cragg Brook, near its junction with the Calder, whilst
ιe south it abuts against and is bounded by the rising
  sides.

ιlthough the Cragg Vale Brook has almost evidently cut
ʒh and worn back the eastern edge of this terrace, the
ι is now seen running against a cliff of grit shales on
ιst bank, and there is no evidence of drift deposits on
ιide.

'o the casual observer this terrace has the appearance of
l-defined alluvial river terrace only.

'or many years, however, local geologists have known that
·posits forming the terrace contained more than the ordinary
alluvium and local rocks, and it appears somewhat strange
the Memoirs of the Geological Survey of the Burnley Coal
  which include this district, make no allusion to the dis-
ɣ of erratics here.

ιlthough these Memoirs were published so late as A.D. 1874,
uthors say in reference to this area, "The only part of
listrict in which drift has been observed is the northern
ɔf Boulsworth."*

ᵊᵉ Geology of the Burnley Coalfield, p. 120.

And again, "An observer cannot fail to be struck ʋ
contrast between the western and eastern sides of the
Chain as regards the glacial deposits. All along the
flanks of the chain and extending into the plains these
are spread in masses, often attaining a thickness of
200 feet, but on the Yorkshire side these deposits are
the strata everywhere appearing at the surface, or only
by soil.*

It would appear, therefore, that these observers did ɩ
the presence of drift with erratics at Mytholmroyd.

So far back as about A.D. 1840, however, according
late James Spencer † (who latterly somewhat boldly d
these deposits as the "Mytholmroyd Moraine"), a con
number of boulders of granite and other foreign rocks,
them half a ton in weight, were found in making the
between Mytholmroyd and Hebden Bridge, the cutting ŧ
traverses the length of the terrace.

In later years many observers have recorded errat
found at the surface, exposed in the footpaths, or at
of the terrace almost throughout its length.‡

The occurrence of glacial boulders in the Calde
originating from Cumberland and the Lake District, i-
course, well established by extensive finds at Millw
Todmorden, Mytholmroyd, Luddenden Foot, Sowerby
Elland, Brookfoot, Mirfield, and elsewhere, and it is not
that these glacial deposits occupy the bed of the va
Todmorden to Sowerby Bridge, covered by more local
alluvium.

In order more thoroughly to examine the natuɩ
Mytholmroyd drift and its contained erratics, a sub-coɩ
the Yorkshire Geological and Polytechnic Society was

* Ibid, p. 133.

† See Halifax Naturalist, Vol. I., p. 21; also Proc. Yorkɩ
Poly. Soc., Vol. XIII., p. 375, section illustrating paper.

‡ See reports of Yorkshire Boulder Committee for 1893, 18

d a visit of the members arranged to be made to view the
alts obtained on the day of the Annual Meeting, November
d, 1899, when we were instructed to submit a report later.

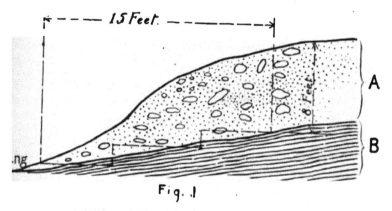

Fig. 1

THE SOUTH SECTION.

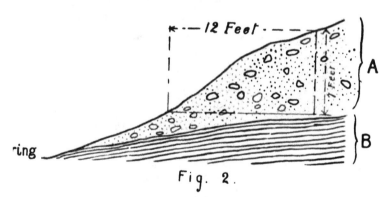

Fig. 2.

THE NORTH SECTION.

A. Drift with erratics.   B. Shales.

Sections cut into edge of terrace of drift deposits at Mytholmroyd
n a field between Scar Bottom and the south side of railway embank-
1ent by Mytholmroyd Station. The parts excavated are indicated by
he dotted lines.

At a preliminary investigation of the area, acting upon the
vice of Mr. P. F. Kendall it was decided to run two cut-
gs or drifts into the edge of the terrace as it rises in the

field between the Cragg Vale Brook and the south
railway embankment, this being thought likely to
satisfactory results than would be obtained by s
the surface.

By enlisting the aid of the tenant of the fa
short drifts were duly cut, running well into the r
which in this field presents two tongues of escarps
embayment between; a section was cut into ea
tongues.

In one cutting (Fig. 1), the southerly one, the
were found to rest upon the undisturbed shale,
junction being traceable by springs; the shales
almost linable with the terrace edge contour, and
therefore was made in a series of three steps for a
in, where a clear face of 8 feet deep by about
wide of drift was exposed.

In the north cutting (Fig. 2) the shales were
commencing a little higher up, and a section was e
12 feet into the terrace ending in a 7 feet face.

Two or three tons of boulders were got out
size from about 12 in. long to pebbles of 1 in.

The great bulk were local grit stones and
some 4 or 5 cwt. were laid aside for further exan
75 boulders were later brought away for careful
these varied in size from about 6 in. by 4 in.
pebbles of 1½ in. long.

A careful examination of the boulders brough
been made, with the help of Messrs. J. H. Howart
Muff, and they were found to consist of and be in
numerical proportion:-

Galliard or Gannister ...   .
Granophyres   ...   .
Andesites   ...   .
Rhyolites   ...   .
Silurian Grits ...   .

Vein Quartz	...	...	...	5
Eskdale Granite	...	...	...	2
Volcanic Ash ...	...	...	...	2
Carboniferous Sandstones	...	...		2
Chert ...	...	...	...	1
Quartz Felsite ...	...	...	...	1

They were imbedded in a loamy, sandy matrix, and were
equally throughout the length and depth of the section,
diately after the grass and a few inches of surface soil
)een removed.

They lay at all angles and positions, and were so separated
he gravelly and sandy matrix that it was quite evident
were not stream laid.

Some of the harder stones possessed flat soles, such as
cterise glaciated stones, but no unmistakable striations
exhibited by any specimen.

The far-travelled stones were well rounded and apparently
: worn, the local galliards were more subangular.

Mr. Simpson has not been able, on subsequent visits, to
any erratics exposed in the western part of the area or in
stream bed of Stubb Clough.   Mr. Law, however, is quite
that he and Mr. Saltonstall have found them all round the
of the terrace.

A careful search was made for erratics on the same contour
as the terrace, but a little further up Cragg Valley, but
out success.

It may be noted that to the west of the terrace beyond
b Clough the valley contracts considerably from both sides,
it is highly probable that it is to the projection thus of
hill sides we owe the preservation of these drift deposits.

It is extremely difficult to account for the phenomena pre-
>d by this drift, with its enclosed erratics, by any other
ry than that it is the product of land ice which travelled
brought its *débris* from the western side of the Pennine
n.   The absence of bedding, the way in which the boulders

were laid and separated by the gravelly matrix, possibility of the deposit being that of running the presence of the erratics throughout the ma lower layer of more local stones, does not lend s theory of local ice being over-ridden by spurs down the valley from over the Pennine watershed.

There are superficial indications further down the southern side between Mytholmroyd and Lu of hummocky drift patches, which offer scope fo repay further investigation.

NOTES ON EAST YORKSHIRE BOULDERS.

BY JOHN W. STATHER, F.G.S.

(*Read July* 14*th*, 1900.)

Our knowledge of the distribution and source of the boulders
East Yorkshire has perceptibly increased during the last few
s, and an attempt is made in the following notes to point out
group the more interesting of the facts, both old and new.

BOULDERS TWELVE INCHES AND UPWARDS IN DIAMETER.

Some ten years ago Mr. G. W. Lamplugh counted and roughly
ified the larger boulders of Flamborough Head and other
ted localities on the Yorkshire coast, and published his results
ie Proceedings of this Society. This work has been continued
members of the Hull Geological Society, who have, up to the
ent time, recorded nearly 4,000 boulders of twelve inches and
ards in diameter. To avoid possible error arising from the
ing beach and other causes, only the boulders actually in place
he clays were noted, or such as had recently and obviously
n from the cliffs. The whole of the coast-line from Spurn
Flamborough has been surveyed in this way, and also
ions of the coast north of Flamborough as far as Saltburn.
lists thus compiled have been published from time to time
the Hull Geological Society and by the Erratic Blocks'
mittee of the British Association.
The cliffs of Holderness, with the exception of certain post-
al deposits, consist entirely of glacial accumulations, and
fore afford exceptional opportunities for the study of East
:shire boulders; and the following table gives the particulars
ned at four localities where the cliff sections were clear
boulders plentiful :—

TABLE I.

	BOULDERS TWELVE INCHES AND UPWARDS IN DIAMETER.	Out Newton (1 Mile of Cliff).	Tunstall (1 Mile of Cliff).	Aldbr (1 Mi Cl
	Origin.	Per cent.	Per cent.	Per c
East Yorkshire Rocks.	Lias...	5·6	13·9	1
	Chalk	1·9	2·9	
	Other Mesozoic rocks, chiefly sandstones ...	3·7	1·8	
Rocks foreign to East Yorkshire.	Carboniferous limestone...	37·8	37·6	2
	Sandstones, grits, &c., chiefly from Carboniferous sources ..	12·4	17·9	1
	Basalts	28·8	23·4	3
	Granites, gneiss, schists, &c.	9·8	2·5	
		100·0	100·0	10
	Actual number of boulders noted	267	274	8

From the above table it will be seen that th
East Yorkshire can be divided into two well-define
first division consisting of rocks from comparatively
and the second division comprising rocks from
localities.

(1) LOCAL ROCKS. The coast of Yorkshire n
lington presents continuous sections of the Jurassic a
strata, and, as might be expected, these rocks are
sented in the glacial beds to the southward.

Lias. In south Holderness hard nodular cor
this formation are plentiful, but large boulders
shales, so characteristic of the lower part of the
places, are rare. Further north, in Filey Bay, bet
Valley and Hunmanby Gap, many masses of Lia
embedded in boulder-clay, both in the cliffs and
shore, and were formerly mistaken for Kimmeridge
Several of the masses in the base of the cliffs

soni beds of the Lower Lias, while on the beach patches
pper Lias occur. One of these patches was observed under
lly favourable conditions during the summer of 1893. The
.er occurred on the beach forty yards from the foot of the
and consisted of a patch of black shale twenty yards long
en yards wide, surrounded by boulder-clay. The shale
ed few, if any, signs of crushing, and contained numerous
preserved fossils, including *Ammonites communis* and *Leda*
. Mr. G. Lether, of Scarborough, also informs me that he
seen similar large masses of shale containing Upper Lias
s in the boulder-clay cliffs situated in the Cliff Bridge
rany's grounds, south of Scarborough.

Oolite. Boulders of Oolite are comparatively rare in south
erness ; but as we proceed northwards and approach the
ibourhood of Filey and Scarborough, where these rocks occur
itu, Oolitic boulders become exceedingly numerous in the
rent glacial clays.

Speeton Clay. Mr. Lamplugh has also pointed out that the
r part of the drifts resting on the Chalk around Selwick's
Flamborough Head, is largely composed of re-arranged
ton Clay.

Chalk. As far as I am aware no boulders of Chalk twelve
es and upwards in diameter have been noted in the drifts
1 of Flamborough Head. In Holderness they occur in fair
bers, though somewhat unequally distributed, as the following
shows :—

t Barmston ...	... 17	per cent. of the boulders are Chalk.		
Skipsea ...	... 14	„	..	..
Atwick ...	... 36	„		
Hornsea ...	... 10	„		
Mappleton...	... 12	„		
North of Aldbrough	24	„		
South of Aldbrough	9	„		
Thorp Garth ...	4	„		
Hilston ...	... 3	„		
Tunstall ...	...: 3	„		

At Withernsea   ...   3 per cent. of the boulders a
., Hollym   9   .,   ..
., Holmpton 12   .,
., Out Newton   . .   2   .,
., Dimlington   ...   3   .,
., Easington 14   .,
., Kilnsea   7   ,,   ,,

This inequality in the distribution of Chalk boulders
different places along the Holderness coast probably a
the fact (pointed out by Wood and Rome) that the l
of the glacial series of Holderness contains more Cl
the upper, and that these basement beds with their h
centage of Chalk boulders rise only occasionally above

(2) FAR-TRAVELLED BOULDERS.   It will be seen 1
No. 1 that, among the far-travelled boulders of the E
shire drift deposits, Carboniferous rocks take numer
leading position; and the Carboniferous area west
of the Tees is generally regarded as their place of or
group of boulders next in numerical importance to
going is the basalts, the source of which is undou
Whin Sill. The next, and last, group is that of th
gneisses, &c., comprising rocks of widely diversified type
variety. Phillips showed long ago that some of these
the English Lake District, and we now know that th
Cheviot and Scandinavian rocks among them; but the
a large number of these rocks have yet to be dete

When the lists of boulders obtained in the sout
shire are compared with the lists obtained in the
bring to light several interesting facts with regard
travelled boulders. Compare, for instance, the lists
Dimlington and Redcliff, in south Yorkshire, with th
Upgang and Saltburn in the north. Before, however
the boulders it is advisable to give a brief descript
localities where the lists were compiled.

(i.) Dimlington is situated on the sea-coast
southern extremity of Holderness. The cliffs ave

feet in height for upwards of two miles, and are entirely
:d of glacial material, chiefly boulder-clay.   Here were
134 boulders of twelve inches and upwards in diameter.
) Redcliff is on the north shore of the Humber, near
 Ferriby, and is twenty-four miles west-north-west of
ton.   The cliff continues along the Humber side for
·ds of a mile with an average height of eighteen feet,
gether with the adjacent beach is composed of boulder-
The boulders recorded here were 373 in number.
.) Upgang is one-and-a-half miles north of Whitby; the
:tions are one hundred feet or more in height, and con-
gely of boulder-clay.   In this neighbourhood Mr. Lamplugh
  and classified two hundred boulders of twelve inches and
s in diameter, the majority of which were of local origin ;
·centages given in the table below are based on his list.
.) The cliffs between Saltburn and Redcar present the
orthern exposure of boulder-clay on the Yorkshire coast.
sections yielded 133 boulders of twelve inches and up-
in diameter.
lter eliminating all the *local* boulders from the lists, at
)ve-mentioned localities, the relative proportion between the
groups of *far-travelled* boulders is as follows :—

TABLE II.

GROUPS.	I. Dimlington.	II. Redcliff.	III. Upgang.	IV. Saltburn.
	Per cent.	Per cent.	Per cent.	Per cent.
rboniferous limestones and sandstones ...	55	59	70	73
salt (Whin Sill) ...	32	30	24	20
ignesian limestone ...	0	0	5	7
anite, gneiss, &c. ...	13	11	1	0*
	100	100	100	100

·veral large boulders of Shap granite were seen in the gardens
)ut the town, which had probably been derived from the neigh-
drifts.

It will be seen from **Table II.** that, althougl
bsoniferous limestones and sandstones, and the ba
travelled into our district from practically the sam
relative proportions of the boulders from the two g
considerably from point to point.   Thus, while it
both groups decrease numerically southwards, the
show that the basaltic group increases relatively fro
southwards.   The explanation of this seems to be tha
boulders of basalt bear transport better than similar i
the Carboniferous sedimentary rocks.

In south Holderness the Magnesian Limestone
is rarely found excepting as pebbles, but these grow
and size northwards.   Large boulders begin to appe
Scarborough, and at **Whitby** and **Saltburn**, as the t
they form from 5 to 7 per cent. of the non-local boul
in the clays.   This rock is matched by the Magn
stone found *in situ* at **Roker, near Sunderland.**

We now come to the boulders of igneous ro
in group 4, which are shown by the table to de
numerically and proportionately northwards.   This
decrease is all the more noteworthy when we rem
boulders of **Shap granite** and other Lake District r
the **Cheviot porphyrites**, all included in this gro
rapidly in the same direction.   This seeming anol
I think, from the influence of the boulders from
Among the boulders of south Holderness occur ver
types which agree with certain Scandinavian rocl
known of these being the **augite-syenite** (*laurriki*
rhomb-porphyry.   These types, although not by any
known in the drifts of North Yorkshire, are much
than in the south.   For instance, at **Dimlingto**
Holderness, we should find at least one hundred bo
above-named Scandinavian rocks to one of **Shap g**
on the other hand at **Robin Hood's Bay** or **Runswi**
near Whitby) the **Shap** boulders outnumber the S
twenty to one.   Seeing then that the boulders

...se both in number and variety southwards, and that among
... are certain Scandinavian types which are *known* to be
... more plentiful in the south of the county than in the
..., I think it may be fairly inferred that the unidentified
... of the group are probably largely from Scandinavia also.

SMALLER BOULDERS, PEBBLES, AND GRAVELS.

Up to the present we are not in possession of even a rough
... of the smaller boulders and gravels of the East York-
... drifts; but the few notes that have been made are of
... interest and suggest a profitable field for further investi-
...n. Among the smaller stones and pebbles we find derived
... of wide range, Carboniferous corals being particularly
...picuous. From the Secondary rocks Lias fossils are perhaps
...most common, though in Holderness specimens from the
...ton clay and Chalk are not rare. North of Flamborough
..., fossils and pebbles from the Cretaceous rocks though rare
...not unknown, and the writer has seen striated pebbles of
... chalk (six inches in diameter) in the glacial clays at Scalby
..., two miles north of Scarborough, and smaller pebbles of
...same character as far north as Kettleness. It has also long
... known that the East Yorkshire boulder-clays, both north
south of Flamborough Head, contain large numbers of black,
..., and green-coated flints, which cannot be matched from
...kshire rocks *in situ*. In addition to this, a well-preserved
...er Cretaceous belemnite, which Mr. Jukes-Brown recognises
Belemnitella lanceolata, is frequently found in the Holderness
...ts, yet is unknown to collectors from the Yorkshire Chalk.
the other hand, *Belemnitella quadrata*, which is common in
Upper Chalk of Yorkshire and Flamborough Head, has not
been noted in the Holderness clays, though sought with
...

It is also worthy of note that among the smaller boulders
pebbles from the boulder-clays and gravels of East York-
...e the percentage of the far-travelled rocks is much higher
...n among the larger boulders. There are certain types also

among the smaller specimens which seldom appe
Among these is a fairly definite group of rock
among East Yorkshire collectors as *porphyrit*
with some confidence to the Cheviot Hills.   The
of this conclusion may be briefly stated as follow
seem to match the descriptions of the Chev
by Mr. J. J. H. Teall and others.   (2) Peb
increase, both in numbers and in size, as
Cheviot district.*

The distribution of both boulders and
inexplicable under the supposition that the drif
in the sea during submergence of the land, b
naturally into place if we acknowledge the
ice-sheets covering not only the land but the
by the sea.

REFERENCES.

S. V. Wood, jun., and J. L. Rome.   "On the G
    Structure of Lincolnshire and South-east York
    Geol. Soc., Vol. XXIV., p. 146.

J. Phillips.   "Illustrations of the Geology of Yor
    Yorkshire Coast," 3rd edit.

C. Reid.   Geological Survey Memoir.   "The Geolo

G. W. Lamplugh.   "The Larger Boulders of Flamb
    Proc. Yorks. Geol. and Polyt. Soc., Vol. IX., p.
    Vol. XI., p. 231; Part IV., Vol. XI., p. 397.

G. W. Lamplugh.   "Drifts of Flamborough Head
    Soc., Vol. XLVII., p. 384.

T. Sheppard and H. Muff.   "Notes on the Glac

ON SECTIONS EXHIBITED DURING THE EXCAVATION OF THE ALEXANDRA DOCK EXTENSION, HULL.

BY W. H. CROFTS.

(*Read November 8th*, 1900.)

This extension is really a small dock, connected with the
Alexandra Dock by a short channel opening into the
‑n side of latter. The extension has an area of about
acres, its centre being about Lat. 53° 44′ 40″, and W. Long.
′ 15″. It is situated on the foreshore forming the northern
of the river Humber, immediately to the east of the
ndra Dock, and near to but west of the eastern boundary
e city of Hull.

The site was originally a mud cranch, the high side of which
es to a height of 12 ft. above O.D., being the level of
.O.S.T. L.W.O.S.T. being 10 ft. below O.D., that, of
e, is the height of the low side at its apparent limit. The
side of the cranch abuts against an artificial bank, formed
rotect the low lying land north of the Humber (in many
s only 8 ft. above O.D., and even less in a few instances),
st the estuarine tides which have gradually warped up the
h to at least 4 ft. above the adjacent country. The date
he first formation of this bank is unknown, but it must
been many hundred years ago; in fact, it is quite certain
here were banks here before the Norman Conquest, for there
llages mentioned in Domesday Book which were on the east
of Hull, and which therefore must have been protected by
k or banks.* On the original surface of the site, part of
xcavated material from the Alexandra Dock itself had been
ited, roughly levelling it up to about 17 ft. above O.D.

The general excavation was carried down to 21 ft. 6 in.
O.D., the floor of the dock being formed at this level, and
hes for the foundations of the dock walls were excavated to
30 ft. to 34 ft. below O.D. The formations exhibited by

---

* King Edward I. and Kingston-upon-Hull. J. R. Boyle, F.S.A.

these works are glacial and post-glacial. The strata
after the removal of the débris from the Alexan
being as follows: Humber warp, shell bed, marl,
stoneless glacial clay, gravel and boulder clay.

The eastern wall trench was most favourable f
observation. At the southern end the strata were as

				ft.
Laminated warp	...	...	...	12
Shell bed	...	...	...	0
Silt	...	...	...	1
Shell bed	...	...	...	1
Glacial gravels	...	...	...	11
Compact boulder clay ...	...	...	**3**

Towards the north the surface of the boulder
a bed of stoneless red clay intrudes into the upper po
gravels, a peat bed makes its appearance, one of th
disappears, and the toe of a sand bank is introduce
warp. At the north end of this trench we find tl
distinct beds of boulder clay separated by grave
stoneless clay having died out, otherwise the sectic
vary from that last described about half way towards
end of the trench.

The floor of the dock itself is formed in the g
over a large part of its area, but towards the nortl
upper boulder clay rises above this level. This gra
largely contributed to the ease with which these
completed, the flood water availing itself to a lar
the course of its glacial predecessors.

The boulder clay forming the bottom of the se
compact stiff grey clay, with a large number of st
boulders in addition to the prevalent basalt, Cheviot
Mountain Limestone, and the usual rocks of the b
Holderness, except that no specimens of Scandinavia
noted. This fact is the more remarkable as in the c
of Holderness the lower the bed the more numero
dinavian rocks become. This clay may be either

NO. 1.—GENERAL VIEW, LOOKING EAST FROM FLOOR OF DOCK.

STERN SIDE OF DOCK, LOOKING NORTH, SHOWING BOULDER CLAY WITH OAK ROOTS,
PEAT, MARL, AND WARP.

Yorks. Geol. and Polyt. Soc. Vol. XIV. Plate XXXVIII.

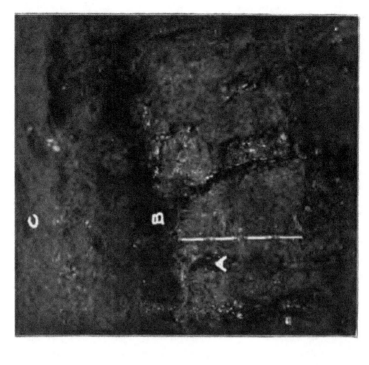

No. 4.—N.W. SIDE OF DOCK LOOKING EAST. A. RED STONELESS CLAY. B. PEAT
WITH ROOTS. C. SAND AND WARP.

3.—PORTION OF VIEW NO. 1 IN DETAIL. A. GLACIAL GRAVEL. B. RED
STONELESS CLAY. C. UPPER GRAVEL WITH INDICATIONS OF PEAT. D. WARP.

wn as the "purple" or the "basement," but as there is
iderable difficulty in differentiating these clays, in the absence
hells or other undoubted feature, in so favourable a situation
observation as the cliffs of the Holderness coast, it is not
sable to be too definite here.

The gravels overlying the lowermost boulder clay are composed
cks of similar character to the boulders in the clay, sometimes
led, sometimes angular, often striated and of distinctly glacial
1, and at the southern end, particularly where most developed,
.fied and divided into thin beds by bands of clay, silt, sand,
These beds dip about 6° south by east, and at the southern
are capped by the lower shell bed. Towards the north the
.ng of the gravel becomes more regular, the constituent
les finer and more compact, and the upper portion gradually
s through bedded sand into a red stoneless clay. Nearer
iorthern end of the trench the glacial beds above the lower
ler clay are as follows :—

					ft.	in.
Red clay	...	...	...	...	5	0
Gravel ...	...	...	...	...	3	6
Stiff grey chalky boulder clay			...		4	0
Grey stoneless clay	...	...	...		0	9
Purple stoneless clay ...	...	...		0	9
Rippled sandy clay	...	...	...		2	0
Stratified sand ...	...	...	...		2	0

The sand is bedded and passes upwards into an interesting
ed sandy clay with carbonaceous markings in the ripples.
ripples are about 10 in., crest to crest, the trough between
being about $2\frac{1}{2}$ in. deep, the axes being about N.N.E. ;
incidence of each overlying crest is not perfect, the
sition of each group of layers developing on the eastern side
he crest, the large number of layers forming each group
; to a considerable amount of time being occupied by their
sition. This bed gradually passes upwards, the junction not
g of a definite character, into the fine purple stoneless clay,
wed by a bed of similar clay and thickness, but of a grey

colour. Overlying these beds there is boulder clay of co
thickness, stiff, grey in colour, and containing a large r
Chalk boulders.

Next in the order of succession is the gra
reduced to 3 ft. 6 in. in thickness, very coarse at its l
fine pebbles at the top, stained black in narrow band
this gravel is 5 ft. of the red stoneless clay with pas
of bedded sand between them. Along the middle of
there is a laminated band formed of extremely fine lay
base of the red clay rises towards the north until it i
supplanted by the gravels. At its southern end the
intrude into it, splitting it into two beds before they
it altogether.

In the south-western portion of the dock area, the l
warp rested on the red clay with the shell bed only int
further to the north, the clay is capped with gravel, a
of glacial origin, with intrusions into the clay, as well a
lenticular patches. This gravel has patches of a v
gravel with peat fragments overlying it in places. T
surface of the clay is in some places quite level, in
undulated as if the agency of water had assisted in
the surface. The clay is apparently quite free from
is of an extremely fine nature, and was no doubt
contemporaneously with the gravels into which it is c
It very commonly forms the upper member of the gla
in this locality, but I do not ever remember seeing
under a peat bed, or where a peat bed has existed at s
as if it was of such an impersistent nature as to be
stand sub-aerial denudation without the protection of a
such as the peat bed.

The upper boulder clay which underlies the red
clay and gravel, although appearing in some places c
colour and very compact, generally follows the descriptio
is known as the Hessle clay, containing a large number
boulders, having ashy coloured joints. Where the red
clay is non-existent the upper surface is of a somewhat

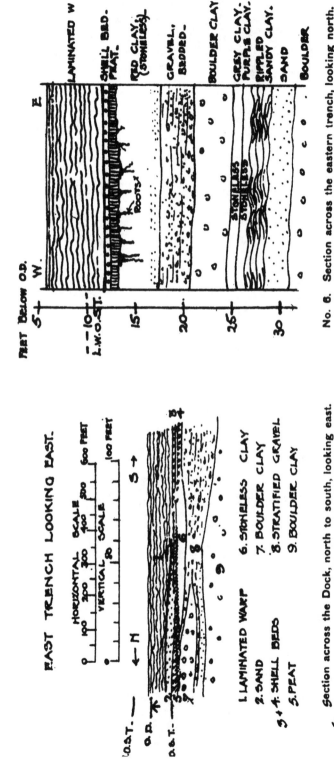

FEET BELOW O.D.

LAMINATED W
SHELL BED. PEAT.
RED CLAY. (STONELESS)
GRAVEL, BEDDED-
BOULDER CLAY
GREY CLAY. PURPLE CLAY.
RIPPLED SANDY CLAY.
SAND
BOULDER

ROOTS

STONELESS STONELESS

No. 8. Section across the eastern trench, looking north.

EAST TRENCH LOOKING EAST.

HORIZONTAL SCALE
0   100   200   300   400   500   600 FEET

VERTICAL SCALE
0   50   100 FEET

1. LAMINATED WARP
2. SAND
3,4. SHELL BEDS
5. PEAT

6. STONELESS CLAY
7. BOULDER CLAY
8. STRATIFIED GRAVEL
9. BOULDER CLAY

No. 6. Section across the Dock, north to south, looking east.

Geol. and Polytec. Soc., Vol. XIV., Plate XL.

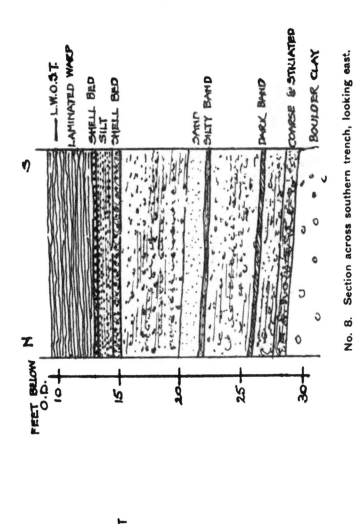

FEET BELOW O.D.

N

S

L.W.O.S.T.

LAMINATED WARP

SHELL BED

SILT

SHELL BED

SAND

SAND SILTY BAND

DARK BAND

COARSE & STRIATED

BOULDER CLAY

No. 8. Section across southern trench, looking east, showing gravels.

SECTION

PLAN

No. 7. Detail of rippled sandy clay in No. 6.

e, on which the peat bed rests with the roots of trees in
running down into the boulder clay itself. This gravelly
ig appears to be the original upper portion of this clay,
onsists of sand and rounded and angular fragments of the
ers common to this clay, some of which are so angular that
can hardly have been subjected to any attrition at all, and,
ey are often quite on the top amongst the peaty fragments,
    cannot have subjected this capping very much to its
bing influence, though the close of the glacial period is
illy associated with flooding.

'he peat bed, which in the northern portion of the dock
s the boulder clay, further towards the south rests upon
xd stoneless clay, sometimes having a gravelly capping as in
ise of the boulder clay.

'his peat bed is similar in some respects to that of the
it meres of Holderness. It covered the whole of the district
Hull, reaching to Hessle on the west, where it gradually
away until at last nothing but a fine line marks the
on of the boulder clay and warp. The peat here did not
i further south than about 200 ft. north of the southern
of the dock, and beyond this there were no indications of
oots having at any time penetrated the underlying beds,
he red stoneless clay at the south-east corner of the dock
i considerable number of roots of reeds running through
d in some places its sandy base had traces of the same
although where the peat itself existed there were but few
of this character.

'he peat consisted of leaves, bark, wood, stumps of trees
in some cases, the roots going down several feet into the
i beds below. No prostrate trunks of any size were noted,
gh this is a common feature of this bed in this neighbour-
    The oak (*Quercus pedunculata*), cherry (*Prunus padus*),
and hazel were represented. Beetles' wing cases were also

ine of the most instructive features of this bed is the fact
its surface 13 ft. below O.D. is 25 ft. below the level of

H.W.O.S.T. The peat was covered with a stifl
generally serving as a matrix, filling up all
upper surface almost perfectly horizontal and le
were cherry stones in considerable quantity.
perfect cherries (*Prunus padus*) were found prese
At a point near the north-western angle of tl
filled up a depression in the underlying bed:
surface, and towards the northern edge of the d
of small pieces of charcoal was found, but an
search revealed no traces of human agency.

The two shell beds, as before stated, at the
the dock, rest on the gravels, and are separate
1 ft. 6 ins. in thickness : but towards the nort
gradually dies out, and generally there is but or
of the marl over the peat bed, the shells formin
on its surface, but in some places penetrating i
cracks. This suggests that the clay had been
dried before the waters of the estuary formed
which is indicated by the large number of sma
the fact that both valves are often intac
represented were *Cardium edule, Tellina solid
piperata, Utriculus obtusus, Rissoa ulva, Littorin
obtusata, Mytilus edulis, Pholas candida,* and .
the five latter have not been recorded before
this bed. The lower bed in places consisted
Scrobicularia piperata packed closely together.

Over the shell bed the Humber warp cr
but from the centre of the dock, thickening t
a laminated sand separates them, the uppe
gradually passes into the warp. The warp itse
alternate rise and fall of the tidal river Hum
colour, dark when damp and lighter when dry
tremely fine layers, each consisting of the sand a
the sand of course being heavier and forming tl

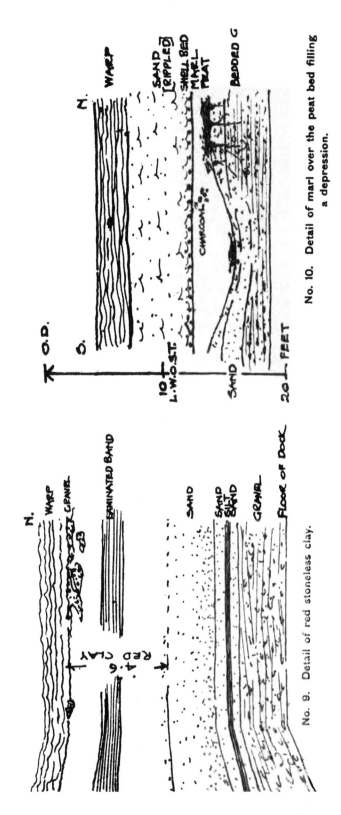

No. 10. Detail of marl over the peat bed filling a depression.

No. 9. Detail of red stoneless clay.

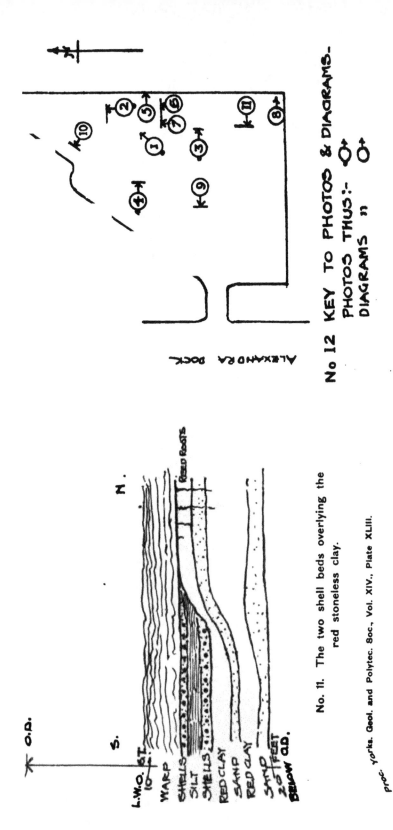

No. 11. The two shell beds overlying the red stoneless clay.

No 12 KEY TO PHOTOS & DIAGRAMS.

PHOTOS THUS:-

DIAGRAMS ,,

[olderness Cliffs, together with detritus brought down the
flowing into the Humber, it is advisable to suspend judgment
a series of exhaustive experiments put the matter beyond
e. This warp is of course identical in composition with the
ow in course of formation along the shores of the Humber.
here is evidence of a dip in the Chalk underlying the
t, corresponding with the basin of the river Hull, and the
t chalk cliffs may possibly be east of the river. The glacial
:s extend from the sea on the east, on to the edge of the
on the west, the depression in the valley of the Hull
filled with the warp. The glacial gravels bear a strong
·lance to those adjoining the ancient meres of Holderness,
do not contain fossil shells, and have the appearance of
formed in a similar manner.
here is a gravelly bed extending over a great part of
·ness between the boulder clays known as the "purple"
'Hessle," filling channels of varying depth in the so-
"purple" clay, the ancient mere beds being apparently
irge channels along which at one time strong currents have
, the percolation water deepening the channel and letting
the gravels. The "purple" clay where the gravels are
developed is generally attenuated, and sometimes missing
her, so that this bed may fairly be connected with that
n. I am inclined to think that as the Kelsey Hill gravels
· a similar position between the "Hessle" and "purple"
they belong to the same period as this great gravel bed,
iat the presence or absence of shells is due to the position
old beach from which they are probably derived. These
s are well exhibited in the excellent coast sections.
he peat bed, although now 13 ft. below high water level
inary spring tides, must at one time have been above the
vater level, when the physical appearance of the district
have considerably differed from its present aspect, but
teresting field this opens out can scarcely be touched upon
Take one point alone: the Humber channel at the present
s of no great depth in places, and I am inclined to think

that from the shape of its preglacial mouth its course past
is not extremely ancient. Off the Ferribys there is a (
boulder clay which appears to be the remains of a moraine
no doubt at one time dammed the Humber waters
a considerable task at the present time, but granted a high
level of only 30 ft. lower than at present, the section:
exhibited suggest a bank more than equal to such a task

The perfect preservation of the cherries found in the
overlying the peat bed seems to suggest that a flood oc
when the cherries were ripening on the trees. The level s
of the marl with the shells from the bed above penetratin
small crevices may suggest that the surface was dried
the subsequent final inundation, which took place, so far a
be read by the method of deposition of the warp, th
a gradual subsidence.

With regard to the pieces of charcoal, an occurrence
had a parallel in the Albert Docks on the west of the
(in fact the two sections strongly resemble one another·
traces of human agency were exhibited, and Dr. Jessen, c
Danish Geological Survey, informs me that charcoal
common occurrence in the peat beds of Denmark, and d
natural ignition.

The North Eastern and Hull and Barnsley Railway:
template the construction of a large dock to the east c
works described in this paper, and it is to be hoped tha
sections will, in connection with the Albert Dock and
works, assist to complete the record of the strata under th
of Hull.

I have to thank Mr. R. Pawley, C.E., the engineer
Hull and Barnsley Railway and Dock Co., and Mr. ·
Stather, F.G.S., for assistance in getting these notes togethe
Mr. W. S. Parrish, for assistance with the photographs.

---

* Notes on the Drifts of the Humber Gap. J. W. Stather,
Proc. Yorks. Geol. and Polytec. Soc., Vol. xiii., Part II. **1896.**

† Albert Dock, Hull. J. C. Hawkshaw, F.G.S., Q.J.G.S., Vo
1871.

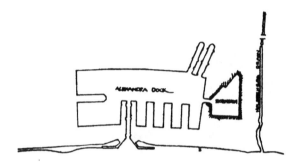

ALEXANDRA DOCK

SITE OF
PROPOSED JOINT DOCK

FILEY BRIGG

FLAMBOROUGH
HEAD
+ BRIDLINGTON

NORTH SEA

GREAT DRIFFELD
+

+ DRIFFIELD

+ NAFFERTON

+ BEVERLEY

+ HULL

+ WITHERNSEA

RIVER HUMBER

SCALE

SPURN

PLAN OF DISTRICT

NOTES ON THE OCCURRENCE OF THE ADWALTON STONE COAL AND THE HALIFAX HARD COAL.

BY THEODORE ASHLEY.

(Read November 8th, 1900.)

At the meeting of the British Association at Bradford in September, 1900, a discussion was held with regard to the conditions under which coal was formed. It was suggested to the writer by Mr. Percy F. Kendall, F.G.S., that it would be useful to put on record all the facts which could be ascertained with reference to some definite seam or seams. In accordance with his suggestion the following details have been compiled, partly from information collected by the writer and partly from the records published by the Geological Survey.

The seams of which a description is given are those known as

(1) The Adwalton Stone or Cannel Coal.

(2) The Halifax Hard Bed Coal.

These two seams were chosen as being the most different from each other in character of any in the north of the Yorkshire Coal Field, and at the same time as being thoroughly representative of their respective types. It is not proposed to deal in any way with their chemical composition or their commercial value, but simply to give such facts regarding their occurrence as bear on the conditions of their formation.

(1) THE ADWALTON STONE OR CANNEL COAL is found, as its name suggests, at the village of Adwalton, which lies about $\frac{1}{2}$ miles to the south-west of Leeds. It is not, however, here that it attains its greatest thickness. It occurs in the Middle Coal Measures lying about 130 yards above the Blocking or Silkstone seam, which is recognised as the dividing line between the Lower and Middle Coal Measures, and underlies an area bounded on the north by the village of Middleton, situated about a couple of miles to the south of Leeds, on the south by

a line a few miles south of Thornhill, and on the west by its outcrop near Birstall. Its boundary on the east is referred to later.

Upon reference to Plate XLV., which is taken from the Geological Survey Map on the scale of one inch to the mile, it is first seen on the north, just north of Middleton: it crops out along the hill-side. Going west from Middleton, a pair of faults is met with throwing the seam down, and on the far side a small oultier occurs. A large fault, running in a south-easterly direction, half-a-mile to the north of Morley Main Collieries, throws the Middle Coal Measures against the Lower series. Further west, it again crops out on the hillsides in the neighbourhood of the village of Adwalton. To the south small oultiers occur west of the main body, formed by a complicated series of faults. A large fault, running east and west in the vicinity of Batley, again throws the seam out, and it is next met with on the eastern side of the valley between Batley and Dewsbury. To the south a complicated system of faulting occurs, making the outcrop very difficult to trace.

Thickness of the Seam. At Middleton, on the northern outcrop, it is only $1\frac{1}{2}$ inches thick, at Morley 5 inches, at Gildersome $9\frac{1}{2}$ inches, thinning down to 6 inches at Adwalton. To the east of Middleton, there is evidence that the cannel in the seam dies out, its place being taken by an inferior coal: for instance, at Rothwell Haigh Colliery no cannel is found, and still further to the north-east the ordinary coal in the seam seems to have deteriorated, for at Woodlesford a seam of "bad coal" is found on the same horizon, after which it disappears altogether.

In the neighbourhood of Batley various thicknesses are met with, all greater than are found in the north, the greatest thickness being at Staincliffe and at Clark Green, where it is proved 16 inches thick.

Passing southwards, it is 11 inches thick at Dewsbury Bank Colliery and at Briestfield, varying a few inches between these points.

In the extreme south, as in the north, it gradually thins out, the cannel being replaced by ordinary coal. This is proved at various collieries — Overton, Grange Moor, and Prince of Wales'—where the seam is known as the Flockton Thick Coal, having a thickness of about 3 feet 6 inches, divided by a bed of shale of varying thickness.

On the east. unfortunately, it has only been proved to a small extent, owing to the dip of the measures in this direction, whereby the seam is overlaid by newer beds and runs deeper than any existing pits ; but at the two most easterly points where it has been proved, viz., at East Ardsley Colliery on the north, and on the south at Hartley Bank Colliery (west of Horbury) it is respectively 3 inches and 4 inches thick.

It will thus be seen that both on the north and south there is actual evidence of the cannel having altogether died away, and on the east the evidence is strongly in favour of the same thing having taken place.

Plate XLVI. represents horizontal sections of the seam, which show very clearly the thinning out and the variable thickness of the bed. The horizontal scale is a half inch to the mile, the vertical scale being a half inch to the foot.

Fig. 1 is a section running north and south, starting at Briestfield with the cannel 11 inches thick, from which point it gradually thins down to 5 inches at Morley, the furthest point north on this line that I have any information about. The coal is then cut off by the large fault bringing in the Lower Coal Measures.

Fig. 2 runs from Adwalton on the west, showing a gradual thickening until Tingley is reached, from which point it again thins down to East Ardsley, where it is only 3 inches thick.

Fig. 3, section running approximately east and west and showing the greatest recorded thickness at Staincliffe (16 inches). From Staincliffe it gradually thins down to East Ardsley, where it is only 3 inches.

The extent of the coal on the south is so small that it has not been possible to draw any section east and west through it.

Of course, as will be seen, the vertical scale of the sectio
is very much exaggerated; this is done in order to show t
variation in the thicknesses. Only the top bed of the se
which is invariably cannel coal, is shown.

Underlying Strata. It is interesting to note the str
underlying this bed of cannel.

The area over which this coal extends, as shown on the pl
does not indicate that the seam does not exist further to
south and east, but only that the cannel bed found over
area ceases. The Flockton Thick Coal into which, as I h
already stated, it gradually passes, is supposed to exist on
east, and is known to do so along its outcrop on the south,
owing to the dip of the measures it has never been pr
further east than at the two points East Ardsley and Ha
Bank Colliery.

On the south the section of the whole seam is, at Bunl
Hill Colliery, Stone Coal 10 inches, Brown Stone 2 inches,
Coal 10 inches, Dirt 1 foot 10 inches, Coal 1 foot.

At Thornhill the section is:—Stone Coal 1 foot, Meas
6 feet, Coal 1 foot.

At Dewsbury Bank the shale between the two coals
decreased to about 2 feet.

It will thus be seen that, except on the south, there is
Middle Coal ; further north, however, there are, as a rule, tl
beds of coal with thin partings between ; the top one is invari
cannel, the other two bright coal, although at Howley P
and adjacent collieries underneath the Middle Bed, which is
2 inches thick, there is found a bed of inferior cannel knowr
" Hubbs."

Thus it would appear that, as the cannel increases in th
ness, the so-called Middle Coal thins out, and the writer sugg
that the cannel thickens at the expense of the bright c
At Morley Main Colliery the Bottom Coal is composed of can
and in some parts of the mine at Tingley all three seams
found lying together without any dirt partings, both top
bottom coals being cannel. The top coal is 5 inches thick

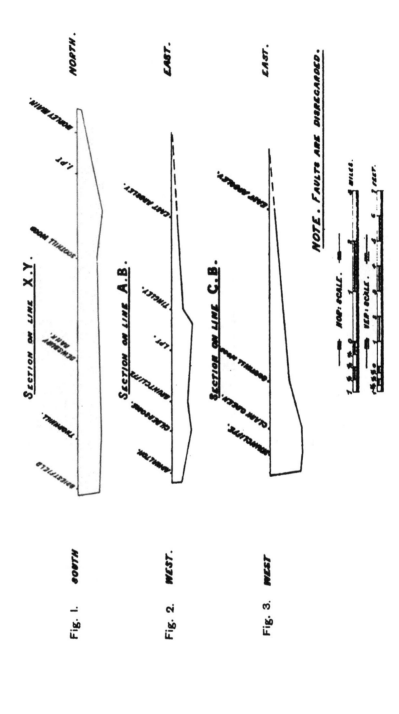

SECTION ON LINE X.Y.

NORTH.

MORLEY MAIN.

L.P.T.

SOOTHILL WOOD.

BEESTONPARK MAIN.

THORNHILL.

BRESTFIELD.

SOUTH

Fig. 1.

SECTION ON LINE A.B.

EAST.

EAST ARDSLEY.

TINGLEY.

L.P.T.

ARNTCLIFFE.

CLECKSTONE.

ARNULTON.

WEST.

Fig. 2.

SECTION ON LINE C.B.

EAST.

EAST ARDSLEY.

SOOTHILL WOOD.

CLAN GREEN.

ARNCLIFFE.

WEST

Fig. 3.

NOTE. FAULTS ARE DISREGARDED.

HOR: SCALE.

MILES.

VER: SCALE.

FEET.

bottom 3 inches. In other parts of this mine the cannel
is found 10 inches thick.

At East Ardsley the cannel is underlaid by bright coal 2 feet
k, there being no parting between them, the bottom half of
ordinary coal being inferior.   This corresponds very nearly
he section found at Hartley Bank Colliery, the bright coal
e being 1 foot 10 inches thick.

Isolated patches of cannel are found in the south as well as
he north, as at Kirby, where a cannel is met with on the top
he Flockton Thick Coal 4 inches to 5 inches thick, and further
h still at Hoyland Silkstone a cannel coal is found above
Flockton 10 inches thick.   Other instances could be given.

Overlying Strata. The roof over the cannel coal is always of
same character, and is composed of from 1 to 2 feet of black
containing few traces of animal life.

Above this shale is a stratum 10 inches to 1 foot thick,
times running thicker, which is crowded with the shells of
racosia.   Above this shell bed are about 20 feet of bluish-
shales containing several layers of ironstone nodules.   Shells
nthracosia are common in the ironstone but do not occur in
shale.

This band of ironstone nodules is found overlying the Flockton
Coal, north of Sheffield, where it has been extensively
ed.   It has also been worked at West Ardsley, where it
ly resembles that of the south.   To the north, however,
with the disappearance of both the cannel and the under-
coals, the ironstone measures die away.

The cannel itself contains very numerous remains of fishes
he form of scales, plates, and spines ; also very numerous
ins of a small animal belonging to the group Ostracoda.
is in marked contrast to any case of bright coal, and indeed
ins of marine animals have never been found in any true coal.

(2) THE HALIFAX HARD BED COAL.—The Halifax Hard Bed
along with the Halifax Soft Bed Coal lying a short distance
w, and more generally known as the "Ganister Coals," are
lowest seams of the Lower Coal Measures.

The Hard Coal extends from the neighbourhood of Shipley right along the western edge of the coalfield through Halifax, Huddersfield, Penistone, and, although not shown on the map (Plate XLVII.), the seam is known to exist right to the southern extremity of the coalfield near Nottingham.

Thickness of the Seam.   On the plan (Plate XLVII.), the points where the bed has been proved are shown, the thickness of the seam being given.   It has only been worked at and near its outcrop on account of the increasing depth to the east, the measures dipping in that direction.

Starting at a point about six miles north of Sheffield, the seam is found 2 feet 3 inches thick.   Going north, it varies in thickness a few inches only, until New Mills is reached, where it is found only 1 foot 6 inches thick.   It, however, soon thickens out at Huddersfield, averaging about 2 feet 3 inches, this thickness being maintained until Queensbury, on the north side of Halifax, is reached.

It thins out somewhat to the north and is found in outlying patches brought in by faults, as at Baildon and Rawdon.

North of Leeds it is found only a few inches thick.

Underlying Strata.   The floor underlying this seam is, except in the neighbourhood of Queensbury, a hard ganister, consisting almost entirely of silica.

Underneath this ganister a bed of fireclay is generally found, in some places worked extensively.   North of Queensbury, however, the fireclay is found immediately under the coal.

Overlying Strata.   The roof is a shale full of *Aviculopecten papyraceus, Goniatites,* and other marine fossils.   Plate XLVIII. is a section of the coal along its outcrop, which runs approximately north and south, and is drawn on the same scale as that of the cannel coal (Plate XLVI.).

COMPARISON OF THE TWO SEAMS.

Upon comparing the two seams, it will be noticed the thickness of the Halifax Hard Coal as compared to the cannel coal is extremely uniform.   The length of the section of the Hard

shown is about 36 miles, while the longest section running
about eight miles.

ON

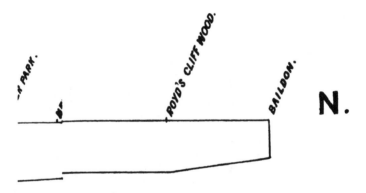

dy stated, of nearly pure silica, there being an almost entire
nce of potash and iron.   It is penetrated in all directions

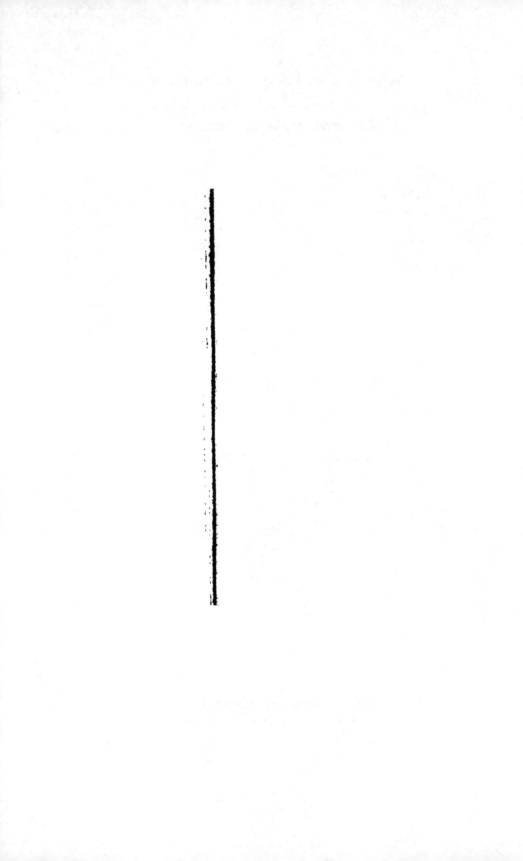

shown is about 36 miles, while the longest section running rough the cannel coal is only about eight miles.

While the roof of the cannel coal and the Hard Coal are alike inasmuch as over the area traced they both contain marine fossils, the floor of the Hard Coal is always practically the same, whereas the floor of the cannel coal varies exceedingly, being sometimes dull and sometimes bright coal. Furthermore, as already indicated, the marine organisms are found actually embedded in the cannel, which is not the case in the Hard Bed.

The cannel coal occurs in the form of a lenticular patch, having its greatest thickness in the neighbourhood of Batley, and thinning out north, south, and east. Unfortunately, it is impossible to say what takes place to the west, as the seam crops out about two miles west of Batley.

The Hard Bed, on the contrary, is not lenticular in form, and is much more constant in thickness.

The cannel coal contains a high percentage of mineral matter, and also numerous remains of marine organisms, all of which are quite consistent with the view that it is a deposit of drifted material laid down in shallow water, and the fact that it never has a definite under-clay or other bed which would have been likely to serve as a floor on which the vegetation of which it is composed could have grown, there seems to be little doubt that it is actually a drift coal.

The Hard Bed, on the other hand, contains a percentage of mineral matter so low as to be quite inconsistent with the drift-wood theory, since the currents which carried the vegetable matter would in all probability have carried also sand, or at all events fine silt. This silt, on being deposited, would of course largely increase the percentage of mineral matter.

The coal is so constant in thickness over such a wide area that it seems impossible to attribute it to deposition in shallow water.

The ganister upon which the Hard Coal rests consists, as already stated, of nearly pure silica, there being an almost entire absence of potash and iron. It is penetrated in all directions by roots and rootlets which are in all probability those of the plants whose dead tissues formed the coal. The ganister is, in fact, a sandy soil which has been exhausted by the removal of its iron and potash salts by the plants which grew upon it.

These two seams furnish examples of coal of widely different origin, and while the "driftwood" theory is probably correct in the case of the cannel it cannot be accepted in that of the Hard Bed.

Some of the sections are taken from the memoir of the Geological Survey on the Yorkshire Coal Field, others being obtained from information kindly given by Mr. Walter Rowley, F.G.S. The writer wishes to express his sincere thanks for the valuable suggestions and help received from Mr. A. R. Dwerryhouse, B.Sc., F.G.S.

In Memoriam.

WALTER PERCY SLADEN, F.L.S., F.G.S., F.Z.S.

BORN AT MEERCLOUGH HOUSE, HALIFAX, YORKSHIRE, 30TH JUNE, 1849;
DIED AT FLORENCE, ITALY, 11TH JUNE, 1900.

n his prime, endowed with rich mental gifts, a refined and
e nature, happy in his family life, beloved of his friends,
ted by all who knew him, bountifully favoured by fortune,
abundant and precious material collected together sufficient
ars of scientific research—suddenly, the bright light of his
; was extinguished at Florence, on 11th June, 1900.

Ve, who held him so dear, have in the death of W. Percy
1 to mourn the loss of the most brilliant and most
sing of our little circle of Yorkshire naturalists.

in appreciative and able memoir appeared in *Nature* of 12th
1900, from the pen of his friend Dr. G. B. Howes. It is
we can add to that memoir; perhaps, however, something
may be said as to his earlier years with which his
guished southern friends were less acquainted.

n the sixties and early seventies, several ardent lovers of
re met together from time to time at the Halifax Museum,
at each other's residences, to compare notes, to exchange
and to discuss Evolution, the origin of species, the work
arwin, Huxley, and Tyndall, which at that day engaged the
ion of the scientific world; but not content with mere
sion, material for study was obtained, and under the skilled
ction of the late A. Campbell (curator to the Halifax
im, and one time assistant to Professor Jamieson, of
urgh), dissections were made and skeletons prepared of
parrots, snakes, crocodiles, fishes, and other animals, and
ctical acquaintance with type-structures acquired. The
iens in the Museum were compared with fossil forms;
ng excursions were made by the friends conjointly or
tely (to investigate marine forms, especially invertebrates)

to Tenby, Milford Haven, the Cumbraes, Portrush, Belfast [?] Weymouth, and the Isle of Man. Some members of the brofraternity possessed first class microscopes, and so histology and micro-organisms were not neglected ; in all this delightful an[d] happy work every month had its discovery where all was new and none was more diligent, none expressed clearer ideas, none studied the foreign literature relative to his work more enthusiastically or more profitably than Percy Sladen.

As a boy Sladen was playfully dubbed by his schoolfellows "the Astronomer Royal ;" indeed he was always an organiser and a leader. For many years his motto might well have been "Something of everything," until he finally added to it "Everything of something." When Mr. Carruthers lectured before the local Literary and Philosophical Society on the microscopic structure of the plants of the Coal Measures, it was our friend who prepared sections, with the aid of an old saw and a slab of stone, to the required thinness, and then demonstrated their vegetable anatomy : it was he who first introduced us to the graptolites and entomostraca of the Palæozoic rocks, and explained the foraminifera and polycystina from deep sea dredgings ; it was he who, along with his intimate friend Mr. John Stubbins, F.R.M.S., first explained to us the niceities of microscope illuminations, and taught us to prepare and mount microscopic objects.

Never to be forgotten is a visit to Tenby, when for the first time the circulation in ascidians (Clavellina) was presented to our wondering eyes. Charmed with Sladen's beautiful sketches made on the spot, and with his description of the structure he demonstrated, we hung over the instrument until the early hours of dawn. It is probable that his observations on the abundant material collected at Tenby and at Milford Haven determined his selection of the Echinodermata as the group about which he determined to know everything of something.

The star fish, sea urchins, and ophiurids were found plentifully both as to species and specimens : many interesting questions were raised and discussed respecting the structure and functions

[*Exemplar*]

Yours most sincerely

W. Percy Sladen

these interesting creatures, and the foundation laid of his
ure extensive and minute knowledge of the group.   The steady,
iet work of this and following years was a real preparation for
ɔ production of his *magnum opus* on the Challenger *Asteroidea*,
iich secured for him a foremost place among echinodermists of
ɔ world.

It has been said that he never attended a regular academic
irse of instruction on the branch of science in which he became
iinent ; that is also true of many of the most eminent men of
ɔry place and age ; but he learned his lesson in the true
iversity, that of Nature, and not only knew the dry bones of
i subject but also, what is not so common, he knew the
ucture, habits, and habitats of the subjects of his loving
idy.   Nor were preliminary and auxiliary studies wanting : his
ictical acquaintance with chemistry, botany, zoology, and
idern languages was more than elementary, and his skill in
croscopic manipulation and drawing was great.

Percy Sladen came of an old Yorkshire family which for
nerations was associated with the Halifax district ; in his
tive town he was held in great respect, and held several
nourable scientific positions.   The Halifax Literary and Philo-
phical Society found in him a most efficient secretary; he was
nnected intimately with the Halifax Scientific Society (mainly
inded through the instrumentality of Dr. Sollas, F.R.S., of
:ford), and rendered valuable scientific service as an honorary
ator of the Halifax Museum.   In all the work he undertook
acquired the character and reputation of being painstaking,
rough, and always reliable ; superficiality and pretence he
ays detested.   A Fellow of the Geological, Zoological, and
rixæan Societies of London, he was for ten years secretary
the last named society, and for some time held the post of
'—president, whilst for eighteen years he acted as the British
sociation Committee's secretary to the Zoological Station at
iles.

In the year 1877 his studies extended further afield, when
paid a visit of some months' duration to the celebrated
M

Zoological Station at Naples, where he was much encouraged by the lively sympathy of its brilliant head, Dr. Anton Dohrn. Diligently he collected the marine fauna of the Mediterranean and perfected his methods of preserving and investigating marine organisms, and thus became specially fitted for the post of secretary of the British Association Committee, and to report on the work done at the station. In this year, too, we find him, in conjunction with his friend Professor P. Martin Duncan, M.B., F.R.S., publishing the results of his examination of the Echinodermata collected by the Arctic expedition 1875-6, followed by further papers on the Echinoderms collected by Captain Sir G. S. Nares' expedition to the Polar Seas, and on the Echinoderms of the Arctic Sea of West Greenland.

In 1878 his first paper appeared in the Proceedings of the Royal Society for that year. It was on *Astrophiura*, a new and remarkable creature, which in the following year was described by him in the Annals and Magazine of Natural History and in the Zoologische Anzeiger. In 1878 his first communication to our own Journal on the Genus Poteriocrinus and allied forms appeared. In 1879 appear papers on the *Asteroidea* and *Echinoidea* of the Korean Sea, his first Palæontological paper to the Geological Society of London on *Lepidodiscus Lebouri*, and his first report for the British Association Committee on the Zoological Station at Naples, the precursor of annual reports which extended to the year 1898. By this time his exact knowledge as a specialist was widely recognised, and collections from various quarters were placed in his hands for description ; in 1880 he describes echinoderms of the Barents Sea, followed by a paper on the occurrence of Pedicellaster in the far north, and another paper to the Yorkshire Geological and Polytechnic Society, the title being " On Traces of Ancient Relations in the Structures of the Asteroidea."

The year 1882 was for him of great educational importance in relation to his echinodermic researches, for he now made a tour of the European Museums, and during his travels he found time to correspond with the little fraternity at Halifax.

ly to an expression of regret at his absence from one of
posed gatherings, and a hope that he might be present in
!) he sympathetically replied, enclosing one of his inimitable
pen and ink sketches, displaying himself and each member
Club preserved in jars of *spirit*, and each characteristically
1 with a suitable droll name of his own invention:
is European visit yielded him a wealth of material and
, notes and drawings, to be in part at least worked up in
moir of the Asteroidea of H.M.S. "Challenger" expedition.
ar of his foreign tour saw the publication of his description
Asteroidea dredged during the course of the "Knight
" as well as the first part of his monograph of the Fossil
idea, which was extended in four more yearly parts, and
ed in conjunction with his friend Dr. Duncan, in the
tographical Society's Volumes, 1881-1885.
, the grief of the Halifax circle, in 1883 Sladen left to
at Ewell, near London. This nearness to the great heart
llectual activity proved of great advantage to him by
g him into more direct intercourse with the leaders of
ic thought, among whom he soon made some stimulating
sting friendships.
gain we find collections committed to him for description,
ne from the Faröe Channel. In the words of Professor
"these reports mostly deal with whole collections, and
reports on those made in the Arctic Region in 1875-1876,
se of the 'Alert,' 'Knight Errant,' and 'Triton,' as also
1ade in the Faröe Channel, the Korean Sea, and the Mergui
alago. In each Sladen produced good results, as in the
ry of genera such as *Micraster* and *Rhegaster*; and what
atural, therefore, than that he should have been entrusted
he working out of the Asteroidea collected by H.M.S.
nger,' the report upon which was the crowning achieve-
f his life."
iring the next ten years he produced the chief work of
: : paper follows paper, each exhibiting trained powers,
work, and original thought. In 1883 he contributed to

the Journal of the Linnæan Society a communication on the Asteroidea of H.M.S. "Challenger;" a description of *Vinaster*, a new genus from the Faröe Channel ; the British Association Report on the Scottish Zoological Station (1881-2-3): and a monograph of the Tertiary Echinoidea of Kachh and Kattywar appear the same year. In 1884 he wrote on the homologies of primary larval plates, of the test of Brachiate Echinoderms, and on the classificatory position of *Hemiaster elongatus*. In 1885 appears in the narrative of the cruise of H.M.S. "Challenger" a general summary of the Asteroidea collected during the expedition, and a paper in the Linnean Society's Journal on the family *Arbaciadæ*. In 1886 he continued bibliographical notices on the Crinoidea, etc., which were commenced the previous year : issued his monograph of the Fossil Echinoidea of the coast of Beluchistan (Palæontologia Indica) ; and a paper on the anatomy of the perignathic girdle and other parts of the test of *Discoidea cylindrica*. In 1887 he published in the Annals and Magazine of Natural History his views on some points in the morphology and classification of *Saleniidæ*. In 1888 were given in the same magazine his objections to the genera *Pseudophygaulus*, *Trachyaster*, and *Ditremaster*. In 1889 appeared the Report on the Asteroidea collected by H.M.S. "Challenger" during the year 1873-1876. We do not pretend to estimate the scientific value of this truly superb work, but content ourselves with quoting from one far better qualified by his critical ability to do so :—" This magnificent work of 900 pages, with its accompanying atlas of 118 plates, ranks among the most masterly and exhaustive of the 'Challenger' volumes. Before taking it seriously in hand, Sladen visited every museum in Europe (with one exception) which was known to contain star fishes of importance ; and, as pointed out by the editor in its preface, it is a monograph of the whole group. The labour involved in its production was prodigious ; and its interest is enhanced by the fact that the bulk of it was written between the hour of 9 p.m. and those of early morning, often after a day's occupation with other affairs. The extension of the family Pterasteridæ

the great addition to our knowledge of the deep sea forms
.ts most salient characters ; but we know not which to
re most, the body of the work, with its laborious descriptions
idividual forms, or the supplemental part, in which there
ven a list of every known species, with a record of its
ymetric distribution.   Elementary student and expert stand
  indebted to him for this monumental work, indispensable
·ogress in the knowledge of the subject with which it deals.
ric names like *Benthaster* and *Marsipaster* are sufficiently
ficant in themselves.   Proceeding to classification, Sladen made
  use of the marginal and ambulacral plates, and his sub-
ion into the sub-classes *Euasteroidea* and *Palœasteroidea*, with
·rdinal divisions to which he was led, has withstood the test
ne, and become the adopted classification of the better text
s, as for example those of Lang and Gregory.   In this his
:nce on the progress of science will live, and it is a matter
·ofound gratification that only a short time before his death
ive expression to the satisfaction this afforded him."

His last memoirs were published in the Palæontographical
ty's volumes for 1890 and 1893, and were on the Cretaceous
·oids.

In the year 1890 he married Constance, elder daughter of
ate Dr. W. C. Anderson, of York, a union of heart and
, yielding a bright and tender sympathy which strengthened
stimulated him in his life's work.   On 14th February, 1898,
incle, Mr. John Dawson, died, leaving him Northbrook, his
:iful home in Devonshire.

The friends of Sladen's youth and early manhood appreciated,
ideed all who were brought in contact with him must have
  his clear and logical powers of mind and refined nature ;
·gst the Yorkshire naturalists who met him again and again
ientific fellowship and communion, his judgment was highly
d ; the late J. W. Davis, F G.S., his friend and neighbour,
. Hick, B.A., B.Sc., John Stubbins, F.R.M.S., Geo. Brooke,
3., G. H. Parke, F.L.S., C. P. Hobkirk, F.L.S., and W. Cash
  amongst this privileged band.

Where once we held debate, a band
  Of youthful friends, on mind and art,
  And labour, and the changing mart,
And all the framework of the land.

When one would aim an arrow fair,
  But send it slackly from the string;
  And one would pierce an outer ring,
And one an inner, here and there;

And last the master bowman, *he*,
  Would clear the mark, a willing ear we lent him.

It has been said that the man of science should have no master, but he may have his admirations; Sladen's were for the best—Darwin, Huxley, Lyell, and Tyndall; in love of truth, devotion to science, and goodness of heart, in honour he was of them. The end and aim of his life was truth.

Somewhat conservative, he had no great love of popular science, yet he gave more than one lecture which was appreciated by the people. Genial and companionable, he yet loved solitary working, and held that "Good work is best done alone."

He was a great lover of good literature and of art; his library is rich in rare books and manuscripts; we remember how he prized the old editions of his books and the first instruments he worked with; nothing could induce him to part with any of them; his specimens were his treasures, and he would often remark that imperfect and broken specimens, and even portions of specimens, illustrated points which might otherwise be overlooked; they were all indispensable to his work.

His charity was genuine and unostentatious; his gift of £2,000 to insure the Yeomanry and Volunteers going to the front in South Africa was only a public instance; many private acts of benevolence are only known to the recipients.

Much material was in his hands awaiting that renewed strength and energy which it was hoped his visit to Italy would give him. Cretaceous Echinoderms and the spoils of the "Albatross" expedition for description among the rest; but it was not to be. That competent naturalists, imbued with Sladen's genius, may be found to describe them, is a consummation devoutly to be wished. **WM. CASH.**

MEMOIRS AND PAPERS

BY THE LATE WALTER PERCY SLADEN, F.Z.S., F.L.S., F.G.S.

Report on the Asteroidea collected by H.M.S. "Challenger"
during the years 1873-1876.

Report on the scientific results of the voyage of H.M.S.
"Challenger": Zoology, Part LI. (Vol. XXX.). Published by order of Her Majesty's Government, 1889,
pp. 935, 118 plates.

General Summary of the Asteroidea collected by H.M.S.
"Challenger" in "Narrative of the Cruise of H.M.S.
'Challenger.'"

Report on the scientific results of the voyage of H.M.S.
"Challenger": Narrative, Vol. I., Second Part. Published
by order of Her Majesty's Government, 1885, pp. 607-617.

The Asteroidea of H.M.S. "Challenger" expedition. Part I.,
Pterasteridæ. Linn. Soc. Journ. Zool., Vol. XVI.,
pp. 189-246, 1882.

The Asteroidea of H.M.S. "Challenger" expedition. Part II,
Astropectinidæ. Linn. Soc. Journ. Zool., Vol. XVII.,
pp. 214-269, 1883.

On the Asteroidea of the Mergui Archipelago, collected for
the Trustees of the Indian Museum, Calcutta, by Dr.
John Anderson, F.R.S., Superintendent of the Museum.
Linn. Soc. Journ. Zool., Vol. XXI., pp. 319-331, 1889.

Asteroidea dredged during the cruise of the "Knight Errant"
in July and August, 1880. Proc. Roy. Soc., Edin., Vol.
XI., pp. 698-707, 1882.

Asteroidea dredged in the Faröe Channel during the cruise
of H.M.S. "Triton" in August, 1882. Trans. Roy. Soc.,
Edin., Vol. XXXII., pp. 153-164, 1883.

(Determination of the Echinoderms) in "The Zoology of
Barents Sea," by W. S. M. D'Urban, F.L.S. Ann. and
Mag. Nat. Hist., ser. 5, Vol. VI., pp. 253-277, Oct., 1880.

On the Asteroidea and Echinoidea of the Korean Seas.
Linn. Soc. Journ. Zool., Vol. XIV., pp. 424-445, 1879.

Description of *Mimaster*, a new genus of Asteroidea from the Faroe Channel. Trans. Roy. Soc., Edin., Vol. XXX. pp. 579-584, 1883.

On *Astrophiura permira*, an Echinoderm-form intermediate between Ophiuroidea and Asteroidea. Proc. Roy. Soc. Vol. XXVII., pp. 456-457, 1878.

Astrophiura permira, an Echinoderm intermediate between Ophiuroidea and Asteroidea. Zoolog. Auzeiger, Jg. II. 1879, pp. 10-11.

On the structure of *Astrophiura*, a new and aberrant genus of Echinodermata. Ann. and Mag. Nat. Hist., ser. 5, Vol. IV., pp. 401-415, Dec., 1879.

On *Lepidodiscus Lebouri*, a new species of Agelacrinitidæ from the Carboniferous series of Northumberland. Abstr. Proc. Geol. Soc., Lond., No. 373, p. 4, 1879; Quart. Journ. Geol. Soc., Vol. XXXV., pp. 744-751, 1879.

Note on the Occurrence of *Pedicellaster* (Sars) in the Far North. Ann. and Mag. Nat Hist., ser. 5, Vol. V., pp. 216-217, March, 1880.

On a remarkable Form of Pedicellaria, and the Functions performed thereby; together with general observations on the allied Forms of this Organ in the Echinidæ. Ann. and Mag. Nat. Hist., ser. 5, Vol. VI., pp. 101-114, Aug., 1880.

On the Homologies of the Primary Larval Plates in the Test of Brachiate Echinoderms. Quart. Journ. Micr. Sci., new ser., Vol. XXIV., pp. 24-42, Jan., 1884.

On Traces of Ancestral Relations in the Structure of the Asteroidea. Proc. York. Geol. and Polytech. Soc., new ser., Vol. VII., pp. 275-284, 1881.

On the Genus Poteriocrinus and allied Forms. Proc. West Riding Geol. and Polytech. Soc., Vol. VI. (new ser., Vol. I.), pp. 242-253, 1877.

Bibliographical Notice: Report upon the Crinoidea collected during the voyage of H.M.S. "Challenger" during the years 1873-76. The Stalked Crinoids. By P. Herbert Carpenter, D.Sc., Assistant Master at Eton College; pp. i-xii., 1-442, 69 plates. (Report on the scientific results of the voyage of H.M.S. "Challenger." Zoology, Part XXXII.) Ann. and Mag. Nat. Hist., ser. 5, Vol. XV., pp. 346-352, April, 1885.

Bibliographical Notice : Report upon the Crinoidea collected during the voyage of H.M.S. "Challenger" during the years 1873-76, Part II. The Comatulæ, by P. Herbert Carpenter, D.Sc., F.R.S., F.L.S., Assistant Master at Eton College, pp. I.-IX., 1-399; 70 plates. (Report on the Scientific Results of the Voyage of H.M.S. "Challenger." Zoology, Part LX.) Ann. and Mag. Nat. Hist., ser. 6, Vol. III., pp. 504-510, June, 1889.

Bibliographical Notice of "Catalogue of the Blastoidea in the Geological Department of the British Museum (Natural History), with an account of the Morphology and systematic position of the group, and a revision of the Genera and Species. By Robert Etheridge, jun., and P. Herbert Carpenter, D.Sc., F.R.S., F.L.S." Ann. and Mag. Nat. Hist., ser. 5, Vol. XVIII., pp. 412-417, Nov., 1886.

Report of the Committee . . . . (on) the Scottish Zoological Station. Rept. Brit. Assocn. Adv. Sci. for the year 1881.

Report of the Committee . . . . (on) the Scottish Zoological Station. Rept. Brit. Assocn. Adv. Sci. for the year 1882.

Report of the Committee . . . . (on) the Scottish Zoological Station. Rept. Brit. Assocn. Adv. Sci. for the year 1883.

Report of the Committee . . . . (on) the Zoological Station at Naples.

    Rep. Brit. Assocn. Adv. Sci. for the year 1879
        ,, ...   1880
                                        1881

Rep. Brit. Assocn. Adv. Sci. for the year 1882
„ „ „ „ 1883
1884
1885
1886
1887
1888
1889
1890
1891
1892
1893
1894
1895
1896
1897
„ „ „ „ 1898

The following works in conjunction with Prof. P. Martin
Duncan, M.B., F.R.S.:—

Report on the Echinodermata collected during the Arctic
Expedition, 1875-76. Ann. and Mag. Nat. Hist., ser.
4, Vol. XX., pp. 449-470, December, 1877.

Appendix No. IX. "Echinodermata" in "Narrative of
Voyage to the Polar Sea, during 1875-6, in H.M. Ship
'Alert' and 'Discovery.'" By Captain Sir G. S. Nares
R.N., K.C.B., F.R.S., London, 1878, op. cit. Vol. l
pp. 260-282.

A Memoir on the Echinodermata of the Arctic Sea to
West of Greenland. London, 1881, pp. I.-VIII., 1-
6 plates.

A Monograph of Fossil Echinoidea of Sind, collected by
Geological Survey of India. "Palæontologia Indi
series 14. Part I. Strata below the Trap, pp. 1-
Plates I.-IV. 1882. Part II. The Ranikot Ser
pp. 21-100; Plates V.-XX. 1882. Part III.
Khirthar Series, pp. 101-246; Plates XXI.-XXXV

1884. Part IV. The Nari or Oligocene Series, pp.
247-272; Plates XXXIX.-XLIII.   1884.   Part V.
The Gáj or Miocene Series, pp. 273-367 ; Plates
XLIV.-LV. 1885.

A Monograph of the Fossil Echinoidea of the Makrán
Series of the Coast of Beluchistan, and of some Islands
in the Persian Gulf, collected by the Geological Survey
of India.   (A supplementary part to the Monographs
of the Fossil Echinoidea of Sind.)   "Palæontologia
Indica," series 14.   Part VI., pp. 369-382 ; Plates
LVI.-LVIII.   1886.

A Monograph of the Tertiary Echinoidea of Kachh and
Kattywar, collected by the Geological Survey of India.
"Palæontologia Indica," series 14, pp. I.-VI., 1-91 ;
Plates I.-XIII. 1883.

On some Points in the Morphology and Classification of
the Salenidæ, Agassiz.   Ann. and Mag. Nat. Hist.,
ser. 5, Vol. XIX., pp. 117-137, February, 1887.

On the Anatomy of the Perignathic Girdle and of other
Parts of the Test of *Discoidea cylindrica* Lamarck. sp.
Linn. Soc. Journ. Zool., Vol. XX., pp. 48-61, Oct., 1886.

A Note upon the Anatomy of the Perignathic Girdle of
Discoi ea cylindrica Lmk. sp., and of a species of
Echinoconus.   Ann. and Mag. Nat. Hist., ser. 6, Vol.
IV., pp. 234-239, Sept., 1889.

On the Family Arbaciadæ, Gray. Part I. The Morphology
of the Test in the Genera *Cœlopleurus* and *Arbacia*.
Linn. Soc. Journ. Zool., Vol. XIX., pp. 25-57, May,
1885.

On the Echinoidea of the Mergui Archipelago, collected for
the Trustees of the Indian Museum, Calcutta, by
Dr. John Anderson, F.R.S., Superintendent of the
Museum.   Linn. Soc. Journ. Zool., Vol. XXI., pp.
316-319, Oct., 1889.

The Classificatory Position of *Hemiaster elongatus* Duncan
and Sladen ; a Reply to a Criticism by Prof. Sven

Loven.   Ann. and Mag. Nat. Hist., ser. 5, Vol. XIV.
    pp. 225-242, Oct., 1884.

Objections to the Genera *Pseudopygaulus* Coquand, *Trachy-
    aster* Powel, and *Ditremaster* Munier-Chalmas. Their
    species restored to *Eolampas* Dunc. and Sladen, and
    Hemiaster Desor.   Ann. and Mag. Nat. Hist., ser. 6,
    Vol. II., pp. 327-336, Oct., 1888.

A Monograph of the British Fossil Echinodermata from the
    Cretaceous Formations (Asteroidea).   Part I., Palæonto-
    graphical Society, Vol. XLIV., pp. 1-28; Plates I.-VIII.
    (issued for 1890). April, 1891.

Do. —Part II. (Asteroidea), Palæontographical Soc. Vol.
    XLVI., pp. 29-66; Plates IX.-XVI., 1893.

SECRETARY'S REPORT, 1900.

The work of the Society during the past twelve months has
of exceptional interest, the General Meetings and Field
irsions have been well attended by the members, and by
ncing beyond our ordinary boundaries opportunities for
ɔendent investigation have been taken advantage of, and
ible knowledge of several Yorkshire geological problems has
gained.   In our May excursion to the borders of Lancashire
ɔ information was gained by the study of knoll-reefs, which
ɔs to make clearer the problem of the Cracoe knolls, whilst
ɡg the visit to the Cheviots the local source of large numbers
ɔe East Riding drift boulders was made certain, and another
ɔer in the history of the great ice-sheets which pressed in
ɔr coast during the Glacial period was disclosed by the fine
ɔnes and overflow valleys of the eastern slopes of Cheviot.
Yorkshire geology is a part of a larger history, and
ently it is only by going beyond the bounds of our county
the requisite evidence can be obtained for settling many
ɔtant points of home geology.
The First General Meeting and Field Excursion were held
litheroe, on Friday and Saturday, May 25th and 26th,
ɔ the presidency and leadership of Mr. Joseph Lomas,
ɔ., of Birkenhead.   The district to be visited included
le Hill, the knoll-reefs of the Clitheroe valley, and the
ɔ of the Hodder at Whitewell.
On the first day wagonettes took the party over Waddington
to Newton-in-Bowland, and thence along the picturesque
ɔ of the Hodder to Whitewell.   Several good examples of
ɔreefs were noticed on the way.   At Whitewell a short
was spent in examining the shales in a little clough opposite
bridge, to which the attention of the members had been
ted by Dr. Wheelton Hind, of Stoke-on-Trent.   These shales
found to contain a very interesting fauna, being crowded
delicate polyzoa, and containing small trilobites and cypris

... The gorge of the Hodder is a pretty wooded cut through ... Here and there springs gush out of the ... masses of porous travertine, on which ferns ... grow in great profusion, forming ... return journey the beautiful scenery and the ... enjoyed, and the deflection of the course ... noted with much interest. This river once ... mile valley which passes south-westwards from ... the Irish Sea, but has been captured by a tributary ... making a turn eastwards near Sandal Holme, ... tributary of the latter river at Great Mitton ... of Clitheroe.

... Meeting was held at the Swan Hotel, Clitheroe ... presidency of Mr. Joseph Lomas, F.G.S. The Chair ... dealt with the physical and glacial conditions ... He alluded to the wonderful contortions of the ... in the knoll-reefs, which gave the observer the ... faults. These he believed to be due to the ... limestone under severe stress, the rock resisting ... in one place and forced into knobs in ... The intermediate shales yielded by sliding layer over ... Mr. Lomas also described the interesting river diversions ... the Ribble having captured both the Hodder ... the Calder. The gorge of the Calder at Whalley ... than that of the Hodder, indicating that the ... the Calder was considerably earlier in geological ... of the Hodder. Mr. Lomas also spoke of the ... of the neighbourhood.

A short paper describing the geology of Pendle Hill ... by the Hon. Secretary. A discussion followed in which Mr. P. F. Kendall, F.G.S., Mr. J. H. Smith, F.G.S., Mr. Cash, F.G.S., and others took part. Mr. Kendall said that ... district of Clitheroe was the battle-ground of two glaciers ... moving from the north over Bowland Fells, and bringing down ... local rocks, and another moving from the Irish Sea, and bringing ... in and patches of shell-marl, Scottish granites, and Lake Distri...

Very few granites had been found in the Clitheroe
t, but a pocket of sand with broken shells had been
ed from Whalley, showing that the Irish Sea glacier had
ced thus far.

'n the Saturday morning the party started early, under
adership of Mr. James Walsh, B.Sc., of Blackburn, to visit
:o quarry. One of the features of the Clitheroe valley con-
f a number of rounded hills, which have been termed knoll-
and which many geologists consider to represent separated
ne and shell reefs, whilst others refer their formation to
movements. The Pimlico quarries have laid bare one of
knolls in a remarkable manner, showing its structure very
fully, and laying open for the fossil-hunter exposures
rfully rich in organic remains. The limestone is, indeed,
ss of crinoid remains, with large numbers of heads in
:nt preservation. After good bags had been secured the
drove by way of Chatburn and Downham to Brash
ı, where sections of the lower strata of Pendle Hill were
   Above the shales-with-limestones comes the Pendleside
:one, which is thick-bedded with chert bands. This is over-
y the Pendleside Grit, which forms the lower ridge of
:, above which comes a hollow groove due to the softer
nd Shales. The summit of Pendle is composed of the
beds of the Millstone Grit series. Leaving Brash Clough
ibden road was followed over the escarpment at Pendle
and the Millstone Grit quarries at the summit were seen,
haze interfered with the full appreciation of the view of
dges and valleys of the Pendle range. The descent to
n is down the dip slope of the Millstone Grit, and at
n the black shales between the Fourth and Third Grits
on. It is in these shales that the longitudinal valley of
n has been excavated. At Whalley the interesting gorge
: Lancashire Calder was seen, cutting right across the
ıent ridges of Millstone Grit to the Burnley Coal-field.
it was paid to the interesting old Parish Church, but,
unately, the Abbey grounds were closed. The party

... ... and, after dinner, separated with many ... ... a very pleasant and successful excursion, ... given to Messrs. Lomas and Wals.. for their ... ...

The ... ... General Meeting was held at Wooler, in N... ... ... on Saturday, July 14th, and was associated ... a F... Ex.... ... under the leadership of Mr. Percy F. ... ... F... ... to investigate the geology of the eastern slopes ... ... Hills from July 13th to 17th.

... ... with this Excursion a preliminary visit was ... ... the Roman Wall, on July 10th, 11th, and 12th, ... the ... ... of the Rev. E. Maule Cole, M.A., F.G.S. ... ... was sent to the members of the East Riding A... ... ... Society to join in this Excursion. The line of ... ... defence known as the Roman Wall, with its accom-... ... ... and associated military camps, extending for a ... ... of ... miles from Wallsend, near Newcastle, to Bowness ... ... Solway, forms the most extensive and interesting remains ... ... Roman occupation of Britain.

... ... place on Tuesday, July 10th, was Gilsland, on ... New ... and Carlisle Railway. A start was made from ... railway station for Birdoswald, the largest camp on the R... ... Wall. On the northern side of the wall is a deep fosse, which is an invariable feature. To the south there runs ... ... its entire length a second line of defence, called the Va... ... which consists of a ditch and two parallel mounds. The wall itself is built of good masonry, but has been plundered by ... ... builders of all sorts, the church of Hexham, the keep of T... wall, and the priory of Lanercost being three notable instances of this mediaeval vandalism in the neighbourhood of Gilsland.

Near Birdoswald a third line of defence has lately been discovered, called the Turf Wall, but it only extends for about three miles. The Birdoswald camp, like that of Chesters, afterwards visited, has the usual north and south gates, but on the east and west sides there are two gates. The chariot ruts t and the grooves for angle irons and pivot holes for the gates are plainly to be seen, being in excellent preservation. At Coome Crag an old Roman quarry was visited, and Roman inscriptions noted. Lanercost Abbey, with its excellent west front and interesting monastic buildings, aroused much interest. In the choir are some fine tombs of the Dacres, Lords of Gilsland.

The return journey was made through Naworth Park to Upper Denton, where an interesting very early Norman church was visited. The chancel arch is possibly Saxon.

On Wednesday morning the party started in the opposite direction (easterly) along the Wall. On the way the leader pointed out how the Roman Wall climbed the highest crags, accompanied invariably by the vallum to the south, at a lower elevation. For some distance the modern road follows the ditch of the vallum. Some 15 miles from Gilsland, Housesteads (the Roman Borcovicus) was reached. Here an hour and a half was spent in thoroughly examining the camp, which has recently been excavated by the Cumberland and Northumberland Antiquarian Societies.

This camp, covering nearly five acres, follows the rule, having four gates. The gates are remarkably perfect, showing the characteristic pivot-holes and ruts from chariot wheels. What appears to have been a forum has been thoroughly excavated, and afforded matter of deep interest and for considerable discussion among the members. This camp is perhaps the most marvellous work of antiquity extant in Great Britain, and many parts of it are preserved in a wonderful way.

A detour was then made to Vindolana, where the only existing complete milestone *in situ* in Great Britain remains on the Stanegate, an old Roman road. A second milestone was found, the base and the lower part of the shaft alone remaining. A portion of this stone was identified in front of the Twicebrewed Inn on the modern road, covered with whitewash.

The camp at Æsica was next visited. This camp has also been excavated by the above-named societies. In the centre is

N

a well. with a fine arch, containing water brought from a distance of six miles by a conduit along the slopes of the hills for the use of the soldiers. The party next proceeded to Caervoran (Magna), which commands a fine view. Here a number of small Roman altars were noticed: also pedestals, querns, and incised stones were found in numbers ornamenting the top of a farmhouse wall. The vallum and wall here are in close proximity.

Following the wall down the steep hill-side the party reached Thirlwall Castle, now in ruins, the mediæval pele or stronghold being built entirely of stones from the Roman Wall. The weather was ideal, and much conduced to the success of the expedition. Every part of the interesting remains visited were seen to perfection and thoroughly examined.

On Thursday the party continued their interesting investigation of the Roman remains by travelling to Hexham and examining the Priory Church, which has been built largely of stones taken from the Roman wall. The Saxon crypt, with its marked and inscribed Roman stones, was very fine, and Mr. Robson, the parish clerk, gave an interesting account of the efforts made to lay bare all its chief features, a work to which he had given and was giving considerable time and care. Much interest was also aroused by the ancient Saxon Frid-stool, or safety seat, in the days when the Priory had enjoyed the privilege of sanctuary. The walls are rich in masons' and other markings, which were pointed out by the leader (Rev. E. Maule Cole, M.A.) During the renovation of the chancel, some curious old chantries were removed, one of which, with its quaint carvings, is placed in the south transept. This transept also contains a bold stone staircase, worn by the feet of generations of Augustinian monks, who descended by it from their dormitory to celebrate midnight service in the church. After a careful examination of this interesting church, which contains so many Roman relics the train was taken to Chollerford to view the Roman bridge over the North Tyne. The eastern abutment of this bridge is in excellent condition, and shows a large wedge-shaped foundation

f solid masonry, which was tightly bound together with iron
ands, the grooves for the reception of which were clearly
marked. This abutment contracts wedge-like towards the river,
nd is defended by a square castellum, which forms the termi-
ation of the Roman wall at that point. The remains of two
iers are to be found in the river bed, and the opposite buttress
t to be seen, but the foundations are now under water, the
iver having worked westward since Roman times.

After lunch a rapid move was made for "The Chesters,"
o view the remains of Cilurnum in the park. It was not one
f the days set apart for visitors, but on the special antiquarian
urpose of the party being explained to Miss Clayton, she kindly
ranted permission for these interesting remains to be viewed.
The camp of Cilurnum is next in size to that at Birdoswald,
isited on Tuesday, and like it, has the unusual number of six
ates, one north, one south, and two each east and west. These
ates were connected by streets running straight across the camp.
Four of these gates are double, and were closed by double doors
which swung on pivots revolving in holes lined with iron, the
emains of which are still seen. Each of these gateways is
lanked by two guardrooms. The two extra gates were single,
and have no flanking guardrooms. The Roman Wall meets the
amp just south of the large east and west gates, and so half
f the camp stands north of the wall. The forum in the centre
f the camp shows many points of interesting coincidence with
hat of Housesteads, but is laid out after a plan of its own.
The north and south street advances to its north gate, and is
ontinued down its centre. The east and west streets pass just
utside its walls. It apparently consisted of an open market-
place surrounded by a covered colonnade supported by square-
ouilt stone pillars. At the southern end are buildings probably
ised for the administration of justice, and one is the treasury.
Under this latter is a vaulted chamber, which was probably used
is a place of safety for the military chest, which could be
ecured to the floor. Several rooms flank the forum on the
ast side, which were probably dwellings, as the rooms are seen

to be supported on pillars allowing of the circulation
air.   The furnace for the production of this heated a
preserved.

Another extensive area of ruined structures exten
the east wall of the camp to the river bank.   These co
a courtyard paved with rough stone slabs, at one end of
are seven arched recesses and several associated chambers
use of the recesses in the courtyard is a matter of curious
troversy, many holding that they are places for the reception
the gods of the family.   The remains of a hypocaust are see
and tiled hot-air flues, part of one being shown in excellent
condition.   The examination of this wonderful excavation had
to be hurried because of the shortness of time, but the party
felt that they had been amply repaid for their visit. and only
longed for time to see all the wonders of Cilurnum. A rapid
march had to be made to catch the return train to Hexham,
and before the party separated a unanimous and enthusiastic
vote of thanks was passed to the genial and learned leader, the
Rev. E. Maule Cole, M.A., F.G.S., for the immense trouble he
had taken to show everything that was possible in the time. and
to make clear to every member the meaning of these complicated
antiquarian relics.

The Cheviots' party was large and representative. and had
as their aim the investigation of the eastern side of the Cheviot
Hills, with the special view of attempting the identification of
certain igneous rocks (porphyrites) which are found in large
quantities as boulders in the drift deposits of Holderness and
along the Yorkshire coast.   The members who arrived on Thur
day utilised Friday morning for a visit to Roddam Dene.
which the basement Carboniferous beds are found resting on
volcanic series.   Roddam Dene is a well-wooded, pictures
little glen, showing numerous good sections of red s
stones and marls, but time did not permit of the thor
investigation of the lower parts of the glen, where the
glomerates are probably best seen.   A pretty little
was found near the head of the stream cutting through a

lava, and showing an interesting diversion of drainage owing
the choking of the upper valley by mounds of moraine. The
noon party drove to Middleton Hall, and then walked to
y Burn, where sections of the volcanic rocks were examined
the stream banks. At Shining Pool a fine lateral moraine,
by a lobe of ice which had forced its way from the north
the eastern spurs of Cheviot, was noted. At the time of
invasion from the north the valleys radiating from Cheviot
have their local glaciers, but were not invaded by the
ice-sheet.

On Saturday, July 14th, the geologists drove to Langleeford,
a rough hill road. The programme laid out a route over
ot Hill (2,676 ft.), descending by Hen Hole, and a long
country walk to Kirknewton; but it was agreed that a
ugh examination of the rocks round the granite area, with
pecial design of noting its altering effect on the rocks into
it had been intruded, was of great importance. The
edge of the granite patch was accordingly carefully investi-
by a party under the direction of Mr. Percy F. Kendall,
S., whilst Mr. G. G. Butler, F.G.S., led another party to
top of Great Cheviot to examine the exposures of the granite
ts sides and summit, and to investigate a patch of porphyrite
ked on its summit. The granite was found to grow finer
ined as the hill was ascended, and the remnants of one of
old lava sheets (porphyrite) was found on the north side
the summit. Good junction sections of the granite and newer
kes with the volcanic series were found by Mr. Kendall's
rty.

The General Meeting was held at the Cottage Hotel,
ooler, after dinner on Saturday evening, July 15th, under the
esidency of Mr. G. G. Butler, M.A., F.G.S., of Ewart Park,
ooler. Mr. J. Norton Dickons (Bradford) and Mr. W. G.
nsfield (Ilkley) were elected life-members of the Society.
ter a brief address by the Chairman, Mr. J. W. Stather,
3.S., of Hull, read a paper on "The Boulders in the Drift
posits of the East Coast of Yorkshire." Mr. Stather briefly

... the .... and distribution of the drift of East
Y...... and pointed out that it contained a remarkable assem-
blage of rocks in the form of boulders. An attempt to tabulate
the larger boulders of the drift, those above one foot in diameter,
... ... by the Hull Geological Society, with interesting
re...... It was found the local rocks of the coast almost always
... the ... ... and the same was true of the Teesdale
... ... boulders. In these cases it was found, taking
... the larger boulders into consideration, that these south-
... ... boulders increased rapidly in numbers in passing from
... ... to the Bridlington sections. On the contrary, the
N... ... boulders increased in numbers as the drift was traced
... ... and the same was true of the numerous granites, most
... have not been identified. Mr. Stather argued that
... ... was indicated by the dispersion of these boulders.
... the conclusion that there was a community of origin
... ... these granites and the known Norwegian rocks, and
... ... came over the North Sea.

... Percy F. Kendall, F.G.S., gave an interesting address
... on facts noted during the excursion. The general
... ... the Cheviots was a floor of denuded Silurian rocks
... ... been covered by a great outflow of lavas in the Old
Red Sandstone period. These lavas, after the cessation of
... ... had been invaded and pushed up by a great
... ... granite, which had considerably altered the beds with
which ... ... come into contact. Then these beds had been subjected ... severe denudation in the Carboniferous sea, and great conglomerates had been formed. A period of quiescence had followed
covering most of the Secondary period. In the Tertiary period
another great volcanic outburst had taken place, producing extensive changes along the broad valley which then extended between
the mainland of Scotland and the outer Hebrides. This outburst
starred the rocks for long distances, and the cracks produced
were filled by basaltic rocks. The glacial period would produce
small local glaciers in the valleys radiating from Cheviot, but
this district was invaded by the great ice-sheet which filled the

North Sea, which has left large morainic mounds, and numbers of beautiful dry valleys at considerable elevations, which were cut by the water flowing from the ice-front when the usual lines of drainage were choked with ice. The lie of these moraines and glacial valleys showed that the Cheviots formed the dividing line in local glacial movements, the local glaciers to the north being deflected northwards, and beaten back on to the coast, whilst those to the south were deflected southwards along the coast, as Yorkshire evidence conclusively proved.

The examination of the Cheviot granite showed it to be very fine-grained at the margin on the hillside above Langleeford and at the faulted junction in Harthope. The fine-grained character of the highest beds on Cheviot and the patch of porphyrite on one of the northern spurs showed that the central mass of granite had not been much denuded, and explained the paucity of granite boulders in the local drift. Strings of grano-phyre (i.e., very fine-grained material of the same composition as the granite) passed through the porphyrites and the granite. One specimen was obtained below Long Crag which showed a small vein of granophyre passing across the junction of the granite with the porphyrite. Veins of tourmaline were found along the edges of the granite, and fragments of granite traversed by tourmaline veins were found in the "cone of dejection" of each of the three small streams descending on the north side of the Harthope valley.

A discussion followed in which several of the members took part, and the meeting concluded with unanimous votes of thanks to the Chairman and the readers of papers.

On Sunday afternoon many of the members accepted the kind invitation of Mr. and Mrs. G. G. Butler to visit Ewart Park, and have tea. The interesting collection of curios in the gallery was shown, and an hour was spent in the spacious gardens.

On Monday, July 16th, the party drove to Roddam to complete the examination of the Dene, and see the fine sections of the basement Carboniferous conglomerates exposed therein. A heavy thunderstorm coming on, the investigation was anything

... ... and by the time Calder Farm ... . the members were in need of a good fire to dry .... After a .... lunch some of the party .... the .... to Ingram, but the major portion ... . to see a fine series of overflow .... by the advance of a glacier into the .... A .... fall of quartz amygdules was .... the valley above the footbridge.

. ... was paid to Chillingham Park .. .. white cattle. This famous herd, which .... was well seen, and lunch was partaken of .. .. On the return journey the party ... ... the moors to the south of the .... glacial overflow valleys. A series of .... was examined continuing across three spurs, ... of cutting by considerable volumes of .... are wide and open at both ends, cut .... drainage : all dip in the same direc- ... . and are related to one another just as a series .. .. be ... a succession of lakes held up by an .... all the easterly valley mouths. In ex- .... of Calder Farm and Akeld abundant boulders .. . . greywacke sandstone were found, ... are also frequently found in the drifts of the Yorkshire .... This was an important identification, as the .... had not been previously known by Yorkshire .... and, though not found *in situ*, the numbers in which it is found as a constituent of the Cheviot morainic gravels gives the direction of its source.

Before the party separated a specially enthusiastic vote of thanks was accorded to Mr. Percy F. Kendall, F.G.S., for the way in which he had put his geological knowledge and experience at the full disposal of the members, and done so much to render the meeting a thorough success. With the exception of Monday morning's thunderstorm the weather was hot and clear, and the excursion was most enjoyable and instructive.

During the year the important series of investigations into the sources of the Aire at Malham have been brought to a successful issue, and a report presented by the Hon. Secretary to the Bradford Meeting of the British Association was received with great interest by the Geological Section. During the year the Malham Sub-Committee. consisting of Messrs. C. W. Fennell, F.G.S., J. A. Bean, W. Ackroyd, F.I.C., F. W. Branson, F.I.C., J. H. Howarth, F.G.S., P. F. Kendall, F.G.S., and W. Lower Carter, M.A., F.G.S. (convener), have held several meetings.

On November 14th, 1899, Mr. Kendall was able to report the excellent reception accorded to the report of the Committee at the Dover Meeting of the British Association. A Committee of the British Association had been appointed to continue these investigations on underground waters, and a grant of £40 had been made for this purpose.

The constitution of the British Association Committee was as follows:—Professor W. W. Watts, M.A., F.G.S. (Chairman), Mr. A. R. Dwerryhouse, F.G.S. (Secretary), Professor A. Smithells, Rev. E. Jones, F.G.S., Mr. Walter Morrison, Rev. W. Lower Carter, M.A., F.G.S., Messrs. G. Bray, Thomas Fairley, P. F. Kendall, F.G.S., and J. E. Marr, F.R.S.

The question of the relations of the British Association Committee to that of our own Society was carefully considered, and, after full discussion, the following resolutions were passed unanimously:—

1. That the Yorkshire Geological and Polytechnic Society having commenced the work of investigating the Underground Waters of Craven, and having completed an important section of that work, desires to continue the investigations and to be credited with the results obtained.

2. That, to this end, the Committee appointed by the Yorkshire Geological and Polytechnic Society desires to carry on the work as heretofore by means of its own members, and to present an abstract of the work done to the British Association Meeting at Bradford,

through the Secretary of the B.A. Committee for the Investigation of the Underground Waters of Craven; and that the complete report, fully illustrated, be published in the Proceedings of the Yorkshire Geological and Polytechnic Society.

In reply to these resolutions the Sub-Committee received very kind communications from Professor W. W. Watts, M.A., F.G.S., the Chairman of the B.A. Committee, agreeing to the suggestions of the Sub-Committee, and promising very cordial co-operation in the carrying out of the scheme of further investigations.

Several meetings of the Sub-Committee were held for the drafting of the final report on the Malham investigations, which was issued with the Proceedings, Vol. XIV., Part I., fully illustrated.

The Council tender their heartiest thanks to the members of this Sub-Committee, who, at a considerable expenditure of time and thought, have so successfully completed the record of these interesting investigations.

At the Spring Council Meeting the correspondence between the Malham Sub-Committee and Professor Watts was reported on, and the grant of £10 by the Dover Meeting of the British Association was announced. The Council resolved that the work of investigating the underground waters of Ingleborough should be commenced forthwith, the problem being tackled first from the Clapham side, and the investigations to commence with Gaping Ghyll.

A large and representative Committee was appointed to have charge of this work in conjunction with the Committee appointed by the British Association, all of whose members were elected on the Yorkshire Geological and Polytechnic Society's Committee. The following gentlemen were elected to serve on this Committee: Messrs. E. Calvert (Buxton), J. W. Tate (Ingleton), J. A. Farrer, J.P. (Clapham), J. A. Bean, C. W. Fennell, F.G.S. (Wakefield), W. Ackroyd, F.I.C., J. H. Howarth, F.G.S., W. Simpson, F.G.S. (Halifax), G. Bingley, F. W. Branson, F.I.C.,

A. Burrell, F.I.C., S. W. Cuttriss (Leeds), R. Law, F.G.S. ¡pperholme), J. J. Wilkinson (Skipton), J. E. Wilson and F. ᵣₐnn, B.Sc. (Ilkley), J. W. Handby (Austwick), Dr. W. ᵤₛhall Watts (Giggleswick), and Rev. G. H. Brown (Settle).

The Committee met at Clapham on April 27th and 28th, d again on June 8th. The following report was presented to ᵤ Geological Section of the British Association at Bradford, Mr. A. R. Dwerryhouse, F.G.S. :—

"The Committee is carrying out the investigation in con- ᵣction with a Committee of the Yorkshire Geological and ᵤlytechnic Society.

"The present is merely an interim report, as the work is ll in progress.

"It was decided that the work should consist of an investi- tion of the underground flow of water in Ingleborough. This ll forms with its neighbour, Simon's Fell, a detached massif, ᵣich is peculiarly suitable for investigations of this nature.

"The summit of the group is formed of Millstone Grit, then llow Yoredale shales and sandstones, the whole resting on a ᵤteau of Carboniferous Limestone.

"Many streams rise on the upper slopes of the hills and ᵢw over the Yoredales, but without exception their waters are ᵣallowed directly they pass on to the Carboniferous Limestone, reappear as springs in the valleys which trench the plateau.

"The Committee first turned its attention to tracing the ᵤter which flows into Gaping Ghyll hole.

"It was generally believed that the water issued at a large ring immediately above the bridge at Clapham Beck Head and ᵢmediately below the entrance to Ingleborough Cavern.

"On April 28th specimens of the water from this spring ᵣre taken for analysis before the introduction of any test.

"Two cwt. of ammonium sulphate was then put into the ᵤter flowing into Gaping Ghyll, and at the same time the ᵣount of the water was gauged and found to be equivalent to ᵢ1,856 gallons per diem. A few hours later a second quantity two cwt. of the same substance was introduced.

"On the same day 1½ lb. of fluorescein in alkaline solution was put into a pot-hole known as Long Kin East, about 1,300 yards north-east of Gaping Ghyll.

"In view of the important influence which the direction of the joints in the limestone had been found to exercise over the flow of underground water, the direction of the joints in the limestone clints in the neighbourhood of Long Kin East was taken, and was found to be N.N.W. to S.S.E., and to run in such a direction as to lead to the probability that the water would reappear at the springs at the head of Austwick Beck, and these were consequently watched.

"The ammonium sulphate put in at Gaping Ghyll reappeared at the large spring at Clapham Beck Head on the morning of May 3rd, and continued to flow until the evening of May 6th, when the water again became normal. Thus the time occupied by the ammonium sulphate in travelling from Gaping Ghyll to Clapham Beck Head, a distance of one mile, was about five days.

"No ammonium sulphate was found in any of the other springs in Clapdale.

"This result proved beyond doubt that Gaping Ghyll was connected with Clapham Beck Head.

"The fluorescein put in at Long Kin East showed itself at Austwick Beck Head, but not at any of the neighbouring springs on May 11th, having taken over thirteen days to travel, the delay being probably due to the small amount of water flowing at the time of the experiments.

"These results are of considerable importance, as they definitely reveal two lines of divergent movement of these underground waters, and indicate a subterranean watershed of much interest. The influence of the master-joints of the Carboniferous Limestone in determining the direction of flow of these underground waters was also, as at Malham, clearly shown.

"The next set of experiments was carried out by the joint Committee on June 8th and following days.

"In order to confirm the results in connection with the Gaping Ghyll to Clapham Beck Head flow, and further

rtain more definitely if there existed any connection between
ing Ghyll and the smaller springs in Clapdale, 10 cwt. of
mon salt was put into the waters of Gaping Ghyll on June
and a further 10 cwt. on June 5th, samples of the water
each of the springs being taken several times a day until
25th.

"One pound of fluorescein in alkaline solution was introduced
the stream flowing through Ingleborough Cave on June 8th,
L 0 p.m., at the point where the water plunges down a hole
the floor of the cave, and marked 'Abyss' on the 6-inch
nance map.

"Five cwt. of ammonium sulphate was introduced into a
on the allotment about 500 yards N.E. of Long Kin East
June 9th, at 3 p.m.; and at 3.15 p.m. on the same day 1 lb.
fluorescein in alkaline solution was poured into the stream
ch flows past the shooting-box on the allotment and sinks
r the Bench Mark 1320·1.

"The fluorescein introduced into the abyss came out of
pham Beck Head, and possibly at Moses Well and other
ings in Clapdale, but this point requires further investigation,
evidence being as yet somewhat unsatisfactory.

"The salt from Gaping Ghyll appeared at Clapham Beck
ad on June 15th, 16th, 17th, 18th, 19th, 20th, and 21st,
ng at its maximum on June 18th, but not at any of the
ler springs.

"The ammonium sulphate put into the sink on the allotment
eared at Austwick Beck Head on June 22nd, the other springs
the neighbourhood being unaffected on that day; but on the
h and 25th there were slight increases in the amount of
monia in two small springs in Clapdale, viz., the small spring
ow Clapdale Farm and Cat Hole Sike. As one of these
eams is close to the farmyard, and the other was at the time
rly dry and flowing through pasture-land, no importance is
ached to these slight increases.

"Of the fluorescein put in below the shooting-box no trace
since been found, and the same is the case with ½ lb. of

...were introduced into Grey Wife Sike, above Newby Cote.

"Several most interesting problems still await solution in this area, one of them being the relations of the Silurian floor which underlies the Carboniferous Limestone of the plateau to the flow of underground water.

"The two sinks, Gaping Ghyll and Long Kin East, are only about 1,300 yards apart, and yet the waters of the one take a direction quite distinct from those of the other, and eventually emerge in a separate valley, the distance between the springs being 1½ miles apart, the great mass of Carboniferous Limestone known as Norber, a hill upwards of 1,300 feet in height, lying between the two valleys.

"In Crummack Dale it is seen that the Silurian rocks form a ridge running in an approximately N.W. and S.E. direction, and unconformably overlain by the Carboniferous Limestone.

"If this line be continued it separates the Gaping Ghyll to Clapham Beck Head flow from that of Long Kin East to Austwick Beck Head.

"Thus it appears that this ridge of Silurian rocks forms an underground water-parting, which the Committee hopes to be able to trace for a considerable distance across the area.

"The magnitude of this undertaking will be to some extent realised when it is stated that upwards of 400 samples of water have been tested for common salt, ammonium, and fluorescein, making in all upwards of 1,200 tests.

"The whole of the grant of £40 has been spent upon the investigation, and a small sum in addition.

"The experiments which have been carried out have indicated which are the most suitable reagents for use in different cases, and it is consequently hoped that future investigations will be carried out at rather less cost than has been the case up to the present.

The Committee was reappointed, with a grant of £50.

At the Meeting of the Committee at Clapham, on April 27th, it was resolved that the following Sub-Committee should

mpowered to make arrangements for further investigations :—
srs. J. H. Howarth, F.G.S., P. F. Kendall, F.G.S., C. W.
nell, F.G.S., J. A. Bean, F. W. Branson, F.I.C., W. Ackroyd,
3., A. R. Dwerryhouse, F.G.S., and W. Lower Carter, M.A.,
S. (convener).

During the year 1899 reports had been received of the
isfactory way in which the finds obtained from the Grass-
n Explorations, organised by this Society in 1892, were
d, and the great danger there was of their serious injury.
Council resolved that, if no suitable and safe locality could
ovided in Grassington for their reception, the collection
l be removed to Leeds and placed in the care of the Leeds
ophical and Literary Society. It was subsequently found
the Grassington Parish Council were willing to provide a
le place in their hall for the case, and accordingly your
il offered the finds to the Parish Council as a gift on the
tions that they would have them locked up and properly
for, and would give the Yorkshire Geological and Poly-
ic Society ready access to them. In reply the Clerk of the
ington Parish Council wrote stating that the case and its
nts were in a satisfactory condition and under the control
le caretaker, and that they would be at our Society's
re for inspection at any time.

During the year there have been elected three life members
16 subscribing members, making a total of 19 new
ers for the year.

On the other hand, we much regret to report the loss of
al members, some of whom have had a long and honourable
ection with our Society. At the Clitheroe meeting we had
cord the decease of Mr. Richard Reynolds, F.G.S., of Leeds,
was elected a member of our Society in 1864, had been a
ber of its Council for thirty years, and a Vice-President
the year 1894. At the same meeting we received the news
he death of Mr. J. McLandsborough, F.G.S., F.R.A.S., of
lford, who had been a member of our Society since 1853.
es of deep regret and sincere sympathy were passed, and the

Secretary sent a suitable letter to the relatives of each of the deceased gentlemen.

Of those who have been connected with the Society for a shorter period, but whose names we shall miss from our roll with deep regret, are the late Messrs. E. Slater, of Farsley, and John Young Short, of Thirsk. These losses together with four resignations leave the present roll of members at 182, a net gain of 11 during the year.

The Rev. W. Lower Carter, M.A., F.G.S. was appointed the Representative Governor of the Yorkshire College, and Mr. William Gregson, F.G.S. the Delegate to the Corresponding Societies Committee of the British Association.

The Hon. Secretary had the honour of being appointed one of the Secretaries of Section C (Geology) at the Bradford meeting of the British Association: and valuable papers on local geology were contributed by several of our members.

The loan of £350 to the Halifax Corporation fell due for repayment on 1st April, 1900, but an offer received for its renewal for five years at 3 per cent. was accepted by the Council.

The Council recommend the following arrangements for the General Meetings and Field Excursions of 1901 :—

Leyburn—for Wensleydale.

Keswick—to examine the rocks *in situ* from which numerous Yorkshire erratics have been derived.

Annual Meeting—Bradford.

Our Proceedings as usual have been forwarded to leading Scientific Societies in various parts of the world, and publications in exchange have been received from the following Societies :—

British Association.
Royal Dublin Society.
Royal Geographical Society.
Royal Society of Edinburgh.
Royal Physical Society of Edinburgh.

Royal Society of New South Wales.
Department of Mines, Sydney, N.S.W.
Department of Mines, Adelaide, S. Australia.
Nova Scotian Institute of Science.
Royal Institution of Cornwall, Truro.
Bristol Naturalists' Society.
Cambridge Philosophical Society.
Essex Naturalists' Field Club.
Edinburgh Geological Society.
Geological Association, London.
Geological Society of London.
Leeds Philosophical and Literary Society.
Liverpool Geological Society.
Liverpool Geological Association.
Hampshire Field Club.
Hull Geological Society.
Herefordshire Natural History Society.
Manchester Geological Society.
Manchester Geographical Society.
Manchester Literary and Philosophical Society.
Nottingham Naturalists' Society.
University Library, Cambridge.
Yorkshire Naturalists' Union.
Yorkshire Philosophical Society, York.
American Philosophical Society, Philadelphia, U.S.A.
American Museum of Natural History, New York, U.S.A.
Academy of Natural Sciences, Philadelphia, U.S.A.
Brooklyn Institute of Arts and Sciences.
Boston Society of Natural History, Boston, U.S.A.
Kansas University, Lawrence, Kansas.
Wisconsin Geological and Natural History Survey, Madison, Wis.,
    U.S.A.
Geological Survey of Minnesota, Minneapolis, Minn., U.S.A.
Chicago Academy of Sciences.
Museum of Comparative Zoology at Harvard College, Cambridge, Mass.
New York Academy of Sciences, New York.
United States Geological Survey, Washington, D.C.
Elisha Mitchell Scientific Society, University of N. Carolina, Chapel
    Hill, U.S.A.
New York State Library, Albany, U.S.A.
Wisconsin Academy of Sciences, Arts, and Letters.
Smithsonian Institution, Washington, D.C.

o

L'Academie Royale Suedoise des Sciences, Stockholm.

Societe Imperiale Mineralogique de St. Petersburg.

Societe Imperiale des Naturalistes, Moscow.

Comite Geologique de la Russie, St. Petersburg.

Instituto Geologico de Mexico.

Sociedad Cientifica "Antonio Alzate," Mexico City.

Australian Museum, Sydney.

Australian Association for the Advancement of Science, Sydney.

Natural History Society of New Brunswick.

L'Academie Royale des Sciences et des Lettres de Danemark, Copenhague.

Kaiserliche Leopold-Carol. Deutsche Akademie der Naturforscher, Halle-a-Saale.

Geological Institution, Royal University Library, Upsala.

Imperial University of Tokyo, Japan.

REVENUE ACCOUNT.

1899-1900. Receipts.	£	s.	d.	1899-1900. Expenditure.	£	s.	d.
To Subscriptions	68	5	0	By Balance due to Treasurer 1st Nov., 1899	18	18	2
,, Sale of Proceedings	4	5	0	,, Expenses of Meetings	11	16	3
,, Transfer from Capital Account	18	18	0	,, Year Book of Societies	0	6	0
,, Halifax Corporation Interest	10	1	3	,, Maps for Wooler Excursion ...	1	7	10
				,, Postcards, Envelopes, Stationery, &c. ...	3	11	0
				,, Postages and Petty Cash	13	11	10
				,, F. Carter, Stationery, Circulars, &c. ..	13	6	7
				,, Chorley & Pickersgill, Printing Proceedings	16	12	8
				,, Maull & Fox, Block for Portrait ...	1	15	0
				,, G. West & Sons, Drawing and Lithographing Plates	14	18	0
				,, Balance in hands of Treasurer Nov. 1st, 1900	5	5	11
	£101	9	3		£101	9	3

CAPITAL ACCOUNT.

To Life Members' Subscriptions	... £18 18 0	By Transfer to General Account	... £18 18 0
To Halifax Corporation Bond .	£350 0 0	Examined and found correct, 8th November, 1900,	

J. H. HOWARTH, Halifax.

RECORDS OF MEETINGS.

Meeting of the Underground Waters Sub-Committee, Halifax 14th November, 1899.

Chairman :—Mr. P. F. Kendall, F.G.S.

Present :—Messrs. F. W. Branson, W. Ackroyd, J. B Howarth, W. L. Carter (Secretary), and Mr. A. R. Dwerryhouse (Secretary of the B.A. Committee).

The minutes of the previous Committee Meeting were read and confirmed.

The Chairman reported that an excellent reception was accorded to the interim report at the Dover meeting of the British Association, and that a grant of £40 had been voted towards the expenses of continuing the investigations.

The following B.A. Committee had been appointed superintend the expenditure of this grant :—

Chairman : Professor W. W. Watts, M.A., F.G.S.

Secretary : Mr. A. R. Dwerryhouse, F.G.S.

Professor A. Smithells, B.Sc., F.I.C.

Rev. E. Jones, F.G.S.

Mr. Walter Morrison, J.P.

Mr. George Bray.

Rev. W. Lower Carter, M.A., F.G.S.

Mr. W. Fairley, F.I.C.

Mr. P. F. Kendall, F.G.S.

Mr. J. E. Marr, M.A., F.R.S., F.G.S.

A lengthy conversation was then held as to the prosecution of the investigations, and as to the relation of B.A. Committee to the Committee of the Yorks. Geol Polytec. Soc. It was unanimously felt that the memb the Yorks. Geol. and Polytec. Soc. would expect the i gations to be carried on as heretofore in their name a their Committee. It was represented that it would be possible for the Yorks. Geol. and Polytec. Committee to

the work and present a report to the Bradford meeting
the B.A., through Mr. Dwerryhouse, the Secretary of the
⬐. Committee; reserving to the Yorks. Geol. and Polytec.
₋ the right to publish the full illustrated report in their
◼ Proceedings.

In order to avoid any misunderstanding on this matter the
⬐wing resolutions were proposed by Mr. W. Ackroyd and
⬐nded by Mr. J. H. Howarth, and carried unanimously. The
⬐retary was instructed to forward copies of them to Professor
⬐tts through Mr. A. R. Dwerryhouse.

Resolved—

(1) That the Yorkshire Geological and Polytechnic Society
having commenced the work of investigating the
underground waters of Craven, and having completed
an important section of that work, desires to continue
the investigations and to be credited with the results
obtained.

(2) That to this end the Committee appointed by the Yorkshire Geological and Polytechnic Society desires to
carry on the work as heretofore by means of its own
members and to present an abstract of the work done
to the B.A. Meeting at Bradford, through the Secretary of the B.A. Committee for the Investigation of
the Underground Waters of Craven; and that the
complete report, fully illustrated, be published in the
Proceedings of the Yorkshire Geological and Polytechnic Society.

The final report of the investigations at Malham was then
⬐sidered. Mr. J. H. Howarth read the Introduction which he
d prepared, which was adopted with slight modifications.

Mr. P. F. Kendall was requested to amplify the geological
⬐ort for publication, and to draft general conclusions, with the
operation of Messrs. J. H. Howarth and A. R. Dwerryhouse,
⬐ to submit the complete report to another meeting of the
⬐mmittee.

Meeting of the Underground Waters Sub-Committee, Halifax. 22nd March, 1900.

Chairman:—Mr. J. H. Howarth.

Present:—Messrs. P. F. Kendall, W. Ackroyd, W. L. Carter (Secretary), and Mr. A. R. Dwerryhouse.

The minutes of the previous meeting of the Committee were read and confirmed.

Letters of regret for non-attendance were read from Messrs. J. A. Bean, C. W. Fennell, and F. W. Branson.

Mr. P. F. Kendall reported that he had not yet been able to complete the geological section of the Report, but hoped to have it ready for the next meeting. He desired the rainfall results at Malham Tarn for the two months preceding Messrs. Morrison and Tate's experiments in 1879.

The Secretary presented an account (£2 10s.) from Mr. Townsend, of Malham, for expenses in collecting and forwarding samples of water for analysis. This was passed for payment.

Mr. A. R. Dwerryhouse and the Secretary read letters from Professor W. W. Watts offering very cordial co-operation in the carrying out of further investigations, and making suggestions for the conduct of the work.

After careful consideration of these suggestions it was unanimously resolved:—

(1) That the investigations round Ingleborough should be carried on under the supervision of a Committee to be appointed by the Yorks. Geol. and Polytec. Soc. and that the B.A. Committee supervise the expenditure and present a report to the Bradford Meeting The full report to be published in the Proceedings of the Yorks. Geol. and Polytec. Soc.

(2) That only one set of accounts be kept, and that the B.A. grant be supplemented by local subscriptions be raised by the Yorks. Geol. and Polytec. Soc. found necessary.

(3) That all the members of the B.A. Committee for the Investigation of the Underground Waters of Craven be appointed members of the Yorks. Geol. and Polytec. Committee.

(4) That the investigation of the Ingleborough area be commenced by the examination of the stream flowing into Gaping Ghyll.

(5) That Mr. E. Calvert (Buxton), Mr. J. E. Tate (Ingleton), and Mr. J. A. Farrer (Clapham) be nominated as members of the Committee.

(6) That the Secretary of the Yorks. Geol. and Polytec. Soc. be the Secretary of the Investigation Committee.

Council Meeting, Philosophical Hall, Leeds, 29th March, 1900. Chairman :—Mr. F. W. Branson.

Present :—Messrs. W. Ackroyd, C. W. Fennell, J. E. Wilson, W. Stather, G. Bingley, J. J. Wilkinson, P. F. Kendall, E. D. Tellburn, J. T. Atkinson, W. Cash, and W. L. Carter (Hon. Sec.).

The minutes of the previous Council Meeting were read and confirmed.

Letters of regret for non-attendance were read from the Rev. E. M. Cole and Messrs. T. H. Mitchell, H. Crowther, R. Law, . Reynolds, and W. Gregson.

Meetings and Field Excursions.—The question of the General Meetings and Field Excursions in 1900 was discussed at some length.

Resolved :—

(1) That the first General Meeting be at Clitheroe on May 25th and 26th.

(2) That the second General Meeting and Field Excursion be to the Cheviots, on July 13th to 16th. Mr. P. F. Kendall to be invited to be the leader with the co-operation of a local geologist.

(3) That the Annual Meeting be held at Selby early in
November, the local arrangements to be in the hands
of Mr. J. T. Atkinson.   A morning excursion to be
arranged to Brayton Barff and Hambleton Hough.

Underground Waters Investigation.—The Hon. Secretary read
the minutes of the Sub-Committee, and explained the methods
suggested by them for co-operation. with the B.A. Committee.

These minutes and arrangements were confirmed.

A Committee was then appointed for the Investigation of
the Underground Waters of Ingleborough, to consist of the fol-
lowing gentlemen :—Messrs. Edward Calvert (Buxton), J. W. Tate
(Ingleton), J. A. Farrer (Clapham), J. A. Bean (Wakefield), W.
Ackroyd (Halifax), G. Bingley (Leeds), F. W. Branson (Leeds),
C. W. Fennell (Wakefield), J. H. Howarth (Halifax), R. Law
(Hipperholme), W. Simpson (Halifax), J. J. Wilkinson (Skipton),
J. E. Wilson (Ilkley), B. A. Burrell (Leeds), S. W. Cuttriss (Leeds),
F. Swann (Ilkley), J. W. Handby (Austwick), Dr. Watts (Gigle-
wick), and Rev. G. H. Brown (Settle), together with the members
of the B.A. Committee for the Investigation of the Underground
Waters of Craven :—Professor W. W. Watts (Birmingham), Mr.
A. R. Dwerryhouse (Leeds), Professor Smithells (Leeds), Mr. Walter
Morrison (Malham), Mr. G. Bray (Leeds), Rev. E. Jones (Embsay),
Mr. T. Fairley (Leeds), Mr. P. F. Kendall (Leeds), Mr. J. E. Marr
(Cambridge), and the Rev. W. L. Carter (Hon. Secretary).

British Association.—The report of the delegate to the Corresponding Societies Committee at the Dover Meeting (Mr. W.
Gregson, F.G.S.) was read by the Hon. Secretary.

The Rev. W. L. Carter reported that he had been appointed
one of the Secretaries of Section C at the Bradford Meeting.

The Secretary read a letter he had received from Mr. G. W
Lamplugh, F.G.S., the Recorder of Section C, with reference to
the preliminary arrangements, and asking for suggestions for
papers, &c.

After considering the matter a number of suggestions were
made as to papers to be read before the Geological Section, and
as to suitable places for geological excursions.

Vice-Presidents and the Council.—The Secretary brought forward the recommendation of the Annual Meeting that the Vice-Presidents should be summoned to the Council Meetings. He pointed out that the rules of the Society only provided for one Vice-President who is a member of Council, but that several noblemen and gentlemen prominent in the county and in the scientific world had been elected honorary Vice-Presidents.

The Council approved of this explanation and appointed Mr. Richard Reynolds, F.C.S., as the Vice-President to have a seat in the Council and be summoned to its meetings.

Grassington Finds.—The Secretary read a letter from the Clerk of the Grassington Parish Council notifying the safe custody of the Grassington Finds, and promising free access to them by the Yorks. Geol. and Polytec. Soc. at any time.

Resolved.—That the Secretary write to the Parish Council expressing the satisfaction of the Council of the Yorks. Geol. and Polytec. Soc. at their acceptance of the custody and care of the finds.

Appointments—The Rev. W. Lower Carter, M.A., F.G.S., was elected the representative Governor of the Yorkshire College. Mr. W. Gregson, F.G.S., was elected the delegate to the British Association Corresponding Societies' Committee.

Geological Photographs.—The Secretary read a circular from the Geological Photographs Committee of the British Association, proposing a scheme of circulation of photographs of geological interest for educational purposes. He reported that there were duplicates of several of the large photographs issued by our Society, and suggested that they might be of considerable use to the Photographs Committee if a grant of some copies were made.

Mr. Kendall inquired whether these photographs were for sale, as he should like to procure a set for the Yorkshire College.

Resolved:—That a set of the large photographs in stock be presented to the Yorkshire College, and that an offer of several sets be made to Professor Watts for the use of the Photographs Committee.

Loan to the Halifax Corporation.—The Hon. Secretary reported that he had received a letter from the Borough Accountant of Halifax, offering the renewal of the mortgage No. 3,635, £55 due April 1st, 1900, for another five years at 3 per cent.

The Secretary reported that he had consulted the Treasurer, the Auditor, and some other members of Council, and on their advice had accepted the offer of the Halifax Corporation.

This action was approved.

Meeting of the Underground Waters Sub-Committee, Leeds, 29th March, 1900.

Chairman:—Mr. F. W. Branson.

Present:—Messrs. P. F. Kendall, W. Ackroyd, and W. L. Carter (Secretary).

The minutes of the previous Committee Meeting were read and confirmed.

Mr. P. F. Kendall reported that the geological report had become considerably extended, and would be completed in the course of a few days.

The question of the Ingleborough investigation was discussed and it was arranged to call a meeting of the full Committee at Clapham on April 27th and 28th.

It was resolved to purchase a copy of " Irelande et Cavern Anglaises,' by Mons. E. A. Martel, for the use of the Committee

It was resolved that two or three six-inch maps of Ingleborough district be purchased, and that a map on 25-inch scale be prepared for the Bradford meeting of the B Association.

Mr. Branson was authorised to consult the chemists o Committee as to the tests to be applied, and to arrange for preparation and conveyance to Clapham.

It was suggested that a supplementary geological should be made at Malham by Mr. Kendall, to ascertain important data with regard to the rock-structure, &c., district.

Meeting of the Committee for the Investigation of the Under-ground Waters of Ingleborough, New Inn, Clapham, April 27th and 28th, 1900.

Chairman:—Mr. J. A. Farrer, J.P.

Present:—Messrs. F. W. Branson, J. H. Howarth, P. F. Kendall, R. Law, W. Simpson, J. W. Tate, J. J. Wilkinson, W. Cuttriss, E. Calvert, W. L. Carter (Secretary), and several gentlemen as visitors.

April 27th.—The members dined together at the New Inn, under the presidency of Mr. J. A. Farrer. After dinner a meeting was held, at which the conditions of the problem were discussed, and the best methods of carrying on the investigations considered. It was resolved to test Gaping Ghyll by means of ammonium sulphate, and to put fluorescein into a water-sink east of Gaping Ghyll.

April 28th.—Specimens of the water for analysis were taken from the principal springs in Clapdale, and the tests were introduced into Fell Beck, just above Gaping Ghyll, and into Long Kin East.

At a meeting at the dinner table it was resolved that Messrs. C. W. Fennell, J. A. Bean, F. W. Branson, W. Ackroyd, P. F. Kendall, J. H. Howarth, A. R. Dwerryhouse, and W. L. Carter (Secretary) should form a Sub-Committee to arrange for further investigations.

———

Meeting of the Underground Waters Sub-Committee, Bradford, May 17th, 1900.

Chairman:—Mr. J. H. Howarth.

Present:—Messrs. F. W. Branson, P. F. Kendall, A. R. Dwerryhouse, C. W. Fennell, and W. L. Carter (Secretary).

The minutes of the previous Sub-Committee Meeting were read and confirmed.

Mr. P. F. Kendall brought up the geological report on the Malham investigations, which was read and adopted. Mr. J. H. Howarth was requested to draft the general conclusions of the whole investigation for adoption by the Sub-Committee.

Mr. Branson reported on the analyses of the samples of water received from Clapham.

Reports were received as to the results of the tests introduced into Gaping Ghyll and Long Kin East, and the procedure in future investigations was considered.

Messrs. F. W. Branson and W. L. Carter were authorised to prepare a brief statement of the Clapham experiments to date, to be issued to the members of the Committee and to the Press.

General Meeting and Field Excursion, Clitheroe, May 25th and 26th, 1900.

Chairman :—Mr. Joseph **Lomas**, F.G.S., of Liverpool.

May 25th.—An early party visited the noted Crinoid bed in the Pimlico quarry. After luncheon the members proceeded by waggonette over Waddington Fell to Newton-in-Bowland. Thence the valley of the Hodder was traversed to Whitewell. An examination was made of fossiliferous black shales in a little clough near the Suspension Bridge. The beautiful gorge of the Hodder, with its pretty, grotto-like tufa formations, was visited. The return journey was taken by Bashall, where a section in the great esker was examined.

The members dined together at the Swan Hotel. After dinner the General Meeting was held under the presidency of Mr. Joseph Lomas, F.G.S.

The following new members were elected :—

Mr. Frederick Justen, F.L.S., London.

Mr. S. W. Cuttriss, Leeds.

Mr. John Hoyle Ashworth, Bradford.

Mr. Arthur W. Cooke, F.C.S., Kirkstall.

Mr. Richard Murray, Leeds.

Mr. R. M. Robson, Filey.

Mr. J. Crowther, B.Sc., Halifax.

Dr. Wheelton Hind, F.G.S., Stoke-on-Trent.

The meeting heard with the deepest regret of the decease of two old and valued members of the Society—Mr. Richard

leynolds, F.C S., who had been for many years a valued member
f Council and a Vice-President, and Mr. J. McLandsborough,
'.G.S., who had been a member of the Society since 1853.
'otes of deep regret and sincere condolence with the relatives
! these gentlemen were passed.

An address was delivered by the Chairman on "The Physical
ad Glacial Geology of the Clitheroe District."

A paper was read by the Rev. W. Lower Carter, M.A.,
.G.S., on "The Stratigraphical Geology of Pendle Hill."

A paper was communicated by Mr. J. R. Mortimer, entitled
Notes on the History of the Driffield Museum of Antiquities
ad Geological Specimens."

A discussion followed in which Messrs. J. H. Smith, F.G.S.
Padiham), P. F. Kendall, F.G.S., W. Cash, F.G.S., J. Walsh,
t.Sc. (Blackburn), and J. Weekes (Clitl eroe) took part, and the
!hairman replied.

May 26th.—Pimlico quarry was visited and good bags of
xsils were obtained.

The party joined the wagonettes at the Workhouse and rode
y Chatburn and Downham to Brast Clough, where a section of
he Pendleside Limestone covered by drift was seen.

The conveyances were rejoined at Worston, Pendle Nick
ras crossed to Sabden, and the road taken along the valley to
Vhalley. The party returned by conveyance to Clitheroe, and
fter dinner separated with heartiest thanks to the leaders,
fessrs. J. Lomas, F.G.S., and J. Walsh, B.Sc., for their able
onduct of the excursion.

---

*Meeting of the Committee for the Investigation of the Under-
round Waters of Ingleborough*, New Inn, Clapham, June 8th,
900.

Chairman :—Dr. D. Forsyth.

Present :—Messrs. F. W. Branson, W. Ackroyd, C. W.
Pennell, A. R. Dwerryhouse, G. Bingley, P. F. Kendall, T.
Pairley, F. Swann, R. Law, and W. L. Carter (Hon. Sec.).

The report of the results obtained from the tests introduced at the previous meeting was considered and, with one or two alterations, adopted, and ordered to be printed and circulated.

The Secretary reported the arrangements made by Mr. Branson and himself for the conveyance of the tests to Clapham. He also reported on the state of the potholes as revealed by a preliminary survey during the earlier part of the day. One ton of common salt had been put into Fell Beck just above Gaping Ghyll on June 4th and 5th.

It was then resolved :—

(1) That 5 cwt. of ammonium sulphate be put into the stream between Long Kin East and the Hunting Box.

(2) That 1 lb. of fluorescein be put into the "abyss" in the Ingleborough Cave.

(3) That 1 lb. of fluorescein be put into the Shooting Box stream.

(4) That ½ lb. of methylene blue be put into Grey Wife Sike.

Mr. Godfrey Bingley then exhibited a number of photographs of Malham, and a selection of eight views was made to illustrate the Malham report.

General Meeting and Field Excursion to Gilsland and Wooler July 10th to 17th.

I. Excursion to the Roman Wall. Leader :—Rev. E. Mac Cole, M.A., F.G.S.

July 10th.—The party met at Gilsland Spa and went wagonette to view the camp at Birdoswald (Amboglanna). Coome Crag old quarries with Roman inscriptions were amined, and a visit was paid to Lanercost Abbey, which built of materials taken from the Roman Wall. The r journey was made by way of Naworth Park to Gilsland.

July 11th.—The party drove to Housesteads (Borco and examined the notable ruins. A detour was mad lolana, on the Stanegate, to see the erect Roman milestone. the return journey the camps of Æsica and Magna were ed, and the line of the Wall followed to Thirlwall Keep, h was built of stones from the Wall.

July 12th.—Train was taken to Hexham and the Priory ch visited. Train was then taken to Chollerford, and the ment of the Roman bridge over the North Tyne was ex- ed. After luncheon the party visited the ruins of Cilurnum Chesters. On returning to Hexham a very enthusiastic vote nanks was given to the leader for his interesting descriptions excellent arrangements.

II. Excursion to the Cheviots.

Leader:—Mr. Percy F. Kendall, F.G.S.

July 13th.—The party met at Wooler and visited Carey a in the afternoon.

July 14th.—The members drove to Langleeford, in Harthope ., where they divided into two divisions. One division ex- ed the stream sections and found excellent junctions of the ite and porphyrite. The other division ascended Cheviot and ained the granite exposures and the patch of porphyrite at north side of the summit.

The General Meeting was held at the Cottage Hotel, Wooler, er the presidency of Mr. G. G. Butler, M.A., F.G.S., of Ewart k, Wooler.

The following new members were elected :—

    Mr. W. G. Stansfield, Ilkley.

    Mr. J. Norton Dickons, Bradford.

    Dr. D. Carmichael, Gosforth.

An address was delivered by the Chairman.

A paper was read by Mr. J. W. Stather, F.G.S., on "The ulders in the Drift Deposits of East Yorkshire and their tribution."

An address was given by Mr. P. F. Kendall, F.G.S., on nts noted during the excursion in connection with the geology the Cheviot Hills.

A discussion followed, and the meeting concluded with vote of thanks to the Chairman and the readers of papers.

July 16th.—The party drove to Roddam Dene and worked up the stream to see the basement Carboniferous conglomerate, in a heavy thunderstorm. After luncheon at Calder Farm some of the party went direct to Ingram, whilst others crossed the moors to the Breamish valley to examine some fine overflow valleys.

July 17th.—A visit was paid to Chillingham Park to see the herd of wild cattle. After luncheon the party returned to Akeld, where an ascent was made to the moors to see some overflow valleys. Before the party separated a hearty vote of thanks was given to Mr. Kendall for his careful and instructive leadership.

———

Meeting of the Underground Waters Sub-Committee. Wakefield, August 3rd, 1900.

Chairman :—Mr. C. W. Fennell.

Present :—Messrs. J. H. Howarth, F. W. Branson, W. Ackroyd, A. R. Dwerryhouse, and W. L. Carter (Secretary).

Letters of regret for non-attendance were read from Messrs. P. F. Kendall and J. A. Bean.

The minutes of the previous meeting of the Committee were read and confirmed.

Mr. W. Ackroyd read the report of the second series of tests at Clapham. The ammonia put in at the pothole P10 had appeared at Austwick Beck Head. The fluorescein put in at the Cave "abyss" had come out at Beck Head. No indications had been obtained of the methylene blue put in at Grey Wife Sike or of the fluorescein introduced into the Shooting Box stream

The report was adopted with the best thanks of the Committee to the chemists for their arduous labours in the analysis and the papers were handed to Mr. A. R. Dwerryhouse incorporation in the British Association Report.

The Malham report was finally revised and adopted.

Council Meeting, Philosophical Hall, Leeds, Oct. 18th, 1900.
Chairman :—Mr. J. H. Howarth.
Present :—Messrs. P. F. Kendall, F. W. Branson, J. W.
ner, F. F. Walton, W· Ackroyd, E. D. Wellburn, J. E.
ord, G. Bingley, W. Cash, J. T. Atkinson, W. Rowley,
!rowther, J. J. Wilkinson, and W. L. Carter (Hon. Secretary).
The minutes of the previous Council Meeting were read and
rmed.
Letters of regret for non-attendance were received from
rs. W. Simpson, R. Law, W. Gregson, and T. H. Mitchell.
Annual Meeting. A letter was read from Lord Ripon agree-
:o the holding of the Annual Meeting at Selby, but regretting
nability to be present owing to the doctor's regulations.
The Secretary announced that Mr. J. T. Atkinson, F.G.S.,
had been a member of Council for 21 years, had kindly
anted to preside in Lord Ripon's place.
Arrangements were then approved for the holding of the
ual General Meeting at the Museum, and for an excursion
!rayton Barff and Hambledon Hough. The dinner to be at
Londesborough Arms Hotel, Selby.
Officers and Council. The present officers and members of
ncil were nominated for re-election with the following altera-
s:—Mr. Walter Rowley, F.G.S., F.S.A., to be a Vice-President in
e of the late Mr. Richard Reynolds, F.C.S., and to be summoned
ll meetings of the Council; Mr. A. R. Dwerryhouse, F.G.S.,
e a member of Council in the place of Mr. W. Rowley.
The Hon. Secretary read the Report and submitted a list
apers for the Annual Meeting. These were approved.
The Treasurer made a financial statement. The following
unts were presented for payment and passed :—

	£	s.	d.
G. West & Son (plates)	14	18	0
F. Carter (circulars and stationery) ...	13	6	7
Maull & Fox (block)	1	15	0
Chorley & Pickersgill (Proceedings) ...	16	12	8
	£46	12	3

The arrangements for the meetings for 1901 were considered and it was resolved that General Meetings and Field Excursions be held at Leyburn (for Wensleydale) and at Keswick (for Lake District Igneous rocks), and that the Annual Meeting be at Bradford.

It was suggested that an evening meeting, for the reading and discussion of papers, should be held at Leeds early in the year.

The Secretary read a letter from Mr. Fred Reynolds in reply to a letter of sympathy and deep regret which he had forwarded to the family on the death of Mr. Richard Reynolds, F.C.S., in accordance with the vote passed at the Clitheroe Meeting of the Society.

It was resolved that a set of large photographs be presented to the Oxford and Cambridge Museums, and a set of the proceedings to the Radcliffe Library, Oxford.

Meeting of the Underground Waters Sub-Committee. Leeds, October 18th, 1900.

Chairman :--Mr. J. H. Howarth.

Present : --Messrs. F. W. Branson, W. Ackroyd, P. F. Kendall, A. R. Dwerryhouse, and W. L. Carter (Secretary).

The minutes of the previous Committee Meeting were read and confirmed.

The report of the B.A. Underground Waters Committee, presented to the Bradford Meeting, was read by Mr. A. R. Dwerryhouse.

It was resolved to continue the experiments on the Chapel-le-Dale side, and Mr. Dwerryhouse and Mr. Howarth were asked to undertake the arrangements, with the co-operation of Mr. J. W. Tate, of Ingleton.

The question of the complete report was considered, and it was decided that Mr. A. R. Dwerryhouse should keep the records of the various experiments until the time came for drawing up the report.

Annual General Meeting, The Museum, Selby, November 8th, '00.

Chairman :—Mr. J. T. Atkinson, F.G.S.

The Hon. Secretary read letters regretting non-attendance >m the Marquis of Ripon, Dr. Sorby, Messrs. W. Rowley, G. ngley, F. W. Branson, and W. Gregson.

The Hon. Secretary read an abstract of the Annual Report.

The Treasurer presented the Financial Statement.

Resolved.—That the Report and Financial Statement, as sented, be adopted and printed in the Proceedings.

Proposed by Mr. J. E. Bedford, F.G.S., seconded by Mr. J. H. fthouse, F.G.S., and carried.

The following new members were elected :—

Mr. Theodore Ashley, Leeds.

Professor J. W. Carr, F.L.S., F.G.S., Nottingham.

Election of Officers and Council.—The following names were minated by the retiring Council :—

President :—

The Marquis of Ripon, K.G.

Vice-Presidents :

Earl Fitzwilliam, K.G.

Earl of Wharncliffe.

Earl of Crewe.

Viscount Halifax.

H. Clifton Sorby, LL.D., F.R.S.

Walter Morrison, J.P.

W. T. W. S. Stanhope, J.P.

James Booth, J.P., F.G.S.

F. H. Bowman, D.Sc., F.R.S.E.

W. H. Hudleston, F.R.S.

J. Ray Eddy, F.G.S.

David Forsyth, D.Sc., M.A.

Walter Rowley, F.G.S., F.S.A.

Treasurer :

William Cash, F.G.S.

Hon. Secretary:
William Lower Carter, M.A., F.G.S.
Hon. Librarian:
Henry Crowther, F.R.M.S.
Auditor:
J. H. Howarth, F.G.S.
Council:

W. Ackroyd, F.I.C.

J. F. Atkinson, F.G.S.

J. E. Bedford, F.G.S.

Godfrey Bingley.

F. W. Branson, F.I.C.

A. R. Dwerryhouse, F.G.S.

J. H. Howarth, F.G.S.

P. F. Kendall, F.G.S.

R. Law, F.G.S.

G. H. Parke, F.L.S., F.G.S.

F. F. Walton, F.G.S.

E. D. Wellburn, F.G.S.

Local Secretaries:

Barnsley—T. W. H. Mitchell.

Bradford—J. E. Wilson.

Driffield—Rev. E. M. Cole, M.A., F.G.S.

Halifax—W. Simpson, F.G.S.

Harrogate—Robert Peach.

Huddersfield—Samuel Jury.

Hull—John W. Stather, F.G.S.

Leeds—H. Crowther, F.R.M.S.

Middlesbrough—Rev. J. Hawell, M.A., F.G.S.

Skipton—J. J. Wilkinson.

Thirsk—W. Gregson, F.G.S.

Wakefield—C. W. Fennell, F.G.S.

Wensleydale—W. Horne, F.G.S.

In addition to the above, Mr. Wm. Gregson, F.G.S., nominated Lord Masham, Swinton Castle, and Sir Christopher Furness, M.P., D.L., Grantley Hall, Ripon, as Vice-Presidents of the Society.

Resolved.—That the Officers and Council as nominated be elected to serve for the ensuing twelve months; and that the best thanks of the Society be given to the retiring Officers and Council for their conduct of the affairs of the Society during the past year.

'roposed by Mr. E. Hawkesworth, seconded by Mr. G. ngs, and carried.

λ paper was read by the Chairman on " Our Society— spect and Prospect."

λ paper was read by Mr. Theodore Ashley on " Notes on 'ccurrence of certain Coal Seams in Leeds and the neighbour- "

λ paper was read by Mr. Percy F. Kendall, F.G.S., "On the lacial Contour of the Vale of York."

A paper was read by Mr. W. H. Crofts, on " New Sections led during the Excavations for the Alexandra Dock Ex-)n, Hull."

A paper was read by Mr. Robert Kidston, F.G.S., "On the)niferous Flora of Yorkshire."

λ series of lantern slides by Mr. Godfrey Bingley, illustrating ºield Excursions to Clitheroe and Wooler, was exhibited by λ. R. Dwerryhouse, F.G.S.

Resolved.—That the best thanks of this Annual Meeting be to the Chairman for presiding; to Jonathan Hutchinson, F.R.S., for kind permission to hold the meeting in the ₁m; to the Earl of Londesborough for allowing the members it Brayton Barff and Hambledon Hough; to the Vicar of for permission to visit the Abbey Church, and for copies e guide thereto; to the readers of the papers; to Mr. ₂y Bingley, for his exhibition of lantern slides; to Messrs. ›lds & Branson, Ltd., for providing a lantern; and to V. N. Cheeseman, for kindly lending his lantern screen.

'roposed by Mr. F. F. Walton, F.G.S., seconded by Mr. Dickons, and carried.

ºhe members and their friends dined together at the ₃sborough Arms Hotel, under the presidency of Mr. J. T. ₁son, F.G.S.

HONORARY MEMBERS.

1887 BODINGTON, Principal N., Litt.D., The Yorkshire College, Leeds.

1892 DE RANCE, CHAS. E., C.E., F.R.M.S., 32, Carshalton Road, Blackpool.

1887 HUGHES, Prof. T. McK., M.A., F.R.S., F.G.S., 18, Hills Road, Cambridge.

1887 JUDD, Prof. JNO. W., C.B., LL.D., F.R.S., F.G.S., Science Department, South Kensington, London.

1887 WOODWARD, HENRY, LL.D., F.R.S., British Museum (Natural History), Cromwell Road, London, S.W.

1898 WHITAKER, WILLIAM, B.A., F.G.S., F.R.S., Freda, Campden Road, Croydon.

LIST OF MEMBERS.

Life members who have compounded for the annual subscriptions are indicated by asterisks (*).

Elected.

1899 ABBEY & HANSON, Surveyors, Huddersfield.

1883*ABBOTT, R. T. G., Whitley House, Malton.

1890*ACKROYD, W., F.I.C., Borough Analyst, Halifax.

1901*ANDERSON, TEMPEST, M.D., F.G.S., 17, Stonegate, York.

1899 APPLEYARD, J. H. R., 5, Willow Lane East, Huddersfield.

1900 ASHLEY, THEODORE, 10, Blenheim Avenue, Leeds.

1900 ASHWORTH, JOHN HOYLE, 28, Silverhill Road, Thornbury Bradford.

1899 ANNIS, ERNEST G., L.R.C.P., Medical Officer of Health Huddersfield.

1875 ATKINSON, J. T., F.G.S., Hayesthorpe, Holgate Hill, York

1879*BARTHOLOMEW, C. W., Blakesley Hall, near Towcaster.

1899 BEAN, J. A., F.G.S., Moot Hall, Newcastle-on-Tyne.

1875 BEDFORD, JAMES, Woodhouse Cliff, Leeds.

1878 BEDFORD, J. E., F.G.S., Arncliffe, Shire Oak Road, Headingley Leeds.

BERRY, RAYMOND, Surveyor, Hipperholme.

BINGLEY, GODFREY, Thorniehurst, Shaw Lane, Headingley, Leeds.

BOOTH, JAMES, J.P., F.G.S., Spring Hall, Halifax.

BOULD, CHAS. H., 5, Wrangthorn Place, Hyde Park, Leeds.

BOWMAN, F. H., D.Sc., F.R.A.S., F.C.S., F.G.S., Mayfield, Knutsford, Cheshire.

BRADLEY, F. L., Bel Air, Alderley Edge, Cheshire.

BRAY, GEORGE, Belmont, Headingley, Leeds.

BRANSON, F. W., F.I.C., 14, Commercial Street, Leeds.

BRIGG, JOHN, M.P., F.G.S., Keighley.

BRIGGS, ARTHUR, J.P., Cragg Royd, Rawdon, Leeds.

BROOKE, Sir THOS., Bart, J.P., Armitage Bridge, Huddersfield.

BROWNRIDGE, C., C.E., F.G.S., 26, North Road, Devonshire Park, Birkenhead.

BUCKLEY, GEORGE, Waterhouse Street, Halifax.

BURRELL, B.A., F.I.C., F.C.S., 5, Mount Preston, Leeds.

CANHAM, Rev. HENRY, F.G.S., Leathley Rectory, Otley.

CARR, Professor J. W., F.L.S., F.G.S., University College, Nottingham.

CARTER, W· LOWER, M.A., F.G.S., Hopton, Mirfield.

CASH, W., F.G.S., 35, Commercial Street, Halifax.

CHAMBERS, J. C., 7, Cardigan Road, Headingley, Leeds.

CHARLESWORTH, J. B., J.P., Hurts Hall, Saxmundham.

CHILD, HY. SLADE, F.G.S., Mining Engineer, Wakefield.

CLARK, J. E., B.A., B.Sc., Lilegarth, Ashburton Road, Croydon.

COLE, Rev. E. MAULE, M.A., F.G.S., Wetwang Vicarage, near York.

COLLINSON, EDWARD, Linden Road, Halifax.

COOKE, ARTHUR W., F.C.S., Kirkstall Lane, Kirkstall, Leeds.

CREWE, Earl of, Crewe Hall, Crewe.

CROFTS, WILLIAM HASTINGS, 60, Freehold Street, Hull.

CROSSLEY, ABRAHAM, 62, Wellington Road, Todmorden.

CROWTHER, HENRY, F.R.M.S., The Museum, Leeds.

1900 CROWTHER, J., B.Sc., Technical School, Halifax.

1899 CRUMP, W. B., M.A., 90, King Cross Street, Halifax.

1900 CUTTRISS, S. W., Prudential Buildings, Park Row, Leeds

1899 DALZELL, A. E., 3, Wesley Court, Crossley Street, Halifax

1879*DAKYNS, J. R., M.A., of H.M Geological Survey, 28, Jermyn Street, London, W.

1883 DALTON, THOS., 65, Albion Street, Leeds.

1894*DAVIS, JAMES PERCY A., Chevinedge, Halifax.

1879 DEWHURST, J. B., Aireville, Skipton.

1900*DICKONS, J. NORTON, 12, Oak Villas, Bradford.

1886*DOBINSON, LAUNCELOT, Park View, Stanley, near Wakefield

1891 DODSWORTH, SIR MATTHEW, Bart., Sunningdale, Bournem'th

1887*DUNCAN, SURR W., Horsforth Hall, Horsforth, near Leeds

1896 DWERRYHOUSE, ARTHUR R., F.G.S., 5, Oakfield Terrace, Headingley, Leeds.

1879 EDDY, J. RAY, F.G.S., Carleton Grange, Skipton.

1894 EMBLETON, HENRY C., Central Bank Chambers, Leeds

1895 FARRAH, JOHN, F.R.Met.S., Jefferies Coate, York Road, Harrogate.

1887*FENNELL, CHAS. W., F.G.S., 82, Westgate, Wakefield

1839 FITZWILLIAM, Earl, K.G., Wentworth Woodhouse, near Rotherham.

1883*FLEMING, FRANCIS, Elm Grove, Halifax.

1893 FORSYTH, DAVID, D.Sc., M.A., 2, Lifton Place, Leeds

1894 FOX, C. E., P.A.S.I., 22, George Street, Halifax.

1891*FURNESS, Sir CHRISTOPHER, M.P., Brantford, West Hartlep

1890 GARFORTH, W. E., C.E., F.G.S., Snydale Hall, Pontef

1875 GASCOIGNE, Col. F. C. T., Parlington Park, Aberford, Leeds.

1881 GLEADOW, F., 38, Ladbroke Grove, London, W.

1899 GREAVES, A. E., St. John's, Wakefield.

1882 GREGSON, W., F.G.S., Baldersby, S.O., Yorkshire.

1898 HALDANE, GEORGE W., M.E., Old Corn Exchange, Wal

1843 HALIFAX, Viscount, Hickleton Hall, Doncaster.

1895 HARKER, ALFRED, M.A., F.G.S., St. John's College, Cam

1887 HASTINGS, GEOFFREY, 15, Oak Lane, Bradford.

187 HAWELL, Rev. JNO., M.A., F.G.S., Ingleby Greenhow, Middlesbrough.

196 HAWKESWORTH, EDWIN, Nursery Mount, Hunslet, Leeds.

400*HIND, WHEELTON, M.D., F.G.S., Roxeth House, Stoke-on-Trent.

181 HORNE, WM., F.G.S., Leyburn.

199 HORSFALL, RICHARD EDGAR, 22A, Commercial Street, Halifax.

190*HOWARTH, J. H., F.G.S., Somerley, Rawson Avenue, Halifax.

182*HUDLESTON, W. H., F.R.S., F.G.S., 8, Stanhope Gardens, South Kensington, S.W.

195*IMBERY, JOHN, Hyde Park Road, Halifax.

401 JOHNSON, Rev. W., Archbishop Holgate's School, York.

187*JONES, J. E., Solicitor, Halifax.

397 JOWETT, ALBERT, B.Sc., The Technical School, Bury.

385 JURY, SAMUEL, 6, Eleanor Street, Fartown, Huddersfield.

400 JUSTEN, FREDERICK, F.L.S., 37, Soho Square, London, W.

888*KERR, R. MOFFAT, Solicitor, Halifax.

892 KENDALL, PERCY F., F.G.S., 5, Woodland Terrace, Chapel-Allerton, Leeds

1899 KING, W. N., P.A.S.I., 34, York Street, Wakefield.

1897 KITSON, R. H., F.G.S., Elmet Hall, Leeds.

1897 LAMBERT, ABRAHAM, 25, Great George Street, Harrogate.

1877*LAMPLUGH, G. W., F.G.S., of H.M. Geol. Survey, 28, Jermyn Street, London, W.

1894 LAW, ARTHUR W., Greetland, Halifax.

1898 LAW, ROBERT, F.G.S., Fennyroyd Hall, Hipperholme.

1897 LOFTHOUSE, J. H., Lyell House, Dragon Parade, Harrogate.

1877*LUPTON, ARNOLD, F.G.S., M.Inst.C.E., 6, De Grey Road, Leeds.

1891*MASHAM, Lord, Swinton Castle, Bedale.

1897 MILLWARD, ADAM, 5, Cambridge Crescent, Harrogate.

1889 MITCHELL, T. W. H., Barnsley.

1896 MORETON, HENRY JAMES, Station Hotel, Lofthouse, Wakefield.

1879*MORRISON, WALTER, J.P., Malham Tarn, Settle.

1878 MORTIMER, J. R., Driffield.

1894 MUFF, HERBERT B., B.A., F.G.S., Christ's College, Cambridge.

Q

1895 MULLER, HARRY, 12, West Chislehurst Park, Eltham, Kent.
1900 MURRAY, RICHARD, Laurel Bank, Potternewton Lane, Chapel Allerton, Leeds.
1875*MYERS-BESWICK, W. B., Gristhorpe Manor, Filey.
1898 NEVIN, JOHN, J.P., F.G.S., Littlemoor, Mirfield.
1889*NICHOLSON, M., J.P., Middleton Hall, near Leeds.
1894 NORTON, A. E., Park Place, Parkinson Lane, Halifax.
1899 OLIVER, HUGH, Solicitor, Brighouse.
1899*ORMEROD, WILLIAM, Scaitcliffe Hall, Todmorden.
1875*PARKE, G. H., F.G.S., F.L.S., St. John's Villas, Wakefield.
1892*PAWSON, A. H., F.L.S., F.G.S., Lawns House, Farnley, Leeds.
1883 PEACH, ROBERT, 28, James Street, Harrogate.
1891 PEARSON, JOHN T., The Hall, Melmerby, S.O., Yorkshire.
1897 PRATT, Rev. CHAS. T., M.A., Cawthorne Vicarage, near Barnsley.
1859*RAMSDEN, Sir J. W., Bart., Byram Hall, nr. Pontefract.
1864*RHODES, JOHN, Bolton Royd, Bradford.
1856*RIPON, The Marquis of, K.G., F.R.S., &c., Studley Royal, Ripon.
1898 ROBINSON, WILLIAM, Greenbank, Sedbergh, R.S.O.
1900 ROBSON, R. M., 21, Belle Vue Street, Filey.
1901 ROEBUCK, W. DENISON, F.L.S., 259, Hyde Park Road, Leeds.
1869*ROWLEY, WALTER, F.G.S., F.S.A., 20, Park Row, Leeds.
1875*RYDER, CHARLES, Westfield, Chapeltown, Leeds.
1901 SAGAR, JOE, The Poplars, Halifax.
1882 Scarborough Philosophical Society, The Museum, Scarborough.
1897 SHEPPARD, THOMAS, F.G.S., Eastgate, Hessle, near Hull.
1892*SIMPSON, W., F.G.S., The Gables, Halifax.
1887 SLATER, M. B., 84, Newbiggin, Malton.
1899 SLATER, A. C., B.Sc., Jesmond House, South Parade, Pudsey.
1879 SLINGSBY, W. C., Carleton, near Skipton.
1876*SMITH, F., The Grange, Lightcliffe.
1851 SORBY, H. C., D.C.L., F.R.S., F.G.S., Broomfield, Sheffield.
1901 SPENCE, EDMUND, Clapham, Lancaster.
1875 STANHOPE, Colonel W. T. W. S., J.P., Cannon Hall, Barnsley.

1881*STANSFELD, ALFRED W., 10, Middleton Villas, Ilkley.

1900*STANSFIELD, W. G., Wilton House, The Grove, Ilkley.

1890 STATHER, J. W., F.G.S., 16, Louis Street, Hull.

1894 STEARS, JOHN, Westholme, Hessle, Yorkshire.

1899 STEWART, W. H., Milnthorp House, Wakefield.

1898*STEWART, WILLIAM, Milnthorp House, Wakefield.

1885*STOCKDALE, T., Spring Lea, Leeds.

1898 STOREY, WILLIAM, Fewston, near Otley.

1878 STRANGWAYS, C. Fox, F.G.S., Abbotsbury House, New Walk, Leicester.

1879*STRICKLAND, Sir CHARLES W., Bart., Hildenly, Malton.

1894 SUTCLIFFE, J. W., The Hollies, Glenroyd, Halifax.

1888 SYKES, Sir TATTON, Bart., Sledmere House, near York.

1898 TATE, JNO. W., M.E., Ingleton, Kirkby Lonsdale.

1897*TEAL, JOSEPH, Banksfield House, Yeadon, near Leeds.

1894 TETLEY, ED. H., B.A., 7, Hillary Place, Leeds.

1880 TETLEY, C. F., M.A., Spring Road, Headingley, Leeds.

1886 THOMPSON, R., Drincote, The Mount, York.

1894 THORNTON, ARTHUR, M.A., The Grammar School, Bridlington.

1875*TIDDEMAN, R. H., M.A., F.G.S., of H.M. Geological Survey, 28, Jermyn Street, London, W.

1897 TINDALL, GEORGE FAWCETT, York Road, Tadcaster.

1899*TURTON, ROBERT B., Kildale Hall, Grosmont, Yorks, R.S.O.

1899 UTTLEY, WILLIAM H., 63, Hollins Grove, Sowerby Bridge.

1884*WALMSLEY, A. T., C.E., F.S.I., F.K.C., 5, Westminster Chambers, Victoria Street, London.

1899 WALSHAW, THOMAS, Lincoln Street, Wakefield.

1894 WALTON, F. FIELDER, F.G.S., L.R.C.P., 19, Charlotte Street, Hull.

1901 WALTON, SIMEON, Ashgrove Cottage, Elland.

1875*WARD, J. WHITELEY, J.P., F.R.M.S., South Royd, Halifax.

1900 WHARNCLIFFE, Earl of, Wortley Hall, Sheffield.

1898 WELLBURN, EDGAR D., L.R.C.P., F.G.S., Beech House, Sowerby Bridge.

1890 WHITE, J. FLETCHER, F.G.S., 15, Wentworth Street, Wakefield.

1881*WHITELEY, FREDK., Clare Road, Halifax.

1898 WHITMELL, CHAS. T., M.A., Invermay, Headingley, Leeds.

1896 WIGHAM, FREDERICK HENRY, F.G.S., The Towers, St. John's, Wakefield.

1889 WILKINSON, J. J., Burnside, Skipton.

1894 WILLIAMS, W. CLEMENT, F.R.I.B.A., George Street, Halifax.

1890 WILSON, Colonel EDMUND, Denison Hall, Leeds.

1894 WILSON, J. E., Yew Bank House, Ilkley.

1897 WILSON, J. MITCHELL, M.D., 4, St. George's Villas, Doncaster.

1901 WILSON, NORMAN McLEOD R., A.M.I.C.E., County Surveyor's Office, Northallerton.

1878 WOODHEAD, JOSEPH, J.P., Longdenholme, Huddersfield.

*** It is requested that any Member changing his residence will communicate with the Secretary.

Vol. XIV.]   [Part III.

PROCEEDINGS

OF THE

YORKSHIRE

GEOLOGICAL AND POLYTECHNIC SOCIETY.

EDITED BY W. LOWER CARTER, M.A., F.G.S.,

AND WILLIAM CASH, F.G.S.

1902.

INGLEBOROUGH.

PART II.* STRATIGRAPHY.

BY PROFESSOR T. McKENNY HUGHES, M.A., F.R.S., F.G.S., WOODWARDIAN

PROFESSOR OF GEOLOGY IN THE UNIVERSITY OF CAMBRIDGE.

The oldest rocks of which we have any representatives in this
area are the great mass of green slates and grit and conglomerate
which form the floor of Chapel-le-dale, and of Ribblesdale for
about half a square mile near Horton.   These so much resemble
the old volcanic series described under the general title of "Green
Slates and Porphyry" in the Lake District, and agree so well
with them in their relation to the rocks with which they are
associated, that, in default of any evidence to the contrary, they
have been referred to the same horizon.

* Continued from Proc. Yorks. Geol. and Polytec. Soc., Vol. XIV., p. 150.

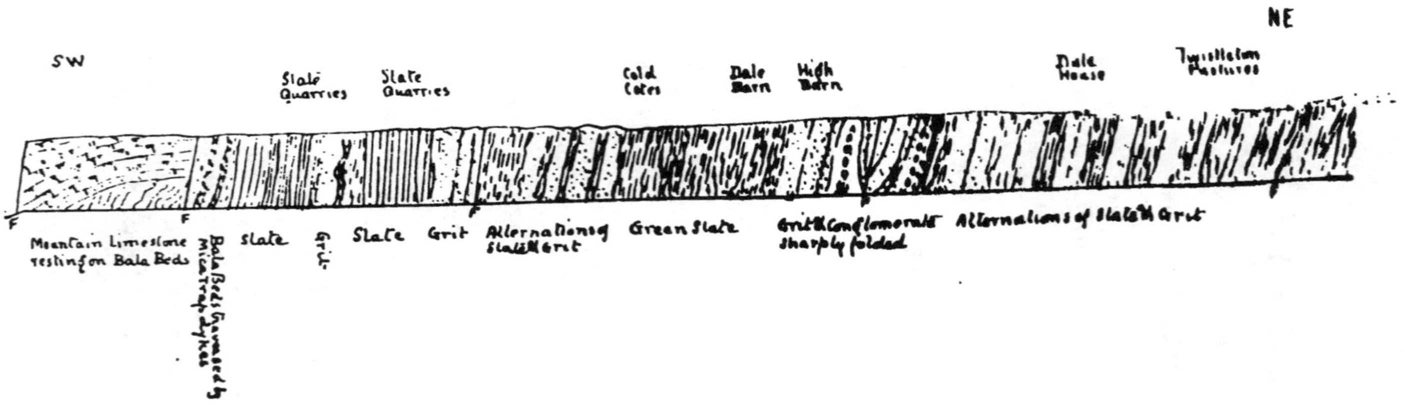

SW   NE

Slate Quarries   Slate Quarries   Cold Cotes   Dale Barn   High Barn   Dale House   Twistleton Scars

Slate Grit   Slate   Grit   Alternations Slate&Grit   Green Slate   Grit&Conglomerate sharply folded   Alternations of Slate&Grit

Mountain Limestone resting on Bala Beds

Bala Beds traversed by Mica Trap dykes

Fig. 9.

SECTION ALONG CHAPEL-LE-DALE.

Scale - 3 inches to 1 mile.

The rock consists, as far as the finer parts are concerned, of
the felspathic dust; the grit is composed chiefly of grains of quartz
and pink orthoclase in a greenish matrix consisting of finer grit
and the same material as the slate. There are also, but rarely,
beds of breccia much like those in the Borrowdale Series. Iron
pyrites occurs in beautifully perfect cubes, and as dendrites in
the slates, while the iron pervades the whole as a green silicate
which is especially conspicuous in some of the more crystalline
quartz veins.

No fossil has ever been found in their equivalents in Cumberland; and, though these Yorkshire beds are more promising than
those of the Lake District, no fossil has yet been recorded from
them. The black marks seen on some faces of rock in the slate
quarry north of Ingleton, which by their outline suggested graptolites, are merely segregations of mineral matter; while in other
cases the combination of concretions and rock crushing have
produced complex arrangements in which imagination has seen
traces of trilobites and other organisms.

The succession as observed or inferred is shown in Section,
Fig. 9. The coarser beds, known as "calliard," indicate the
direction of the dip, and show, as pointed out by Professor
Sedgwick,* that in the slate quarries "the planes of fission are
parallel to the original laminæ of deposit." But in the absence
of such bands it would be very difficult to make out the bedding.
The apparent dips are generally high, from 70° to 90°, some, however, being as low as 50°. The strike is always N.W. and S.E.

The series would appear at first to have a fairly uniform
dip to the S.W., the general differences of material showing a
succession in that direction, and the details of the stratification
at the S.W. end of the section where they can be made out
indicating a coincidence of bedding and cleavage at about 80° to
the S.W. (see Section, Fig. 9). Where, however, marked beds,
such as the grits and conglomerate of Dale Barn and those recently
exposed in the so-called granite quarry on the opposite side of the

* Quart. Journ. Geol. Soc., Vol. VIII., p. 45.

valley allow us to trace the bedding, it is sometimes seen that they are folded over in sharp plications of such a kind as it would be impossible to detect in the fine homogeneous slate. It may be, therefore, that the thickness of the series is not so great as would be inferred from this apparently regular succession at a high angle for 2¼ miles along the valley, but that with a general succession to the S.W. each portion is repeated in a series of sharp folds as shown in the diagram (Fig. 10). The pressure that produced these folds must have been in the same direction as that which produced the cleavage, but whether there is evidence of a cross strain in what looks like double cleavage or cleavage in two directions in the Ingleton quarries, or whether that is due to local readjustments to meet some accidental greater resistance we cannot now make out.

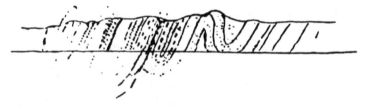

Fig. 10.

In the Ingleton "granite" quarry the coarse grit is excavated for road metal, the word "granite" being frequently used in commerce for any rock with a coarse grain. The more unyielding grit is here seen crushed into the associated soft slaty material, and, as there was much stretching of the mass, especially at the contact surfaces, when the mud was being squeezed out into a cleaved slate and the tough grit had to accommodate itself to the changed relations, there was a good deal of tearing and inclusion of fragments in the adjacent rocks, and it is not always easy to decide whether a given specimen is a part of the brecciated conglomerate or owes its character largely to brecciation in place and drag along the planes of readjustment.

When two consecutive formations of different character occur in a district which has been subjected to great and violent earth movement they are almost sure to give way along the line of contact, and when denudation attacks such a junction there is almost always. a result which obscures the junction. Plenty of examples of this may be found among the overlying rocks in the steep precipices of Ingleborough, but we shall find the difficulty which arises from such conditions still greater when we are trying to make out the line of junction between such ancient deposits as the Green Slate Series and the Coniston Limestone and Shale.

This prepares us to find that junction sections between the two are exceedingly rare and obscure. In the gorge below Thornton Force the two series are seen in contact close to where the great fault leaves the bed of the stream and runs into the hill. Here there are thin papery shales presenting a somewhat intermediate character between the fine ash beds below and the calcareous shales above, but there is always the doubt that we may be here dealing not with true beds somewhat crushed, but with masses which owe their bed-like character to mechanical rearrangement during periods of movement long subsequent to their original deposition.

There is no other junction of the Green Slate Series and the Coniston Limestone and Shale Series seen anywhere in the immediately adjoining area, though the two formations occur very close together in several places and in the same relative position. Some green slaty rock exposed at the surface in a weathered condition near Moughton Sike in the Crummack valley may belong to this series, or to passage beds between it and the Coniston Limestone.

In the northern part of Ribblesdale the green slates are exposed in the bed of Douk Gill to the east of the village of Horton. They dip at a high angle in a southerly direction, and pass up into calcareous slates full of Bala fossils. About 150 yards down stream these turn up again in a synclinal fold broken at its west end by a fault* which cuts off the Bala Beds, bringing the green slates again to the surface.

---

* See Sedgwick, Quart. Journ. Geol. Soc., Vol. VIII., p. 49.

The Green Slate Series is seen in several sections, or on glaciated surfaces near Horton in Ribblesdale, but, except in the railway cutting south of the station, there is no continuous exposure for more than a few yards, so that it is difficult to correlate the horizons with those seen in Chapel-le-dale.   It is, however, probable that it is the upper part of the series that is here represented from the occurrence of the Coniston Limestone Series so close to it on the south, and from the transitional character of the upper part of the series, which here contains much calcareous matter and even beds of limestone.   The main mass, which succeeds in apparently descending section, consists of green slates, grits, and conglomerates not very unlike those which appear to underlie the passage beds in the gorge at the foot of Chapel-le-dale. The beds are generally inclined from 45° to 75° in a south-westerly direction, but they vary to such an extent and in such a manner in the direction and amount of dip, and are in places so crushed and veined, as to suggest that they are much folded and perhaps repeated, a supposition which is strengthened by the fact that all the pre-Carboniferous rocks are more and more intensely plicated as we follow them up Ribblesdale from south to north.

The Green Slate Series of Ingleborough is probably the "geological equivalent"* of the Green Slates and Porphyry of the Lake District and of the Bala Volcanic Series of North Wales. But there is a difference between the results of volcanic activity in these three areas.   In North Wales there are lavas and volcanic breccias and ashes associated with fossiliferous marine deposits.   It is clear that the volcanic masses were here built up on a sea floor, but that most of the lavas and ashes of successive great eruptions remain where they were first thrown out, and are not merely the volcanic material distributed far and wide over the bottom of the sea.

In the Lake District, on the other hand, there are lavas and agglomerates, and ashes, but no traces of fossiliferous sediments associated with the great masses of volcanic ejectamenta, nor is

* Whewell, History of the Inductive Sciences, Vol. III., p. 532

there any sufficient proof that the material was sorted by currents of water. For aught we can say at present the stratification may have been produced by the showering down of the coarser and finer, of the heavier and lighter material over a land surface, or into the sea, where it directly settled to the bottom, beyond the reach of wind, waves, and tides, but no signs of ancient land surfaces or of marine sediment with traces of life have yet been detected.

The Ingleborough district seems to have been still further from the region of volcanic activity. No lava flows appear to have reached it. There is a very large proportion of the finest material, and, where bands of coarser character occur, they consist of well-rolled sand and grit such as might be derived from a coarse granite, and, more rarely, some beds of breccia of doubtful origin. No signs of land surfaces, nor of the fauna or flora of a sea bottom, have so far been detected here, though we may hope that among the grit bands fossils may yet be found. On the whole it seems probable that we have here volcanic material transported from a distance, and quietly setting to the bottom of a lake or sea over which no coarser or heavier material than the grit of Dale Barn was ever strewn. We should not expect the thickness of beds to be as great as that which was attained by the cones nearer the sources of eruption, and this consideration, as well as observation of the nature of the folds where they can be detected, would support the suggestion (p. 325) that the apparent sequence along Chapel-le-dale is deceptive, and that the beds are repeated over and over again as shown in the diagram (Fig. 10).

We cannot attach much importance to the absence of fossils in a formation which has suffered so much deformation as has this Chapel-le-dale Series of green rocks. Almost all traces of organic life would probably have been entirely obliterated or been dragged out beyond all possibility of recognition, and, if they are found, it will probably be where protected by some of the more unyielding beds or on the turn of a fold where the distortion was not so violent.

BALA BEDS.

In ascending section above the Green Slate Series and in descending section below the Silurian flags, grits, and conglomerates, we always find masses of roughly-cleaved shales with calcareous bands and concretions and beds of limestone. Notwithstanding the folded and faulted condition of the older rocks in this area the true sequence can generally be established in these beds, and the newer are found to occur next in succession below the Silurian, and the older next in succession above the green slates, though some members are here and there missing, being probably dropped out by faults. Some bands are highly fossiliferous, and the fossils as well as the position of the series makes it quite clear that this is the equivalent of the Coniston Limestone and Shale of the Lake District and the Bala Limestone and Shale of North Wales. The details, however, vary across this limited area, and, as might be expected, the exact correlation of subdivisions with those of more distant regions is at present very incomplete.

They are seen in the bed of the stream striking parallel to the great fault which traverses the gorge of the Twis or Greet below Thornton Force, as described above.*

The part of the series here exposed consists of a shale with layers of lenticular concretions of limestone. The lower beds are concealed by masses of drift and torrent *débris*—and that which is seen is so much crushed that all traces of fossils have been mostly obliterated, while the few that are found are generally so obscure to be almost unrecognisable.

However, there are fossils in it, and when the zones of the Bala Beds of this district come to be worked out in detail there may be enough palæontological evidence here to determine the exact horizon.

Sedgwick records the following fossils as having been found here near the upper dyke :—*Stenopora* (*Favosites*) *fibrosa* Gold., *Halysites*

_____

* Proc. Yorks. Geol. and Polytec. Soc., Vol. XIV., pp. 146, 147. See tion Fig. 6.

catenularia Linn., *Orthis actoniæ* Sow.; to these I may add a *Petraia* (*Streptelasma*) *æquisulcata*. Near Twistleton Manor House the Coniston Limestone is seen in a roadway very much crushed and traversed by veins of calcite. It has evidently been much altered here and has a brown, earthy character, being probably dolomitized.

. It crosses Dale Beck east of Twistleton Manor, but denudation has cut it down so much that it is generally covered by the alluvial gravel, except after a great flood, when portions are for a time left bare, or where baked portions are preserved in contact with the dykes, which stand out in reefs and buttresses having the same trend as those seen in the Greet.

A few small exposures among the roches moutonnées show that just south of Skirwith the same beds cross the eastern road to Chapel-le-dale and the stream below the farm.

At the lower end of Crina Bottom, whose waters drain into Jenkin Beck, there are shales and bands of concretionary limestone which weather into light, porous, gingerbread-coloured rock by the removal of the lime. These beds apparently dip at a high angle, but this is not clear, as there are only small exposures, and the beds deviate somewhat from the normal strike of the Coniston Limestone Series in the adjoining area, running E.N.E. and W.S.W. They are probably near a crush.

From this point for three miles to the south-west denudation has failed to reach down through the Mountain Limestone, on the upthrow side of the great fault, to the level of the Silurian and Bala. On the downthrow side of the fault the Mountain Limestone has nowhere been cut through.

As we have seen (p. 134) the base of the Carboniferous rocks is reached at different depths dependent upon the character of the underlying rocks, and the Bala shales seem always to have suffered more erosion than any of the other pre-Carboniferous rocks of the district. Therefore they have generally a greater thickness of Mountain Limestone above them. These depressions caused a larger gathering of subterranean waters in the fissured limestone, and thus valleys are apt to be opened up along them

in later times, and this again often results in their being deeply
buried in glacial, alluvial, and terrestrial drifts.   So that we have
to study a very complex series of causes and effects to explain
the present distribution of the exposures of these Bala beds.
The ground-plan (Fig. 11) shows the relation of the Bala Beds and
Silurian (Upper Silurian of Survey) to one another, and represents
what we should see if all the Carboniferous and superficial deposits
were swept off.   Clapham Beck and its tributary torrent, Cat Hole
Sike, have cut down to them, exposing a black shale with calcareous
bands or nodules, and thin beds of limestone.   Traces of fossils have
been found in Cat Hole Sike, and it is probable that a small
excavation in that watercourse would give an opportunity of
collecting a sufficient number of species to determine the horizon.

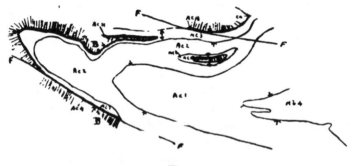

Fig. 11.

GROUND-PLAN SHOWING RELATION OF SILURIAN AND BALA SERIES
BENEATH THE CARBONIFEROUS ROCKS.

The Mountain Limestone is faulted against them on the
south, and, though the actual line of fault is not exposed, its
position is seen above the footbridge at the north end of the
lake in Mr. Farrer's grounds at Ingleborough, near Clapham, and
a line of depression up the hill side probably marks its course.
The Mountain Limestone gets more and more crushed as we
follow it towards the waterfall, which represents very nearly the
north wall of the fault.   Above this the stream tumbles over
a ———— of ledges formed by bands of limestone in shale.
———have yielded a considerable number of ill-preserved

specimens of common Coniston Limestone fossils. The cleavage
is irregular, owing to the irregular texture of the beds, and also,
perhaps, partly to movements subsequent to the cleavage of the
rock, but it is generally from 50° to 70° S.S.W. The joints are
not very pronounced, the most conspicuous being nearly at right
angles to the strike.

Whether it comes to the surface continuously between Spring
Valley Sike and Norber Sike under the heavy masses of drift
below Robin Proctor's Scar, on the north side of Thwaite Lane,
is a matter rather of inference than of direct observation. In
the neighbourhood of Austwick, Crummack, Wharfe, and Woodend, all within an area of about two miles north and south, by
one mile east and west, we have the best sections of the upper
part of the Bala Beds seen anywhere in this district. Their
general distribution is shown on the plan (Fig. 11), and their
relations to one another and to the overlying formations are
shown in sections.

There is no exposure of the underlying Green Slate series in
this area unless we suppose that some of the shales south of Wharfe
may represent it, being muddy sediment deposited further away
from the focus of volcanic eruption; but the occurrence near
Horton in Ribblesdale of Green Slates of the same character as
those of Chapel-le-dale goes against this suggestion, while the beds
of coarse tuff seen in the shale in Wharfe Mill Dam and elsewhere in that neighbourhood show that the products of eruption
still reached this area either directly from the crater or drifted
as ordinary sediment from unconsolidated cones of volcanic ash.

A very instructive traverse may be made by turning into
Norber Sike at the Sheepfold by the bend of the road from
Austwick Town Head to Norber Brow. The Bala shales can
be followed almost continuously along the stream to the spring
thrown out at the base of the Mountain Limestone. They
consist of a strongly-cleaved calcareous mudstone, yet not so
evenly cleaved as to yield slates.

In the upper part of the section bands of flat concretions of
limestone lie with their larger surfaces adjusted to the cleavage

planes. These represent part of the Coniston Limestone. From
this point the basement bed of the Mountain Limestone run
nearly east for about ⅛ mile. If we leave it here where it
turns to the north and cross the field in a south-westerly
direction, a distance of about 150 yards, to where the road to
Crummack takes the corresponding turn to the north, we find
a road-cutting in the shale from which during several excursions
my party have collected a large number of well-preserved fossils.
It was here—about the middle of the cutting—that Mrs. Hughes
found the first specimen of Dindymene recorded from this district.

The fossils* found in this cutting, which we will refer to as
the Norber Road Section, are:—

> *Diplograptus* (like *pristis* His.).
> *Diplograptus truncatus* Lapw.
> *Dicellograptus anceps* Mch.
> *Stenopora* (*Favosites*) *fibrosa.*
> *Tentaculites anglicus* Salt.
> *Ateleocystites* sp.
> *Agnostus trinodus* Salt.
> *Ampyx nudus* Murch.
> *A. rostratus* Sars.
> *Staurocephalus* (*Sphærocoryphe*) *unicus* Thom.
> *Trinucleus seticornis* His.
> *Dindymene Hughesiæ* Roberts.
> *Cybele Löveni* Linnrs.
> *Lichas laxatus* M'Coy.
> *Turrilepus* sp.
> *Phacops* (*Pterygometopus*) sp.
> *Phillipsinella parabola* Barr.
> *Leptæna transversalis* Wahl.
> *Leptæna sericea* Sow.
> *Orthisina* sp.

* Cf. Marr, Rept. Brit. Assoc., 1881, p. 650; Proc. Yorkshire Geol.
Polytec. Soc., N.S., Vol. VII., p. 397; Geol. Mag., Dec. III., Vol. IV.,
No. 1, 1887, p. 35; ib., Vol. IX., No. 333, p. 97. Reynolds, Geol. Mag.
Dec. IV., Vol. I., p. 108.

The Norber Road beds strike in an east-south-easterly direction, but the wide extent of ground from north to south over which the series here extends, and the high angle of dip, suggest that, unless it has a much greater thickness than we are justified in believing, the beds must be repeated by folds. Accordingly we find that instead of the north-north-easterly dips of the Norber Road section and of the Sike head, we have south-south-westerly dips further down the Sike and along Wharfe Gill Sike, near Woodend, where all the fossils found, namely, *Ateleocystites* sp., *Phacops* (*Pterygometopus*) sp., occur also near Staindale, in the Norber Road Section. Further north, also along Crummack Lane, southerly dips prevail. An anticlinal

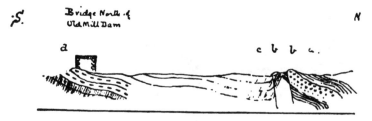

Fig. 12.

a. Striped sandy shale.
b. Conglomerate.
c. Cleaved mudstone.
d. Felspathic ash-like beds, similar to those seen south of the barn in the field on the west.

fold is seen also west of Wharfe Mill Dam between the two barns, and another in the bed of the stream north of the Dam House Bridge Fault (see below). Although, therefore, the true dips are not so easy to detect in the Bala beds as in the overlying Silurian, still it can be made out that the beds are thrown into large spoon-shaped anticlinals and synclinals, with smaller adjustment folds and faults on the margin.

As the result of this the calcareous shales in which *Trinucleus* is fairly common, and which we saw striking in a south-south-easterly direction at Norber Brow, are thrown off to the north and south by the time we reach Crummack Beck, where

older beds of the series are brought up in the anticlinal arch,
as seen between the two barns near Dam House Bridge (Fig. 12)
and in Wharfe Mill Dam (Fig. 13).   These older beds consist
of mudstones which break into larger prism-shaped fragments
and slabs than the overlying calcareous shales.   There are also
beds of fragmental felspathic and other rocks such as are usually
called volcanic ash, but whether they are lapilli showered down into
the sea directly from eruptions or only the waste of distant still unconsolidated beds of volcanic ash distributed over the sea bottom it is
impossible now to tell.

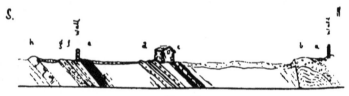

Fig. 13.

a. Striped sandy shale not much cleaved.

b. Conglomerate not well seen in the line of this section.

c. Cleaved shale with subordinate calcareous band.

d. Felspathic ash-like beds and yellow porcellanous rock with subordinate
     black bands, one very conspicuous.   These are probably the beds
     seen under the bridge.

e. Slate strongly cleaved 70° S.S.W.

f. Felspathic, speckly, ash-like beds exposed at the gate for a horizontal
     distance of about 8 feet.

g. Very tough, granular, crystalline rock, with small sago-like grains
     of transparent quartz, about 10 inches seen.

h. Limestone, with a knobbly, irregular surface, about 2 feet seen.

The strike of *e, f, g, h* is clear, but not the dip: indeed, it may be that
between this and the barn we have crossed the axis of the fold, and that we
are already on the southern limb of the anticlinal.

Fossils occur both in the mudstones and in the ash, and especially
in the intermediate rock in which there is a mixture of the ash and
mud.   The fossils found here are *Calymene blumenbachii* Brongn.,
Illænus davisii Salt, *Trinucleus seticornis* His., *Leptæna transversalis* Wahl., *Strophomena.*   The large *Strophomenas* are so con-
spicuously the most abundant form that, although they are not

peculiar to this horizon, they characterise it by the numbers of in-
dividuals in proportion to other species, so that for convenience of
local reference I formerly called these beds the *Strophomena* Shales.
The same remarks apply also to the name *Trinucleus* Shale by which
I referred to the overlying beds. Closer work will doubtless by-and-
by divide the series into many palæontologically distinguishable
zones.

The whole series is nipped out by faults near Wood Lane and is
not seen again to the E.

The section is much complicated by a fault which may be seen
crossing the stream a little above Dam House Bridge. The direction
of this fault is slightly oblique to the strike of the Bala Beds, having
a little more north and south in it, so that it brings the Silurian rocks
against lower and lower beds of the Bala series as we follow it to the
east-south-east. The beds of both series are similarly folded on the
north or downthrow side of the fault, but, owing to the downthrow
being on the north, higher beds of the Bala series are found in the
tops of the anticlinal folds on the north side of the fault than are
exposed in the corresponding and contiguous folds on the south of
it. Thus in the stream close above the small waterfall which marks
the exact position of and is due to the Dam House Bridge fault there
is a small subordinate anticlinal which throws off the basement bed
of the Silurian on either side, so that black shales are exposed between
two bands of conglomerate, the one with a northerly, the other
with a southerly dip (see Figs. 12 and 13). It is probable that this
is a pre-Carboniferous fault.

In these black shales the following fossils have been found :—

> *Orthis testudinaria* Dalm.
> *Strophomena siluriana* Dav.

They therefore represent the highest beds of the Bala seen in
this district, and are the equivalent of the *Strophomena siluriana*
beds so largely developed in the tributaries of the Rawthey, especially
in Sarley Beck, some 15 miles N.N.W. of Wharfe. It was on specimens collected in Sarley Beck that Davidson founded the species
siluriana. In the Lake District their equivalent is seen in the
shales of Ashgill, near Coniston.

About three-quarters of a mile north of the section near White which has just been described, where the principal feeder of Crummack Beck issues from the base of the Carboniferous rocks (see Plate L) the stream has cut down on to the top of another anticlinal fold which brought Bala Beds within reach of recent denudation (Fig. 14. This interesting spot is known as Austwick Beck Head. from the name of the stream into which the water eventually finds its way, though the tributary which immediately receives it is known as Crummack Beck.   Just below the junction of the stream from Austwick Beck Head and Moughton Sike the Coniston or Bala

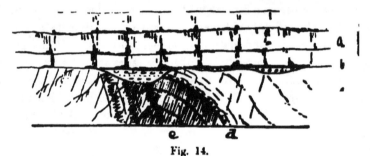

Fig. 14.

DIAGRAM ILLUSTRATING THE MODE OF OCCURRENCE OF THE STRATA
AT AUSTWICK BECK HEAD.

a. Mountain Limestone.
b. Basement bed of the Mountain Limestone.
c. Silurian.
d. Basement bed of Silurian.
e. Bala Shale and Limestone.

Limestone is seen dipping south.   It is at the surface in Moughton Lane with a similar strike and under the Mountain Limestone scar on the east is again seen dipping south at about 50°.

Close to the keld or springhead the basement bed of the Silurian is exposed resting on the Bala Shale.   This section is not very clear owing to the creep-forward of the basement bed of the Carboniferous for some distance down the side of the stream ; in fact, we have here an example of the occurrence of a depression along the outcrop of Bala Beds which I referred to above (p. 331).   It is increased somewhat in this case by a crush and some faulted ground along

Photographed by Godfrey Bingley, Headingley, Leeds. AUSTWICK BECK HEAD, CRUMMACK DALE.

the ant t
and hig ?
Shuriar
of the N
Fi) in v
axis (Fi
Be
clii of
upper
of ch
the
may
a l

ction between them and the more tough and unyielding Silurian.
 depression in the pre-Carboniferous surface was filled with lime-
ne and conglomerate, and, moreover, as usual in the case of the
er hollows, caused the conglomerate to be made up almost
rely of fragments of the underlying rocks, so that it is easily
ounded with the thin line of basement bed of the Silurian which
so composed of fragments of the Bala series. As the subter-
an denudation of the Mountain Limestone went on this depression
rmined that the spring should come out here, and resulted also in
deeper portion of the basement bed of the Carboniferous being
nged further out into the valley. A little care, however, and,
 floods, a little clearing of the torrent *débris*, enables us to dis-
uish and work out the relations between the two basement beds.
In Ribblesdale, the next valley on the east, the Coniston Lime-
e series is exposed at Crag Hill and west of the village of Horton.
section at Crag Hill is full of interest for the stratigraphist and
ontologist, and would well repay careful detailed work. The beds
rought to the surface by a sharp anticlinal fold, the axis of which
about 10° S. of E.
The fall of the ground to the east is greater than the inclination of
nticlinal axis in the same direction, and therefore we find higher
higher beds of Bala series as we ascend the hill, and at last see the
rian beds folding over the Bala series 100 feet or so below the base
ne Mountain Limestone, as shown in the diagrams (Figs. 15, 16,
in which are given a ground plan (Fig. 15), a section along the
 (Fig. 16), and another at right angles to it (Fig. 17).

Behind and above the farm there is a fine section exposing a large
 of limestone and calcareous shale split off so as to expose the
per face of the rock. On this numerous well-preserved specimens
characteristic Bala fossils were to be seen weathered out. Most of
e more obvious of these have been long ago collected, but plenty
y still be found with a little care and trouble. The following is
list of those determined :—

> *Heliolites interstinctus* Wahl.
> *Stenopora* (*Favosites*) *fibrosa* Goldf.
> *Petraia* (*Streptelasma*) *æquisulcata* M'Coy.

B

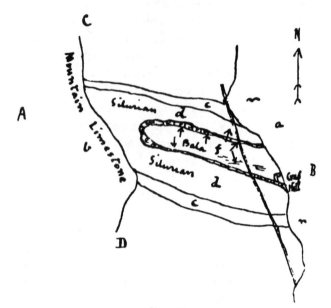

Fig. 15.

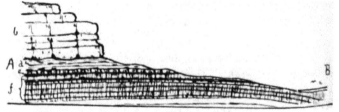

Fig. 16.

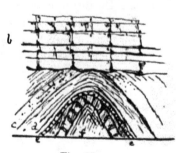

Fig. 17

Cheirurus bimucronatus Murch.

Cybele verrucosa Dalm.

Illænus bowmanni Salter.

Phacops ? obtusicaudatus Salt.

Discina.

Leptæna sericea Sow.

L. transversalis Wahl.

Lingula ovata M'Coy.

Orthis Actoniæ Sow.

Orthis biforata Schloth

Orthis calligramma Dalm.

Orthis elegantula Dalm.

Orthis flabellulum Sow.

Orthis vespertilio Sow.

Rhynchonella.

Strophomena grandis Sow.

Modiolopsis antiqua Sow.

Murchisonia.

grey crystalline limestone (e) is taken for the present at
: as the basement bed of the Silurian, and will be described
'y.

Coniston Limestone and Shale are again seen a short distance
st of Horton-in-Ribblesdale. The stream issuing from the
n Limestone in Dow Gill, about 100 yards below its source,
through the basement beds and exposed the green slates
at a high angle a little west of south. They are followed in
:ly conformable succession by the Coniston Limestone and
hich are here highly fossiliferous. About 150 yards down
hey begin to turn up to a fault thus lying in a broken synclinal.

SUMMARY.

ENIG SERIES.

; forms the base of the group, and in the typical areas of North
nd the Lake District consists of black shales, here and there
ndy or streaked with sandy layers. It has a very distinctive
·ith indications in the upper part of the commencement of

volcanic outbursts somewhere not far off.   At the base there is sometimes a coarse grit.

In the Lake District, where this division is well developed, its base and its top are still not well defined.   Around Ingleborough it is not represented at all, unless it may be among the lowest parts of the volcanic series of Chapel-le-dale.

In all such questions of correlation it must be borne in mind that volcanic accumulations are from the nature of the case of more limited extent than ordinary muddy or sandy sediment, and that among them we must expect to find, though we cannot often hope to identify with certainty, synchronous beds of very different character in closely adjoining areas.   This is what Whewell[*] referred to in his remarks upon "Geological Equivalents," and what was further developed afterwards under the head of "Homotaxis."

THE BALA SERIES. Volcanic Stage.

The great masses of green and grey slates, and grits and brecciated conglomerates of Chapel-le-dale, are referred to this horizon from their similarity to the Borrowdale beds of the Lake District, and because of their constant occurrence below the Calcareous stage of the Bala series, the horizon of which is determined by abundant fossil evidence.

THE BALA SERIES. Calcareous Stage.

This is the horizon of the beds, the distribution and character of which I have been describing in this part of my paper.

There are three places where the fossils are numerous and well-preserved, and the relative position of the beds can be made out—

1. The stream courses north and east of Horton-in-Ribblesdale (see p. 341).
2. The slope west of Crag Hill up to the base of the Mountain Limestone (see p. 339), and, best of all,
3. The area which lies across the southern end of Crummack Dale, between Austwick and Wharfe Gill (see p. 333).

Other areas have been mentioned above, but in them either the exposure is so small or the beds are so crushed that the sequence is obscure and the fossils generally unrecognisable.

_____

[*] Op. cit.

My aim in this paper is to offer some notes which may be helpful to the field geologist on and around Ingleborough, and with that view I will conclude this part of it with a few practical hints.

These are the most easterly exposures of the upper portion of the Bala Beds to be found in this district.   It is a variable series, and some knowledge and experience would be required in order to establish the true sequence in each area and correlate the several zones exposed in the different sections.   But if anyone wants a very interesting and useful, though difficult, piece of work to do let him make out the details of the above three areas, stratigraphically and palæontologically, zone by zone.   It will take a long time to do it properly, but it is not one of those heart-breaking districts where no definite results follow long and careful work.   There are plenty of exposures and plenty of determinable fossils, and it only requires careful observation of the succession and discrimination of exact horizons where fossils have been found.   In such cases all the specimens should be fully labelled in the field.

I gave many years ago in the Proceedings of this Society* and in the Geological Magazine,† a tentative classification, which I think still holds good as far as it goes, but I never visit the district without learning something new about it.

Now as to the best centres to work from.   The village of Horton-in-Ribblesdale is close to the two first-named areas, or, if a more luxurious base of operations is preferred, the train from Settle will set the traveller down and pick him up at Horton Station.

For the third area Austwick or, but little further off, Clapham are most convenient.

If the home-life and folk-lore of the people interest our geologist, let him try to get taken in at one of the farmhouses that lie nestled in the snuggest corners of this margin of the Fells, and I feel sure that, if he is a man of the right sort himself, he will carry away with him a pleasant picture of the best side of old English country life.

(*To be continued.*)

---

* Proc. Geol. Polytech. Soc., W.R. Yorkshire, 1867, p. 565.
† Geol. Mag., Vol. IV., Aug., 1867, p. 346.

THE FLORA OF THE CARBONIFEROUS PERIOD.

BY ROBERT KIDSTON, F.R.S., F.R.S.E., F.G.S.

SECOND PAPER.

In the first paper, the *Ferns, Equisetites,* and *Calamites* were dealt with ; in the present communication the *Lycopodiaceæ, Sphenophylleæ, Cordaiteæ, Coniferæ,* and *Ginkgoaceæ* will be shortly considered.

LYCOPODIACEÆ.

Among the Carboniferous Lycopods a few are found of comparatively small size, perhaps not much larger than some exotic species of *Selaginella,* whilst others, like *Lepidodendron,* attained to arborescent dimensions.

No group of Carboniferous plants rises to the same importance as that of the Lycopodiaceæ. It comprises several genera and some of these contain many species.

The Lycopods must also have supplied much of the material from which our coal seams are formed. Their importance in this respect can be judged by the fact that it is of frequent occurrence to find in coal, bands of over half an inch in thickness formed entirely of Lycopod spores.

We shall now examine shortly the chief genera belonging to this group which occur in Britain.

I.—LYCOPODITEÆ.

Lycopodites Brongniart. Members of the genus *Lycopodites* are very rare in Carboniferous times ; only three species have come under my notice as British, and each has only been represented by a single specimen. One of these, *Lycopodites ciliatus* Kidston, is from Yorkshire and was collected by Mr. Hemingway.

The *Lycopodites* are all small and had whorled or spirally-placed leaves. In *Lycopodites gutbieri* Göpp. (Plate LXIV., fig. 1), for the ise of the stem, the two lateral rows of leaves are large, single-nerved, and sickle-shaped, while the two ventral rows are very small and loosely adpressed to the stem.   Such types have a great resemblance to some forms of *Selaginella.*   In other species the leaves appear to be whorled as in *Lycopodites Stockii* Kidston.

In some cases the fructification is in the form of terminal cones. The *sporangium* is placed at the base of the bract, and in form and position, as far as one can observe from impressions, seems to agree entirely with that of *Lycopodium.*   In the other case the sporangia are borne at the base of the leaves on an ordinary branch, which has apparently undergone little or no modification, as in *Lycopodites ciliatus* Kidston.   Such forms compare with the common *Lycopodium selago* of our hills and moors.

It is not yet known whether *Lycopodites* produced only one kind of spore (*isosporous*) or two kinds, macrospores and microspores (*heterosporous*).

Another genus which is placed in this group is *Archæosigillaria* Kidston, represented by a single species, *Archæosigillaria vanuxemi* Göpp., sp.   It is very rare, and the only examples I have seen are from the Lower Carboniferous of Westmorland.

II.—LEPIDODENDREÆ.

Lepidodendron Sternberg.   This is one of the most common genera of Carboniferous plants, and occurs plentifully throughout the whole formation.

The *Lepidodendra* were of arborescent dimensions, attaining a height of a hundred feet, with trunks two feet and over in diameter. The stem divided dichotomously and formed a much ramified head (Plate LVI., fig. 1).   The outer surface of the bark bears contiguous (*Lepidodendron aculeatum* Sternberg, Plate LI., fig. 1), or more or less distant (*Lepidodendron serpentigerum* König sp., Plate LI., fig. 2) rhomboidal or fusiform *leaf-cushions,* on whose surface, generally above the centre, is situated the *leaf-scar.*   Within the leaf-scar are three punctiform cicatricules, the central being the vascular scar;

... ... which are probably glandular organs, are called the
... ... The leaves are lanceolate, short, or long and grass-like,
with a single nerve.

The leaf-cushions were probably considerably elevated in the
... ... as impressions frequently show them in this state. The
flattened condition in which they are usually
found is in all likelihood the result of pressure
and the collapse through decay of the more
delicate inner tissue. When the leaf-cushions
are distant, the cortex between them is invariably ornamented with fine, irregular,
longitudinal, flexuous lines, as seen on
Lepidodendron serpentigerum Konig. (Plate
LI., fig. 2). The sub-cortical surface is
generally longitudinally striated, and the
single cicatrice here shown is the scar of the
vascular bundle.

Fig. 1.—Leaf cushion
of *Lepidodendra*
... ... Sternb.,
slightly enlarged.
For description see
text.

The following terms have been applied
to the various parts of the leaf-cushion and
scar. Within the cushion is the leaf-scar, Text
Fig. 1. *a*, in form generally rhomboidal or
sub-triangular, and containing the vascular
scar and the two parichnos ; extending both
above and below the leaf-scar is a central
keel which often bears notches on its lower
part. Immediately above the leaf-scar, in
the line of the central keel, is a small cicatricule called the ligule scar, *c*, and just
beyond it is generally the small triangular
notch, *e*. The area surrounding the leaf-
scar is the "field," *d*. In most species of
Lepidodendra immediately below the leaf-scar
are two oval pits, one on each side of the
keel ; these are probably glandular organs, *f*.
The "field" is generally free from any markings, but in *Lepidodendron Wortheni* Lesqx.

×2

Fig. 2.—*Lepidodendron
Wortheni* Lx. Leaf
cushion, showing
ornamentation. (No.
2731.)

Plate LI., fig. 3 and text fig. 2) and *Lepidodendron acuminatum* Göpp. sp., it is ornamented with irregular transverse lines.

The fructification consists of cones (*Lepidostrobus*), and in the great majority of species these terminate the small branches (Plate LII., fig. 2). In a few species, as in *Lepidodendron Veltheimianum* Sternb. (Plate LVII., fig. 1), the cones are sessile and are borne on the large stems in two opposite rows, the cones of one row alternating with those of the other row. It is a peculiar and marked character of these so-called Ulodendroid Lycopods that the fructification is only produced on stems of considerable size and age. When the cones at maturity fall from the stem they leave a small cup-like depression, which, during the subsequent life of the tree, increases in size as the tree increases in girth.

In many species the cones contain macrospores in the lower sporangia and microspores in the upper sporangia, but whether all the *Lepidodendra* possessed heterosporus cones or not, is not yet ascertained.

If a well-developed stem of *Lepidodendron* showing structure is examined, it will be found in most cases to consist of a central pith, surrounded by a zone of primary wood, the component elements of which are arranged without definite order, and to which when once formed no increase takes place. This is succeeded by a zone of secondary wood with medullary rays, in which the elements are arranged in radial order. Immediately outside is a cambium from which additions are made to the secondary wood, by which means its extent of increase in width is only limited by the life of the tree. The whole is surrounded by a very thick cortex, generally separable into three zones, which have been termed the inner, middle, and outer cortex. The outer cortex consists of long, tough, fibrous tissue, which adds strength and hardness to the outer portion of the bark. The leaf-bundles spring from the outer surface of the primary wood and passing upwards and outwards enter the leaves.

Some species have a solid axis and are destitute of a pith, while in a few supposed *Lepidodendron* stems secondary wood has not yet been observed, but this may simply be that sufficiently old stems for the development of secondary wood in these species have not yet been discovered.

Lepidophloios Sternberg.   Plants of arborescent growth with dichotomous ramification (*Lepidophloios Scoticus* Kidston. Plate LV., fig. 1).   Stem and branches bearing well-developed scale-like leaf-cushions, at or near whose summit is placed the leaf-scar.   Leaf-cushions imbricated, pedicel-like (Plate LV., fig. 2), upright or deflexed, exposed portion with slightly curved or straight sides, or rhomboidal in outline (*Lepidophloios laricinus* Sternbg., Plate LVL fig. 2), smooth or keeled, sometimes provided with a small tubercle immediately beneath the leaf-scar.   Leaf-scars rhomboidal or rhomboidal elongate, with lateral angles rounded or acute. Within the leaf-scar are three punctiform cicatricules, of which the central is the vascular scar. Subcortical cicatrice single.

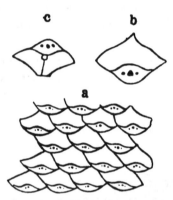

Fig. 3.—*a*, *Lepidophloios Scoticus* Kidston, natural size; *b*, cushion and leaf-scar, enlarged (No. 529); *c*, *Lepidophloios acerosus* L. and H. sp., cushion and leaf-scar, enlarged. (No. 768.)

The fructification is borne on specialised branches, and consists of caducous stalked cones arranged in several spirals (Plate LV., fig. 3).

Lepidophloios is not nearly so common as *Lepidodendron*, and is easily distinguished from that genus by the form of the cushion and the position of the leaf-scar, which is always at the upper end of the pedicel-like cushion, though when the cushion becomes deflexed the leaf-scar appears as if placed at the base. Text Fig. 3.   The exposed portion of cushion and leaf-scar combined is generally of rhomboidal form (Plate LVI., fig. 2).

The mode of fructification is also very different from that of *Lepidodendron*.   The cones (*Lepidostrobus*) are always developed in several spiral series : they are stalked, but the stalks as well as the cones are deciduous, and on falling leave a circle of deflexed leaf-cushions with a small central point (Plate LV., figs. 1 and 2).

Decorticated examples of fruiting branches of *Lepidophloios* were named *Halonia* (Plate LIII., fig. 2) before their true nature was known. In this condition the stem bears spirally-placed rows of mamillæ-like protuberances, but in the corticated condition the depressions between the mamillæ were filled up with the cortex, so that when the bark is present they rise little above the general level of the branch (Plate LV., fig. 2).

The structure of the stem is similar in type to that of *Lepidoden-dron*, though the secondary wood appears to be produced at a later period and in some species it has not yet been observed.

Lepidophloios occurs in both the Upper and Lower Carboniferous.

Lepidostrobus Brongniart. In *Lepidostrobus* are placed the cylindrical, ovoid, or oblong cones with a ligneous axis and single-nerved bracts or sporophylls arranged in spirals (Plate LII., fig. 2, and Plate LV., fig. 3). The bracts consist of two parts, a basal portion or pedicel spring-ing from the axis almost at right angles, and on which is placed the sporangium, and a limb which extends upwards at an acute angle from the extremity of the pedicel, Text Fig. 4. The

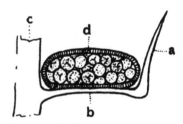

Fig. 4.—*Lepidostrobus*—*c*, axis ; *b*, bract or sporophyll, bearing sporan-gium *d*, containing macrospores ; *a*, limb of bract. (Restored.)

lower bracts bear the macrosporangia, and the upper the micro-sporangia. The macrospores are smooth or apiculate and are provided with a triradiate ridge.

Lepidostrobi are extremely common in Carboniferous strata, but in the majority of cases it is impossible to refer them to the parent plant, hence they are placed in the provisional genus *Lepidostrobus*, in which are included cones belonging to *Lepidodendron, Lepidoph-loios*, and *Bothrodendron*.

Lepidophyllum Brongniart. *Lepidophyllum* embraces the single-nerved, more or less lanceolate leaves so frequently met with, and which it is impossible to refer to the plants which have borne them, Text Fig. 5. They belong in part to *Lepidodendron* and *Lepidophloios*, and also to some *Sigillaria*.

III.—BOTHRODENDREÆ.

Bothrodendron Lindley and Hutton.   The *Bothrodendra* i
one of the most interesting genera of the arborescent Lyco
Their stems ramified by dichotomous div
(*Bothrodendron minutifolium* Boulay (
Plate LIV., fig. 1).   On the mature st
the small distant leaf-scars, surmounted v
a small cicatricule, are oval or oval with m
or less prominent lateral angles, and cont
three punctiform cicatricules.   The b
between the cushionless leaf-scars is or
mented in all cases with perhaps
exception.   In the common Lower
Middle Coal Measure species, *Bothrode*
minutifolium Boulay sp., it is beau
adorned with a series of short, tran

irregular lines and
corrugations, which
divide the outer
surface into very
numerous small,
irregular, vermicu-
lar, oblong shagreen,
each particle of which
bears a row of little
pit-like dots, Text
Fig. 6. This beautiful
structure can only be
seen with a lens, the
leaf-scars themselves
being only about one-
twentieth of an inch in diameter.

Fig. 5. --- *Lepidophyl-*
lum (Lepidostrobus !)
majus Brongt. (No.
2527.) Natural size.

Fig. 6.—*Bo*
minutifoli
sp. *a*, Por
natural si
scar and
tion of bar

On the young branches of *Bothroden-*
dron minutifolium, and possibly on other
species also, we find a most interesting condition of the leaf-scar   The young branches bear finely-c

.ted, rhomboidal cushions (Plate LIV., fig. 2), near whose
is placed the small leaf-scar. As growth proceeds, and at
ly period, these cushions become entirely effaced, and the leaf-
are carried more widely apart (Plate LIV., fig. 1 at *a*, and

Small twigs showing the cushions might easily be mistaken
ung branchlets of *Lepidodendron*.

E two species the mode of fructification is known. In *Bothro-*
n minutifolium it consists of long slender cones (Plate LIX.,
with bracts arranged either in close whorls or very gentle spirals,
hich I cannot at present deter-
These slender cones terminate
branchlets.   In the other
s, *Bothrodendron punctatum*
l H., the fructification consists
ile cones borne in two vertical
which gave rise to cup-like
ssions on the stem in a similar
er to that pointed out when
ing of the fructification of the
ndroid section of *Lepidoden-*
These fruiting portions of
odendron can, however, even
decorticated, be easily dis-
ished from the corresponding
of *Lepidodendron* by the posi-
of the umbilicus of the cone

In *Lepidodendron* it is always
l, in *Bothrodendron* it is eccen-
nd always placed near the
margin of the cup-like depression.

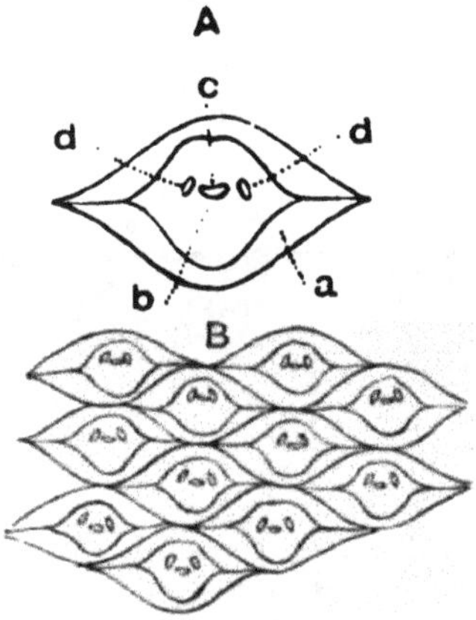

Fig. 7.—*Sigillaria Brardii* Brongt.
Cope's Marl Pit, Longton, Staffordshire. Shale above Peacock
coal, Middle Coal Measures. A,
Leaf cushion (*a*). Leaf scar (*b*);
c, cicatricule of vascular bundle;
d d, parichnos; enlarged. (No
817.)

: leaves of *Bothrodendron minutifolium* and *Bothrodendron*
tum, the only two Carboniferous species of which the foliage
wn, are small, single-nerved, and broadly lanceolate.
he internal structure of the stem is unknown.
he genus is represented both in the Upper and Lower Carbon-
s, but is very rare in the latter division.

The *Bothrodendra*, in the characters they possess, seem to hold an intermediate position between the *Lepidodendra* and the *Sigillaria*.

IV.—SIGILLARIEÆ.

Sigillaria Brongniart. Arborescent Lycopods, with cactus-like or columnar trunks, or very sparingly dichotomously branched stems. The bark is longitudinally ribbed or smooth. The hexagonal leaf-scars are contiguous, Text Fig. 7, or distant with more or less

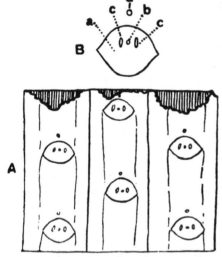

rounded angles, the two lateral angles being most prominently developed. Text Fig. 8. Within the leaf-scar are three cicatricula of which the central is punctiform, transversely elongate, or sub-triangular; the two lateral are upright, and either straight or lunate, and larger than the central vascular scar. The leaf-scar is frequently placed on a more or less prominent elevation, which in the case of the ribbed *Sigillaria* slopes downwards from the leaf-scar. Immediately above the leaf-scar a small cicatricule is frequently present. Sub-

Fig. 8. *Sigillaria principis* Weiss. Old Mills Pit, Farrington-Gurney, Somerset. *Hor.* Lower Series of the Upper Coal Measures. A, natural size; B, leaf-scar enlarged; *a*, area of scar; *b*, cicatrice of vascular bundle; *c c*, parichnos; *d*, "ligule" scar. (No. 421.)

epidermal cicatrices, three; the two lateral vertical, large and lunate, or straight, and united in the centre by the circular or oblong vascular scar; or two through the obliteration of the central scar. They attain considerable size on old stems and, through coalescence, sometimes appear as one large scar. The sub-epidermal surface is longitudinally striated.

The leaves are single-nerved, linear lanceolate, or long and grass-like.

The fructification is in the form of caducous cones (*Sigillarios-trobus*), in some stalked (Plate LV., fig. 4) and placed in the hollows between the ribs (*Sigillaria mamillaris* Brongt., Plate LXI., fig. 3), or between the leaves (*Sigillaria Brardii* Brongt., Plate LVIII., fig. 3 at *a*) in the non-ribbed species. On falling from the stem the stalk leaves a distinct circular, or irregular, shaped scar. The cones form regular verticils of a single row (*Sigillaria elegans* Sternb. sp., Plate LVIII., fig. 3), or of several rows (*Sigillaria tessellata* var. *nodosa* Bowman sp., Plate LVIII., fig. 1), or may be somewhat irregularly placed, especially on the non-ribbed species. In others the cones are sessile and form two opposite alternate rows (Plate LII., fig. 4), leaving, when shed, two vertical rows of cup-like depressions with a central umbilicus (*Sigillaria discophora* König. sp., Plate LX., fig. 1). (*Ulodendron* L. and H. in part.)

The Rhizome of most species of *Sigillariæ* is *Stigmaria* Brongt., in others *Stigmariopsis* Grand 'Eury.

The *Sigillariæ* have been divided into four sections, according to whether the stem is ribbed or smooth, and whether the leaves are close or more or less distant. These four groups, though well characterised in some species, pass into each other, and though they may assist in classifying a very difficult genus, they cannot be regarded as natural divisions or genera as originally supposed. In all, the leaf-scar is of similar structure.

SECTION I.—*Rhytidolepis* Sternberg.

In this section the stems are distinctly ribbed, the ribs straight or slightly flexuous, with surface smooth or variously ornamented. Leaf-scars alternate on neighbouring ribs and occupying the whole or only part of the width of the rib, either close or more or less distant. There is frequently a transverse lunate depression above the leaf-scar. The cone scars are situated in the furrows. Typical form, *Sigillaria mamillaris* Brongt. (Plate LXI., fig. 3).

SECTION II.—*Favularia* Sternberg.

Stem ribbed, ribs flexuous and divided into sub-hexagonal compartments by transverse depressions. Leaf-scars alternate, occupying the whole width of the rib and resting on each other, or only separated by a very narrow space.   The lateral angles of the alternate leaf-scars project slightly and impart to the furrows a more or less zig-zag course (*Sigillaria elegans* Sternb. sp., Plate LVIII. fig. 3).   The cone scars form a verticel round the stem.

The chief distinguishing character of this section is the close leaf-scars and zig-zag furrows.

SECTION III.—*Clathraria* Brongniart.

Stems without ribs with leaves placed on contiguous, slightly elevated, rhomboidal cushions, which are separated by deep oblique furrows.   Cone scars forming irregular verticils and placed in the furrows between the leaf-cushions, or in two opposite vertical rows.

Typical form, *Sigillaria Brardii* Brongt. (Plate LVIII., fig. 2.

SECTION IV.—*Leiodermaria* Goldenberg.

Stems without ribs having distant leaf-scars without cushions. Surface of bark between the leaf-scars variously ornamented, often with fine longitudinal, flexuous striæ, which are frequently cross-hatched with delicate lines (*Sigillaria camptotœnia* Wood sp., Plate LXI., fig. 2).   The cone scars form irregular broad verticils.

The *Rhytidolepis* and *Favularia* sections pass into each other. and only in a very few species can the distinction be observed.

Any interfoliar space on the surface of the ribs is seldom or never entirely free from some ornamentation in the form of transverse lines or small irregular punctations, especially immediately above and below the leaf-scar.   Sometimes these ornamentations are very prominent and form a distinct central band connecting the leaf-scars (*Sigillaria rugosa* Brongt., Plate LXI., fig. 1).   A slightly raised line generally descends from the lateral angles, occasionally extending to the lower leaf-scar and forming the limit of the central band of ornamentation.

The two sections *Clathraria* and *Leiodermaria* also pass into each other ; in fact, they occur on the same specimen (Plate LIX., fig. 1) and thus appear to be only conditions of growth.   The distance or approximation of the leaf-scar cannot always be regarded as a specific mark, for even in the ribbed Sigillariæ, with normally distant leaf-scars, specimens are occasionally found on which, apparently from enfeebled conditions, the leaf-scars become approximated (*Sigillaria Sauveuri* Zeiller, Plate LIV., fig. 4). There are, therefore, really only two sections, those with ribs and those with smooth or unribbed stems.

Those species with the cones arranged in two vertical rows are few in number.   *Sigillaria discophora* König sp. (=*Ulodendron minus* L. and H.) (Plate LX., fig. 1) is frequent in the Upper Carboniferous, while another Ulodendroid form, *Sigillaria Taylori* Carr sp., occurs in the Lower Carboniferous.

The leaves in the great majority of *Sigillariæ* are long and grass-like, with a single nerve, but a few had shorter lanceolate leaves.

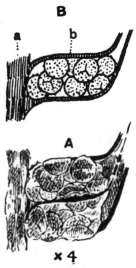

The cones of *Sigillaria* (*Sigillariostrobus*) have hitherto always been discovered separated from their parent trunks though frequently attached to their pedicels, and from one such showing the Sigillarian leaf-scar, M. Zeiller was able to prove their Sigillarian origin.   The Sigillarian origin of these cones had long been suspected, but satisfactory proof was wanting.

Owing to our inability of associating these cones with the species to which they belong they are placed in the genus *Sigillariostrobus* (Plate LV., fig. 4).   These cones differ essentially in their structure from those of *Lepidostrobus*.   In *Lepidostrobus* the sporangia are placed on the pedicel-

Fig. 9.—*Sigilliariostrobus ciliatus* Kidston. A, two sporangia containing *macrospores* (× 4). B, restoration of sporangium—*a*, axis ; *b*, wall of sporangium.

like portion of the bract, whereas in *Sigillariostrobus* the sporangia are developed within the inflated and hollowed-out substance of the base of the bract, Text Fig. 9.   The cones of *Sigillaria* were probably heterosporous ; at least one specimen I possess points to this conclusion, but owing to the imperfection of this example, the opinion requires confirmation before being definitely adopted.

C

The structure of the Sigillarian stem is, in some cases, of the same type as occurs in *Lepidodendron* where the primary wood forms a close ring enclosing the pith ; in other examples the primary wood forms a circle of distinct, but closely-placed, vascular wedges. The two types pass into each other, the passage taking place even in the same specimen, by the lateral union of the wedges among themselves. As in *Lepidodendron*, increase to the vascular system takes place by the addition of a zone of secondary wood from a cambium layer. The whole is enclosed by a thick bark.

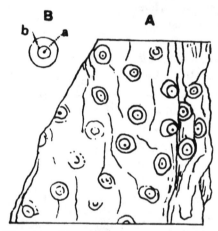

Fig. 10.— *Stigmaria ficoides* Sternb. sp. A, portion of rhizome, natural size : B, rootlet scar—*a*, vascular cicatrice ; *b*, circular depression.

Sigillaria is an extremely distinct genus which comprises many species. It is rare in Lower Carboniferous times, and reaches its maximum development in the Middle Coal Measures, where it is represented by a considerable number of species.

The *Sigillariæ* must also have played a most important part in the formation of coal, and probably the spore-bands, to which reference has already been made, were largely contributed to by spores from Sigillarian cones.

Stigmaria Brongniart. *Stigmaria* (Plate LVI., fig. 3, and Text fig. 10), which is the rhizome of the arborescent Lycopods, diverge from the base of the trunk in four main branches. Shortly after leaving the trunk these again bifurcate, forming eight branches, and again, at a distance of a few feet from the previous fork, these bifurcate into sixteen branches. They do not again divide, or if so only very rarely. Any terminations I have seen end in a blunt point.

The outer surface of the cortex bears quincuncially-arranged rootlet scars, consisting of a slightly raised rim, containing a circular depressed ring placed about midway between the central single vascular cicatrice and the outer margin of the scar, Text Fig. 10.   The rootlets are very long and bifurcate at the extremity, though it is seldom that one finds them in this perfect condition.   It is possible that some of the rootlets did not bifurcate, but remained simple.

Specimens of *Stigmaria* showing structure are not uncommon. The rhizome consists of a large pith surrounded by a zone of exogenously developed scalariform tracheides, which is divided into a number of wedges by the thick primary medullary rays.   These wedges again contain many less prominent secondary medullary rays. The vascular system is enclosed in a thick cortex.   The cast of the pith cavity shows the impression of the netted cylinder of the vascular axis (Plate LII., fig. 3), which is a distinguishing character between *Stigmaria* and *Stigmariopsis*.

Stigmaria is the most common fossil one meets with in Carboniferous rocks.   It is the rhizome of *Lepidodendron*, many *Sigillariæ*, and most probably also of *Lepidophloios*, though it has not been actually found united to the stems of the last-mentioned genus.

Several species and varieties occur, the characters being founded on the size of the rootlet scars and the ornamentation of the outer surface of the bark.

Stigmariopsis Grand 'Eury.   Though this genus is closely related to *Stigmaria*, from the investigations of Solms-Laubach, it must, I think, be separated from it.   It is true that little is known of the organisation of *Stigmariopsis*, but the little known appears sufficient to raise the fossil to generic rank.

Stigmariopsis (*Stigmariopsis anglica* Kidston, Plate LI., fig. 4) is founded on Stigmaria-like rhizomes which are proportionately shorter and thicker.   They spring from the hollow cup-like base of a Sigillarian stem in four primary arms, which again bifurcate, possibly several times.   From the lower surface of the four primary divisions, immediately at the base of the trunk, spring downward directed conical growths—the tap-roots of R. Brown.   The surface of the rhizomes and tap-root-like growths bear quincuncially-arranged rootlet scars, similar in structure to those of *Stigmaria*.   Like *Stig-*

maria. the rootlets bifurcated at the extremity though some may have been simple.

The outer surface of the bark between the rootlet-scars bears an irregular reticulation of slightly raised ridges.   This is probably produced by a sub-epidermal layer of sclerenchymatous tissue, the strands of which uniting amongst themselves form a net-like reticulation (Plate LI., fig. 5).   The dense nature of this tissue would assert

Fig. 11.—*Omphalophloios anglicus* Sternb. sp.   A to E, from different portions of the same specimen (No. 426) ; F and G, portions of another example (No. 433) ; A, C, and F, natural size ; B, D, E, G, enlarged— all from Camerton, Somerset.   For explanation of lettering see text.

itself on the outer surface when the specimen was subjected to pressure and decay.   The cast of the pith cavity is ribbed and somewhat in appearance like the cast of a Calamite, but without joints (Plate LIV. fig. 5).

The genus is rare in Britain, but probably more common than suspected, as it may have been passed over for *Stigmaria.*

The most important characters on which this genus stands are the Calamite-like ribbed cast of the pith cavity, and the irregular mesh-like markings on the outer surface of the bark.   The corresponding cast of the pith cavity of *Stigmaria* would, on the other hand, show the impression of the surrounding netted cylinder, the openings in the mesh being the channels through which the primary medullary rays passed out.   The character of the pith cast can, however, be seldom observed, and in its absence one has only the reticulated or wavy ornamentation of the outer surface of the cortex to direct one, but this seems to be a characteristic feature of the genus.

Omphalophloios White.   The fossil now placed in *Omphalophloios* was originally described by Sternberg as *Lepidodendron anglicum*, and subsequently placed in *Stigmaria* by Brongniart.

In *Omphalophloios* (*Omphalophloios anglicus* Sternb. sp., Plate LXIV., fig. 4) the cortex is divided into clearly-defined rhomboidal areas, Text Fig. 11, A.C.F., within which, and a short distance above the centre, is an elevated sub-cordate, or oval cushion, with a slightly raised, ring-like margin, Text Fig. 11, B *b*, containing a little above its centre an oval scar with a single vascular cicatrice, Text Fig. 11, B *c*.   Immediately below this upraised cushion, and attached as it were to its outer side, is a triangular ridge-like elevation containing a small pit, Text Fig. 11, B *d*.

Omphalophloios is probably the rhizome of one of the arborescent Lycopods, and is easily distinguished from *Stigmaria* on the one hand, and from Lepidodendron on the other, by the form of the vascular scar and cushion.   This fossil is very rare in Britain, and has hitherto only been found in the Radstock series of the Upper Coal Measures of Somerset.

Affinities of the Carboniferous Lycopods.   I can here only summarise the conclusions arrived at as to the affinities of the various Carboniferous genera of Lycopods.   To give all the evidence on which these opinions are based would too far extend the present paper.

The genus *Lycopodites* is very closely related to the existing genera *Lycopodium* and *Selaginella*, and is probably their progenitor.   Some botanists have even placed the members of the genus *Lycopodites* in *Lycopodium*, but we are scarcely warranted to take this course until we possess more definite knowledge of their fructification.

Lepidodendron, *Lepidophloios*, and *Bothrodendron*, though they possess the essential characters of the *Lycopodiaceæ*, appear to have entirely passed away without leaving any descendants.

Sigillaria, in the manner in which the sporangia are immersed in the sporophylls, shows considerable affinity with the existing and diminutive *Isoëtes*. In fact, Goldenberg, many years ago, stated his belief that *Sigillaria* was an arborescent form of *Isoëtes*, and subsequent investigations have added much to strengthen this view.

SPHENOPHYLLACEÆ.

Sphenophyllum Brongniart.   The genus *Sphenophyllum* holds a unique place amongst both fossil and recent plants.

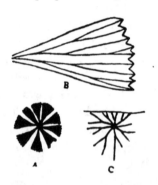

Fig. 12.--A and B, *Sphenophyllum cuneifolium* Sternb. sp.   A, Whorl of leaves, natural size (No. 2706).   B, Leaf, enlarged to show teeth and nervation (No. 1566).   C, *Sphenophyllum trichomatosum* Stur, natural size (No. 1046).

The various species of *Sphenophyllum* have slender ribbed stems with tumid or swollen nodes : the ribs do not alternate at the nodes, but continue in the same line.   The internodes vary considerably in length (*Sphenophyllum myriophyllum* Crépin Plate LXIII., fig. 2), and the branches are given off at irregular intervals, one branch only being given off from a node.

The leaves in the typical form are cuneate, Text Fig. 12, A and B, or wedge-shaped, but they vary much, not only on different species, but even on the different parts of the plant, and this variation can be observed even on the same specimen.   The normal wedge shaped leaf passes into leaves divided into two deeply-cut lobes (*Sphenophyllum majus* Bronn. sp., Pl. LXII., *a* at *a*), which again are bifid, or they may be reduced dichotomously-divided filiform segments which radiate fan-like in the base, Text Fig. 12, C, or may even be simple, narrow, lanceolate leaves forming a whorl round the stem.   These latter seem to restricted to the larger branches.   The nervation repeatedly bifurcate a single veinlet going into each tooth or segment of the leaf.

The fructification, often placed on short lateral branches, is ammonly in the form of terminal cones composed of more or less odified leaves whose basal portions unite to form a saucer-like llar surrounding the axis. The distal portion of the bract is free nd erect. In the cones the whorls of bracts alternate, but on te stems the leaves are superposed.

In *Sphenophyllum majus* Bronn. sp. ae fruiting portion of the plant is little .odified from an ordinary foliage branch, lough the bracts are united into a very arrow collar which surrounds the stem, g. 13.i

Fig. 13. — *Sphenophyllum majus* Bronn. Bract showing four sporangia —sessile and united to each other by their bases.

The arrangement of the sporangia . the cones of the various species varies nsiderably.

In *Sphenophyllum cuneifolium*, the common Middle and Lower al Measure species, the sporangia usually form three whorls placed ι pedicels of varying length which, however, all spring from the act close to the axis, Text Fig. 14. It therefore follows that those the first whorl have shorter pedicels than those of the second, while

those of the third whorl have pedicels longer than those of the two inner circles of sporangia. A small vascular bundle enters each pedicel or sporangiophore, which is therefore most probably a modified segment of the bract.

In *Sphenophyllum trichomatosum* Stur the sporangia are sessile, Fig. 15. In *Sphenophyllum majus* Bronn. sp. the sporangia are also sessile, but united by · their bases into groups of four, Fig. 13.

g. 14. — *Sphenophyllum cuneifolium* Sternb. sp. Arrangement of the sporangia.

The spores of *Sphenophyllum* are very characteristic, the spore membrane being ornamented with spine-like projections connected a series of reticulate ridges.

The stem of *Sphenophyllum* consists of a solid axis, formed of primary, three-rayed, vascular star. developed centripetally. To this is added a zone of secondary wood, formed from a cambium layer. and as this secondary zone increases in width, the star-like form of the primary wood is lost and the bundle becomes circular in form through the external addition of the secondary wood. The whole is enclosed by a firm bark.

Owing to the dimorphic condition of the leaves, it was formerly supposed by some that *Sphenophyllum* was an aquatic plant, but this is not the case.   The whole structure of the plant makes it clear that *Sphenophyllum* was terrestrial, though from their long delicate stems they must have had some support to keep them in an upright position, and this support they probably found by scrambling among the surrounding vegetation.

Sphenophyllum, not only in the structure of its stem, but also that of its cones, exhibits so many peculiarities that it is impossible include it with any other group of plants. With the casts of *Calami* it has a certain superficial resemblar in its noded stems and whorled leaves, the solid axis and non-alternating ribs *Sphenophyllum*, along with its dicho mously-divided leaves as well as structure of its cone, differ so much from the *Calamites* that any systematic re tionship is entirely precluded. With *Asterocalamites* it has a greater rese blance in the ribs of both not alternati

Fig. 15. — *Sphenophyllum trichomatosum* Stur. Arrangement of sporangia (enlarged).

at the nodes and in the leaves being dichotomously divided. But *Sphenophyllum* differs here also in its solid axis and in the structu of the cone, presuming that *Pothocites* is the cone of *Asterocalamite*

Nor with the Lycopodiaceæ does *Sphenophyllum* seem to ha any close connection.   The jointed stem, dichotomously-divid leaves, and structure of the fructification are very different fr anything found amongst the Lycopods.

The genus *Sphenophyllum* must therefore stand alone as a pecul and interesting type of plant which appears to have become extii in early geological times.

GYMNOSPERMS.

We now pass to the Gymnosperms, of which we shall briefly consider the *Cordaiteæ, Coniferæ, Ginkgoaceæ.* Of the *Cycadaceæ* there is no evidence of their occurrence in British Carboniferous rocks.

CORDAITEÆ.

In the *Cordaiteæ* are placed the genus *Cordaites* and certain genera of fossil fruits, some of which belong to *Cordaites*, but which, owing to our ignorance of the species of *Cordaites* to which they belong, necessitates their being placed in separate genera until their parentage has been ascertained.   It is possible, however, that some seeds at present included in the *Cordaiteæ* may not belong to that group, but may be the seeds of other plants, which, with the exception of the fruit, may be quite unknown to us.

Cordaites Unger.   This genus occupies a very prominent and important place amongst the plants of the Carboniferous Period, for though *Cordaites* extends into the Permian Formation it is essentially a Carboniferous genus, for the plant-bearing beds near St. John, New Brunswick, from which the late Sir William Dawson recorded *Cordaites* as Devonian, are, there is very strong reason to believe, really Carboniferous.

Cordaites had thick stems two feet or more in diameter, and which attained a height of 100 feet and terminated in an irregular, much-branched, dense, leafy head.

The branches bore long, lanceolate, spathulate, or linear leaves, which in some cases must have been two feet or more in length.   The leaves have parallel veins (*Cordaites principalis* Germar. sp., Plate LXIV., fig. 3, base of leaf, and Plate LVII., fig. 2, apex of leaf), which increase in number in the leaf by dichotomous division.   In some species the veins are equally strongly marked on the surface ; in others, lying between the veins are one or more finer parallel threads, which appear to represent bands of sclerenchyma rather than weaker veins.

The leaves are spirally arranged, and when shed leave a transversely oval scar on the stem (*Cordaites principalis* Germar. sp., Plate LIII., fig. 1).   Well-preserved stems show on the leaf scars transverse row of little points, the cicatrices of the numerous vascular strands which entered the leaves.   The same character is observable in the bases of the shed leaves.   (Plate LXIV., fig. 3.)

The leaves of *Cordaites* according to their form and shape are placed in the three following genera :—

Cordaites or *Eucordaites*.   Leaves oval-lanceolate, lanceolate or spathulate, with rounded apices (*Cordaites principalis* Germar. sp., Plate LVII., fig. 2).

Dorycordaites.   Leaves lanceolate with sharp points (*Dorycordaites palmæformis* Goppert sp., Plate LVII., fig. 3.

Poacordaites.   Leaves long, narrow, and grass-like.   (*Poacordaites microstachys* Gold. sp., Plate LXIV., fig. 2.)

As a type of *Cordaites*, *C. principalis* Germar. sp. may be mentioned Plate LVII., fig. 2., Plate LXIV., fig. 3).   This is extremely common in the Middle and Lower Coal Measures.

Dorycordaites.   This is represented by *Dorycordaites palmæformis* Gopp. sp., which, however, is rare.   (Plate LVII., fig. 3.

Poacordaites is the most rare in Britain and only occurs in Upper Coal Measures, where *Poacordaites microstachys* Gold. sp. very sparingly found.   (Plate LXIV., fig. 2.)

In the Lower Carboniferous *Cordaites* occurs, but is extremely rare and represented by different species from those found in Upper Carboniferous rocks.

The male and female flowers are borne on different spikes, and the female organs have long been known under the name of *Antholithes Pitcairniæ* L. and H., though their nature was not at first understood.   The male flowers form a spike of distichous or spirally arranged flowers consisting of several whorls of bracts, from among which, or perhaps springing from their centre, arise a number filaments, bearing three or four tubular anthers.

The female infloresence generally consists of a spike of distichous grouped bracts, from the axils of which spring sessile or more or less long-stalked seeds.

These infloresences, both male and female, are now generally included under the name of *Cordaianthus* Grand 'Eury (*Cordaianthus* with *Samaropsis fluitans* Dawson, Plate LXV., fig. 5).

The Structure of the Stems of Cordaites. Little was known about the structure of the stem until M. Grand 'Eury showed a few years ago that the tree described by Witham, under the name of *Pinites Brandlingi*, is really the stem of *Cordaites*. In the young state in this species the pith fills up the whole of the large medulla, but as growth proceeds, the pith not growing with the same rapidity as the surrounding tissue becomes transversely ruptured, resulting in the formation of a series of transverse lenticular cavities, forming a chambered pith. As growth continues the transverse diaphragms of pith get broken up, and eventually the pith cavity becomes an empty tube, except at its margin, to which cling annular rings of pith—the remains of the diaphragms which once extended across the cavity. Inorganic casts of these pith cavities are not uncommon, and are the fossils to which the names of *Artisia* or *Sternbergia* have been given (*Artisia transversa* Artis. sp., Plate LXV., fig. 6). Surrounding the pith in *Cordaites* (*Pinites*) *Brandlingi* is a zone of secondary wood, which consists of radially-arranged tracheides, separated by primary and secondary medullary rays, of one cell in thickness. The first-formed elements of the wood, the protoxylem, formed of narrow spiral tracheides, are followed by larger spiral tracheides; these are succeeded by scalariform tracheides, and then follow tracheides characterised by laterally-placed bordered pits with an oblique opening. It is this latter tissue which forms the bulk of the wood, to which additions are made from a cambium layer. The wood is enclosed by a thick cortex whose outer portion contains bands of dense, thick-walled fibrous tissue. The leaf-bundles which usually appear in pairs in transverse sections of the stem, before entering the leaf split up into numerous small strands, which pursue a parallel course through the leaf as already described. What are probably the roots of *Cordaites* have been described under the name of *Amyelon*.

True, "annual rings" seem to be absent from *Pinites* or *Araucarioxylon* (*Cordaites*) stems, though specimens are occasionally met with where layers of feebly-developed wood are separated by broad bands of wood fibres of normal size.

In *Cordaites* (*Pinites*) *Brandlingi* it is thus seen no *primary* wood occurs, the whole of the tracheides consisting of *secondary* wood.

Some other stems, however, which have been referred to *Araucarioxylon*, *Pinites* or *Dadoxylon*, such as *Araucarioxylon Beinertianum* Goepp. sp. and *Pitus (Araucarioxylon) antiqua* Witham and others, in which the secondary wood agrees in all essential characters with that just described, possess *primary* as well as *secondary* wood. According to the species, the primary wood consists of few or many isolated groups of tracheides, usually of small extent, which are situated in the pith close to the surrounding zone of secondary wood, or resting on it.   To these bundles no additions of new elements are made when once they are fully developed.   It is probable that these stems, with primary and secondary wood, either in whole or in part, also belong to *Cordaites*.   Many of these have been found in the Lower Carboniferous, especially in the Calciferous Sandstone series, where, however, Cordaites leaves are rare.   This may be only an accident of circumstances, for in the beds where the large trees are found, almost invariably no other plant-remains are discovered with them.   In every case which has come under my observation they occur as drift trees, generally embedded in sandstone or other coarse-grained material where delicate fossils could not be preserved.   Their cortex, also, with one exception, has always disappeared through decay or attrition.   It must also be further borne in mind that a natural sorting takes place in all water-carried vegetable material; the smaller coming to rest at one place and the larger, such as tree trunks, at another.

Such is the manner of distribution in which we generally find fossil plants to occur.

The structure of the stems we have just been considering leads to very important conclusions in regard to the affinities of these plants.   It has generally been considered, and as often stated, that these so-called *Dadoxylon* or *Araucarioxylon* stems belong to the *Coniferæ* ; in fact, it has been usual to suppose that the *Coniferæ* occupied a very important place in older palæozoic times : this however, is not the case, for although in the structure of the secondary wood a great similarity is shown to that of *Araucaria*, still other more important structural characters, such as the large chambered pith and, when present, the mesarch structure of the primary bundle, point much more to their affinity being with the *Cycadaceæ*.

The *Cordaiteæ*, however, possess distinctive characters of their own, and though showing more affinity to the Cycads than to the Conifers, form a distinct group which cannot be united with either, but must be regarded as equal in importance with them.

SEEDS.

Many isolated gymnospermous seeds occur in the Carboniferous rocks of which some certainly belong to the *Cordaites*. These, according to their form and structure, have been placed in many genera, as it is seldom ever possible to refer them satisfactorily to their parent stems. The more important of these genera are :—

Samaropsis Göppert. Generally small oval seeds, lenticular in section, pointed at the apex and rounded or slightly cordate at the base, and surrounded by a more or less prominent membranous wing with a notched or acute apex (*Samaropsis bicaudata* Kidston, Plate LVIII., figs. 5-6).

Cardiocarpus Brongniart. Smooth flattened seeds, oval or circular in outline, sharp or obtuse at summit, sometimes slightly cordate at the base and surrounded by a more or less distinct marginal wing (*Cardiocarpus* cf. *emarginatus* Artis. sp., Plate LXI., fig. 5).

Carpolithes Schlotheim. Smooth oval seeds, generally of small size, without any wing and not cordate at base (*Carpolithes perpusillus* Lesquereux, Plate LXV., fig. 4, and *Carpolithes ovoideus* Göppert and Berger, Plate LII., fig. 1).

Rhabdocarpus Göppert and Berger. Oval or cordate seeds containing a hard nucule surrounded by a more or less fleshy envelope in which are numerous hypodermic strands that impart a striated appearance to the outer surface of the compressed fruit (*Rhabdocarpus multistriatus* Sternberg sp., Plate LXI., fig. 4).

Trigonocarpus Brongniart. Hard, nut-like, oval seeds, circular in transverse section and provided with three prominent, and between them three less-prominent, ridges. Before maturity the seed is surrounded by a large pericarp which extends considerably beyond the apex of the nut.

Figs. 1 and 2, Plate LXV., show specimens of *Trigonocarpus Parkinsoni* Brongt. enclosed in the pericarp. The *Carpolithes alata*

L. and H. is founded on such a condition of *Trigonocarpus*, and is quite distinct from the genus *Carpolithes* as defined above. Plate LXV., figs. 3*a*, 3*b*, 3*c*, show three specimens of *Trigonocarpus* removed from their pericarp, with the three prominent ridges.

CONIFERÆ.

From what has been stated when describing the *Cordaiteæ* it is seen that true members of the *Coniferæ* are very rare in British Carboniferous rocks. The only example which has come under my notice is a small specimen of *Walchia imbricata* Schimper (Plate LXIII., fig. 1), which may possibly be only a form of *Walchia piniformis* Sternberg, from the Upper Coal Measures passed through when sinking the shaft of the Hamstead Colliery, Great Barr, near Birmingham.

In *Walchia* Sternberg are placed trees of a very Araucarian-like habit and growth. The branches are regularly pinnate, being arranged in two opposite or alternate rows. The spirally-placed leaves are more or less sickle-shaped, coriaceous, keeled, and wide towards their decurrent base. The fructification consists of small terminal cones, whose detailed structure has not yet been clearly made out. There is reason to believe, however, that *Walchia* holds a close affinity with the recent *Araucaria*.

GINKGOACEÆ.

The type and only existing species of the *Ginkgoaceæ* is the Maidenhair Tree of Japan and China—*Ginkgo biloba* L. = (*Salisburia adiantifolia* Smith). That *Ginkgo* is a very ancient plant type and extends far back in geological times has been clearly shown, but it is doubtful if the *Nœggerathia flabellata* L. and H. from the Lower Coal Measures, Bensham Seam, Jarrow Colliery, which Saporta includes in his genus *Ginkgophyllum* under the name of *Ginkgophyllum flabellatum*, holds any real affinity with the true *Ginkgoaceæ*. The leaves of *Ginkgo* are broadly cuneate, with a long slender petiole. The apex is irregular, and frequently divided into two or more cuneate lobes. The nervation spreads fan-like from the base of the leaf. The *Ginkgoaceæ* are separated from the *Coniferæ* on account of their mode of fertilisation, which takes place through the agency of antherozoids.

In *Nœggerathia flabellata* L. and H. the leaves or leaflets, for I think it is uncertain which term to apply to the ultimate divisions, are cuneate, with a slightly rounded irregular apex, and whose nervation radiates from the wedge-shaped base. Though larger than the leaves of *Ginkgo*, the leaves or leaflets of *Nœggerathia flabellata* L. and H. have a great general resemblance to them, and on account of this resemblance, perhaps only superficial, some authors have placed Lindley and Hutton's species in *Ginkgophyllum*.

Nœggerathia flabellata L. and H. (= *Psygmophyllum flabellatum* Schimper) is very imperfectly known, and of its fructification we are in entire ignorance. The enrolment, therefore, of *Nœggerathia flabellata* in the *Ginkgoaceæ* appears to me to be on insufficient evidence. What its true systematic position is remains to be discovered.

CONCLUSION.

We have now passed in short review the principal genera of Carboniferous Plants. In the time at our disposal it has been impossible to mention many interesting but less common species. Nor has it been possible to enter into any detailed description of the internal organisation of the plants which we have considered. To enter into these points in detail is beyond the scope of the present paper, whose object is only to give a sketch of the more important plant groups of the Carboniferous Formation.

Many excellent Text Books are in existence, and to these I refer the student for a more complete treatment of the subject. *

---

* Schimper, *Traité de Paléont. Végét.*, Vol. I.—III., Paris, 1869-74. Renault, *Cours de Botanique Fossile*, Vol. I.—IV., Paris, 1881-85. Schimper and Schenk in Zittel, *Handbuch der Palæontologie*, II. Abth. *Palæophytologie.* München and Leipzig, 1879-1890, Williamson ; Williamson and Scott ; and Scott. Various papers "*On the Organisation of the Fossil Plants of the Coal Measures.*" *Phil. Trans.*, London. From 1871 to present date. Seward, *Fossil Plants for Students of Botany and Geology*, Vol. I., Cambridge, 1898 (Vol. I. not yet published). Scott, *Studies in Fossil Botany*, London, 1900. Graf zu Solms-Laubach, *Fossil Botany, being an Introduction to Palæophytology from the Standpoint of the Botanist*, English Edition, Oxford, 1891. Schenk, *Die Fossilen Pflanzenreste*, Breslau, 1888. Potonié, *Lehrbuch der Pflanzenpalæontologie mit besonderer Rücksicht auf die Bedürfnisse des Geologen*, Berlin, 1899. Zeiller, *Éléments de Paléobotanique*, Paris, 1900. Kidston, *Carboniferous Lycopods and Sphenophylls, Trans. Nat. Hist. Soc., Glasgow.*, Vol. VI. (new series) part I., pp. 25-140, 1901.

INDEX.

EXPLANATION OF PLATES.

PLATE LI.

1. *Lepidodendron aculeatum* Sternberg. From near Stevenston, Ayrshire. *Hor.* Whistler Seam. Lower Coal Measures [2482].* Specimen received from Rev. D. Landsburgh, D.D. Portion of stem showing contiguous cushions and leaf-scars. Natural size.

2. *Lepidodendron serpentigerum* König. Grange Colliery, Kilmarnock, Ayrshire. *Hor.* Stranger Coal. Lower Coal Measures [2498]. Specimen received from Mr. A. Sinclair. Portion of specimen showing distant cushions and leaf-scars, with interfoliar cortex ornamented with irregular fine ridges. Natural size.

3. *Lepidodendron Wortheni* Lesquereux. Lower Writhlington Pit, Radstock, Somerset. *Hor.* Radstock Series. Upper Coal Measures [374]. Portion of stem showing contiguous leaf-cushions with short, fine, irregular, transverse ridges. Natural size.

4. *Stigmariopsis anglica* Kidston. Monckton Main Colliery, near Barnsley, Yorkshire. *Hor.* Barnsley Thick Coal. Middle Coal Measures [2342]. Collected by Mr. Hemingway. Portion of rhizome showing rootlet scars and ornamentation of surface of cortex. Natural size.

5. *Stigmariopsis anglica* Kidston. Monckton Main Pit, near Barnsley, Yorkshire. *Hor.* Barnsley Thick Coal. Middle Coal Measures [2335]. Collected by Mr. W. Hemingway. Portion of rhizome showing sub-epidermal surface with the characteristic ridges. Enlarged.

---

* The figures enclosed in brackets give the registration numbers of the mens in the Author's collection.

D

y R. Kidston.

Proc. Yorks. Geol. and Polytec. Soc., Vol. XIV., Plate LI.

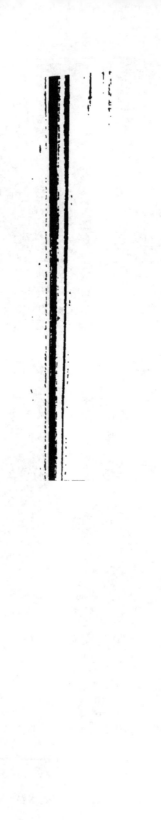

PLATE LII.

Carpolithes ovoidens Göppert an l Berger. Camerton, Somerset. *Hor.* Radstock Series. Upper Coal Measures [3021]. Collected by Mr. G. West. Portion of a slab showing a number of seeds. About natural size.

Lepidodendron lycopodioides Sternberg. Monckton Main Colliery, near Barnsley, Yorkshire. *Hor.* Barnsley Thick Coal. Middle Coal Measures [2718]. Collected by Mr. W. Hemingway. Small branch-bearing terminal cone. About natural size.

Stigmaria ficoides Sternberg sp. Calderbank, near Airdrie, Lanarkshire. *Hor.* Kiltongue Coal. Lower Coal Measures [2599]. Collected by Mr. R. Dunlop. At *a* is seen the cast of the pith cavity, the raised-up fusiform ridges on which are the casts of the openings to the primary medullary rays. The wood plates are seen at *b*. About natural size.

Sigillaria Taylori Carr. sp. From the bituminous Oil Shales, Midlothian. Calciferous Sandstone Series [16]. Collected by Dr. Macfarlane. Showing a vertical row of immature cones. Two-thirds natural size.

1

2

3

b.

a.

4

raphed by R. Kidston.

Proc. Yorks. Geol. and Polytec. Soc., Vol. XIV., Plate LII.

PLATE LIII.

Cordaites principalis Germar. sp.   Bonnington Pit, Kilmarnock, Ayrshire.   *Hor.* Whistler Seam.   Lower Coal Measures [1561].   Specimen from Rev. D. Landsburgh, D.D.   Stem showing leaf-scars.   About natural size.

Halonia tortuosa L. and H.   Smithies, near Barnsley, Yorkshire.   *Hor.* Woolley Edge Rock.   Middle Coal Measures [2176].   Collected by Mr. W. Hemingway.   Decorticated fruiting branch of *Lepidophloios*.   Two-thirds natural size.

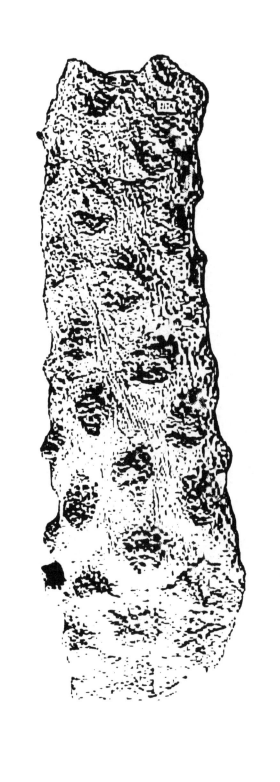

PLATE LIV.

1. *Bothrodendron minutifolium* Boulay sp.   Bonnington Pit, Kilmarnock, Ayrshire. *Hor.* Whistler Seam.   Lower Coal Measures [1568].   Specimen from Rev. D. Landsburgh, D.D.   At *a* is seen the fully-developed stem with distant cushionless leaf-scars ; at *b* is shown the young condition where the stem bears rhomboidal areas in which are placed the leaf-scar ; at *c* an intermediate condition is represented.   About three-quarters natural size.

2. *Bothrodendron minutifolium* Boulay sp.   Part marked *b* in fig. 1 enlarged to show the rhomboidal fields or cushions bearing the leaf-scars.

3. *Bothrodendron minutifolium* Boulay sp.   Part marked *c* on fig. 1 enlarged to show the still slightly elevated leaf-scar and the disappearance of the field.

4. *Sigillaria Sauveuri* Zeiller.   Longton Hall, Longton, Staffordshire. *Hor.* Great Row Rock.   Middle Coal Measures [2199].   Collected by Mr. John Ward, F.G.S. *Sigillaria* of the `Rhytidolepis` Section, showing approximation and reduction in size of the leaf-scars, which probably represents an enfeebled condition of growth. About three-quarters natural size.

5. *Stigmariopsis.*   Cast of the pith cavity.   Specimen communicated by Graf zu Solms-Laubach from one of the examples described in his paper " *Uber Stigmariopsis* Grand 'Eury." *Dames u. Kayser. Palæont. Abhandl.* New Series.   Vol. II., part 5, page 223, 1894 [2601]. About three-quarters natural size.

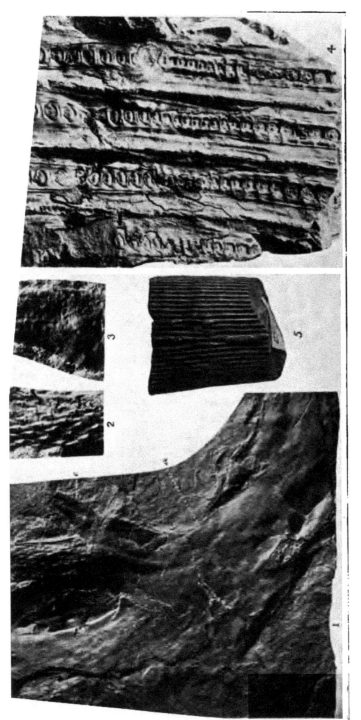

PLATE LV.

Lepidophloios Scoticus Kidston.   West Calder, Midlothian.
Hor. Oil Shales.   Calciferous Sandstone Series [1798].
Halonial branch showing leaf-scars and spirally-placed
scars from which stalked cones have fallen.   About one-
third natural size.

Lepidophloios Scoticus Kidston.   West Calder, Midlothian.
Hor. Oil Shales.   Calciferous Sandstone Series [1810].
Small portion of branch showing leaf-scars and spirally-
placed cone scars.   About natural size.

Lepidophloios Scoticus Kidston.   Water of Leith, below
Canal Bridge, Slateford, Midlothian.   *Hor.* Calciferous
Sandstone Series [1822].   Cone attached to its pedicel.
About natural size.

Sigillariostrobus rhombibractiatus Kidston.   Monckton
Main Colliery, near Barnsley, Yorkshire.   *Hor.* Barnsley
Thick Coal.   Middle Coal Measures [2263].   Collected
by Mr. W. Hemingway.   Lower portion of cone attached
to its pedicel.   About natural size.

r R. Kidston

roc. Yorks. Geol. and Polytec. Soc., *Vol. XIV.*, Plate LV.

PLATE LVI.

1. *Lepidodendron Veltheimianum* Sternberg.   Hailes Quarry, Midlothian.   *Hor.* Calciferous Sandstone Series.   Specimen in the collection of the Geological Survey of Scotland, Edinburgh.   Illustrating the dichotomous ramification of the genus.   Much reduced in size.

2. *Lepidophloios laricinus* Sternberg.   Low Moor, Yorkshire. *Hor.* Black Bed Coal.   Middle Coal Measures [1404]. Collected by the late Mr. J. W. Davis, F.G.S.   Portion of stem showing cushions and leaf-scars.   Three-fifths natural size.

3 *Stigmaria ficoides* Sternberg sp.   Watermill Pit, Clackmannan.   *Hor.* Fakes over Cherry Coal.   Lower Coal Measures [2545].   Rhizome showing rootlet scars and attached rootlets.   Reduced in size.

1

2

3

R. Kidston.

Proc. Yorks. Geol. and Polytec. Soc., Vol. XIV., Plate LVI.

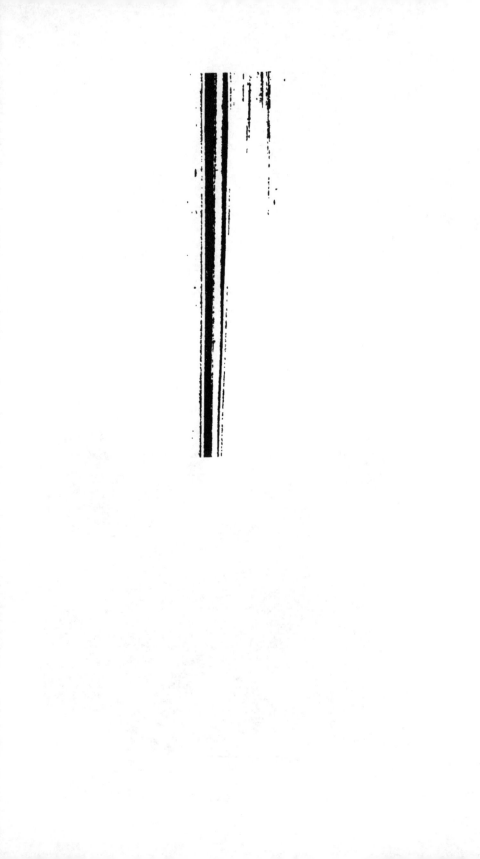

PLATE LVII.

Lepidodendron Veltheimianum Sternberg.   Shore, Wardie, Midlothian.   *Hor.* Calciferous Sandstone Series [2275]. Collected by Mr. J. A. Johnston.   Specimen showing the leaf-scars and four of the cone scars.   Two-fifths natural size.

Cordaites principalis Germar. sp.   Monckton Main Colliery, near Barnsley, Yorkshire.   *Hor.* Barnsley Thick Coal. Middle Coal Measures [1478].   Collected by Mr. W. Hemingway.   Upper portion of leaf showing blunt apex and parallel nervation.   Two-fifths natural size.

Dorycordaites palmæformis Göpp. sp.   Monckton Main Colliery, near Barnsley, Yorkshire.   *Hor.* Barnsley Thick Coal.   Middle Coal Measures [2907].   Collected by Mr. W. Hemingway.   Complete leaf showing pointed apex. Two-fifths natural size.

PLATE LVIII.

Sigillaria tessellata Brongt. var. *nodosa* Bowman sp. Braysdown Colliery, Radstock, Somerset. *Hor.* Radstock Series. Upper Coal Measures [3024]. Portion of stem showing at *a* a verticel of cone scars. About natural size.

Sigillaria Brardii Brongniart. Cope's Marl Pit, Longton, Staffordshire. *Hor.* Shale above Peacock Coal. Middle Coal Measures [817]. Specimen collected by Mr. John Ward, F.G.S. Portion of a stem showing the cone scars at *a*. About natural size.

Sigillaria elegans Sternberg sp. Wombwell Main Colliery, near Barnsley, Yorkshire. *Hor.* Shale over Barnsley Thick Coal. Middle Coal Measures [989]. Specimen collected by Mr. W. Hemingway. Portion of a specimen showing at *a* a verticel of cone scars. About natural size.

Cordaiocarpus Cordai Geinitz sp. Cadeby Colliery, Conisborough, Yorkshire. *Hor.* Shale on the horizon of the Woolley Edge Rock. Middle Coal Measures [1899]. Collected by Mr. W. Hemingway. About natural size.

and 6. *Samaropsis bicaudata* Kidston sp. Long Craig Bay, 1½ miles west of Dunbar, Haddingtonshire. *Hor.* Calciferous Sandstone Series [1940 and 1941]. About natural size.

PLATE LVIII.

. *Sigillaria tessellata* Brongt. var. *nodosa* Bowman sp.   Braysdown Colliery, Radstock, Somerset.   *Hor.* Radstock Series.   Upper Coal Measures [3024].   Portion of stem showing at *a* a verticel of cone scars.   About natural size.

2. *Sigillaria Brardii* Brongniart.   Cope's Marl Pit, Longton, Staffordshire.   *Hor.* Shale above Peacock Coal.   Middle Coal Measures [817].   Specimen collected by Mr. John Ward, F.G.S.   Portion of a stem showing the cone scars at *a*.   About natural size.

3. *Sigillaria elegans* Sternberg sp.   Wombwell Main Colliery, near Barnsley, Yorkshire.   *Hor.* Shale over Barnsley Thick Coal.   Middle Coal Measures [989].   Specimen collected by Mr. W. Hemingway.   Portion of a specimen showing at *a* a verticel of cone scars. About natural size.

4. *Cordaiocarpus Cordai* Geinitz sp.   Cadeby Colliery, Conisborough, Yorkshire.   *Hor.* Shale on the horizon of the Woolley Edge Rock.   Middle Coal Measures [1899].   Collected by Mr. W. Hemingway.   About natural size.

5 and 6. *Samaropsis bicaudata* Kidston sp.   Long Craig Bay, 1½ miles west of Dunbar. Haddingtonshire.   *Hor.* Calciferous Sandstone Series [1940 and 1941]. About natural size.

Photographed by R. Kidston

PLATE LIX.

Fig. 1.   *Sigillaria Brardii* Brongniart.   Railway Cutting, Florence Colliery, Longton, Staffordshire.   *Hor.* Newcastle-under-Lyme Group.   Upper Transition Series [818].   Collected by Mr. F. Barke, F.G.S.   The upper portion of this specimen shows the *Sigillaria Brardii* Brongt. (section *Clathraria*) and the lower part *Sigillaria denudata* Göppert (section *Leiodermaria*) in organic union.   The intermediate portion is the *Sigillaria rhomboidea* Brongt.   About natural size.

Fig. 2.   *Bothrodendron minutifolium* Boulay sp.   Monckton Main Colliery, near Barnsley, Yorkshire.   *Hor.* Barnsley Thick Coal.   Middle Coal Measures [1201].   Collected by Mr. W. Hemingway.   Branchlet-bearing terminal cone.   About natural size.

by R. Kidston.

PLATE LX.

ig. 1. *Sigillaria discophora* König. sp. (= *Ulodendron minus* L. and H.). Cinderford, Bradley. Coal Measures [2136]. Portion of bark showing leaf-scars and part of one of the two vertical rows of cone scars. About natural size.

1

PLATE LXI.

1. *Sigillaria rugosa* Brongniart. Monckton Main Colliery, near Barnsley, Yorkshire. *Hor.* Barnsley Thick Coal. Middle Coal Measures [2852]. Collected by Mr. W. Hemingway. Specimen of *Rhytidolepis* section showing central band of ornamentation. About natural size.

2. *Sigillaria camptotœnia* Wood sp. Gelli, Ystrad, Rhondda, Glamorganshire. *Hor.* No. 2 Rhondda Seam. Upper Transition Series (Lower Pennant Series) [1773]. Collected by Mr. W. O'Connor. Specimen of the *Leiodermaria* section showing ornamentation of the interfoliar bark. The small scars between the leaf-scars on the upper part of the specimen may be cone scars. About natural size.

3. *Sigillaria mamillaris* Brongniart. Monckton Main Colliery, near Barnsley, Yorkshire. *Hor.* Barnsley Thick Coal. Middle Coal Measures [2873]. Collected by Mr. W. Hemingway. Specimen of *Rhytidolepis* section showing leaf-scars on upper portion, and cone and leaf-scars on lower portion. About natural size.

4. *Rhabdocarpus multistriatus* Sternberg sp. Radstock, Somerset. *Hor.* Radstock Series. Upper Coal Measures [352]. About natural size.

5. *Cardiocarpus* cf. *emarginatus* Artis sp. Monckton Main Colliery, near Barnsley, Yorkshire. *Hor.* Barnsley Thick Coal. Middle Coal Measures [1437]. Collected by Mr. W. Hemingway. About natural size.

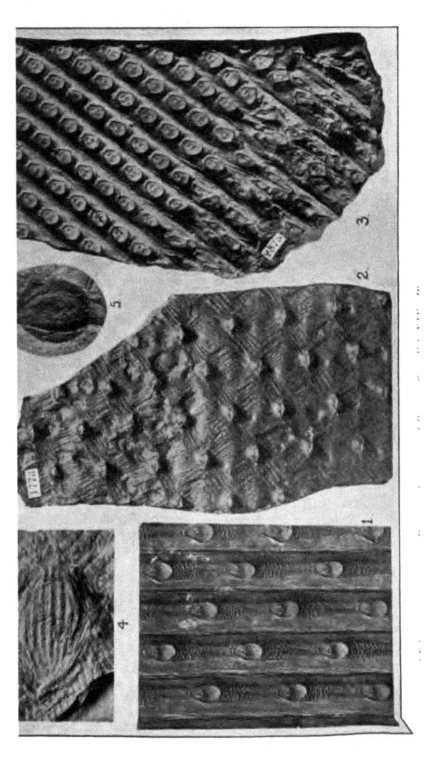

PLATE LXII.

Sphenophyllum majus Bronn. sp.   Monckton Main Colliery,
near Barnsley, Yorkshire.   *Hor.* Barnsley Thick Coal.
Middle Coal Measures [2701].   Collected by Mr. W.
Hemingway.   *a* and *b*. Different parts of the same slab.
The leaves on fig. *a* at *a*, *b*, *c*, exhibit different degrees of
leaf division.   About natural size.

PLATE LXIII.

'ig. 1. *Walchia imbricata* Schimper.   From shaft of Hamstead
Colliery, Great Barr, Staffordshire.   *Hor.* At depth of
350 feet from surface.   Upper Coal Measures.   About
natural size (see Trans. Roy. Soc. Edin., Vol. XXXV.,
Part 6, page 324).

Fig. 2. *Sphenophyllum myriophyllum* Crépin.   Oaks Colliery,
Barnsley, Yorkshire.   *Hor.* Woolley Edge Rock.   Middle
Coal Measures [2206].   Collected by Mr. W. Hemingway.
Specimen showing jointed and ribbed stem and lateral
branch.   About natural size.

G

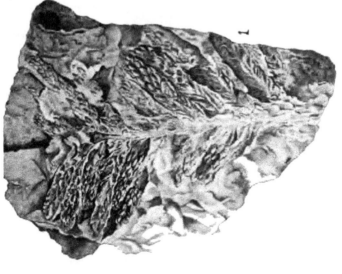

PLATE LXIV.

ig. 1. *Lycopodites Gutbieri* Göppert.   Camerton, Somerset. *Hor.* Radstock Series.   Upper Coal Measures [1501].   Collected by Mr. W. Hemingway.   About natural size.

ig. 2. *Poacordaites microstachys* Goldenberg sp.   Camerton, Somerset. *Hor.* Radstock Series.   Upper Coal Measures [3022].   Small stem showing leaves attached.   About natural size.

'ig. 3. *Cordaites principalis* Germar. sp.   Monckton Main Colliery, near Barnsley, Yorkshire. *Hor.* Barnsley Thick Coal.   Middle Coal Measures [1479].   Collected by Mr. W. Hemingway.   Between the letters *a* and *a* is seen the row of small cicatrices of the vascular strands which enter the leaf.   About natural size.

?ig. 4. *Omphalophloios anglicus* Sternberg sp.   Camerton, Somerset. *Hor.* Radstock Series.   Upper Coal Measures [433].   Collected by Mr. G. West.   Portion of specimen showing the scars placed in rhomboidal areas.   *Note.*— Owing to the direction in which the light has struck the specimen the rhomboidal areas are not so distinctly seen as they would be were the light coming from a different direction. The rhomboidal character of the area is, however, well seen at *a*.   About natural size.

PLATE LXV.

ig. 1.   *Trigonocarpus Parkinsoni* Brongniart.   Bonnington Pit, Kilmarnock, Ayrshire.   *Hor.* Whistler Seam.   Lower Coal Measures [1580].   Specimen received from the Rev. D. Landsburgh, D.D., showing nut enclosed in its pericarp.   In this condition the fossil is the *Carpolithes alata* L. and H.   About natural size.

ig. 2.   *Trigonocarpus Parkinsoni* Brongniart.   Bonnington Pit, Kilmarnock, Ayrshire.   *Hor.* Whistler Seam.   Lower Coal Measures [591].   Specimen received from the Rev. D. Landsburgh, D.D.   This example, though not so far developed as that shown at Fig. 1, exhibits the enclosed nut more clearly.   About natural size.

ig. 3.   *Trigonocarpus Parkinsoni* Brongniart.   Peel Quarry, near Bolton, Lancashire.   Middle Coal Measures.   Collected by Mr. R. Law.   *a* [2656], *b* [2663] side views.   *c* [2658], base of nut showing point of attachment.   About natural size.

Fig. 4.   *Carpolithes perpusillus* Lesquereux.   Pit near Kirkwood, Lanarkshire.   Lower Coal Measures [163 and 164].   Collected by Mr. R. Dunlop.   About natural size.

Fig. 5.   *Cordaianthus* with *Samaropsis fluitans* Dawson.   Monckton Main Colliery, near Barnsley, Yorkshire.   *Hor.* Barnsley Thick Coal.   Middle Coal Measures [2374].   Collected by Mr. W. Hemingway.   The *Samaropsis fluitans* Dawson, one of which is seen at *a*, are probably the seeds of the *Cordaianthus* with which they occur.   About natural size.

Fig. 6.   *Artisia transversa* Artis sp.   Brierley Tunnel, near Barnsley, Yorkshire.   *Hor.* Houghton Common Rock.   Middle Coal Measures [1261].   Collected by Mr. W. Hemingway. The pith cast of *Cordaites*.   About natural size.

H

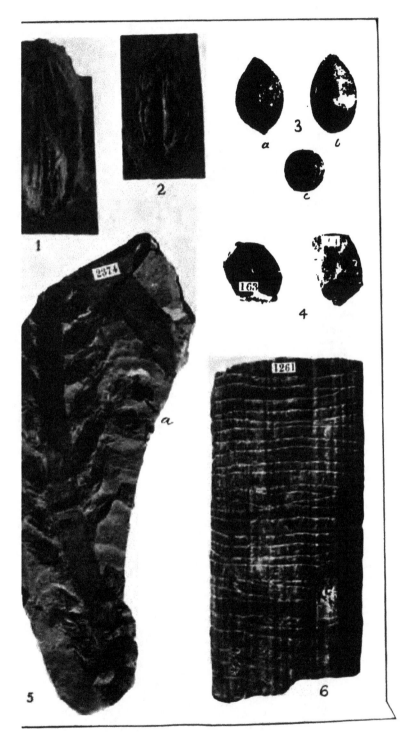

by R. Kidston

Proc. Yorks. Geol. and Polytec. Soc., Vol. XIV., Plate LXV.

ON THE CIRCULATION OF SALT AND ITS BEARING ON
GEOLOGICAL PROBLEMS, MORE PARTICULARLY THAT OF THE GEOLOGICAL
AGE OF THE EARTH.

BY WILLIAM ACKROYD, F.I.C., F.C.S., PUBLIC ANALYST FOR HALIFAX.

(Read April 25th, 1901.)

INTRODUCTION.

Common Salt or Sodium Chloride is probably the most widely distributed of all sodium compounds, as it is one of the most soluble. During our investigation of the underground waters of N.W. Yorkshire, a fruitful aspect of its solubility was presented to us in the disappearance of three tons of salt in a runnel of 19,800 gallons per day. It went down the Smelt Mill Water Sink and reappeared after 10 days at Malham Cove, where the amount of combined chlorine in the water suddenly rose from one up to six parts per 100,000, and in the course of another eight days slowly fell to the normal unit.* We had a similar experience at Clapham. It is apparent, therefore, that any salt in the soil, or in underground channels is carried quickly away by the water, and such soil or underground channels would be thenceforth free from the compound, unless in the course of disintegration fresh quantities were exposed for solution. This, however, is so slow a process that it quite fails to account for the enormous amounts of chloride annually conveyed by rivers to the sea.

The position is, perhaps, better seen in the following statement of fact :—The Widdop Reservoir, belonging to the Halifax Corporation, contains some 640·5 millions of gallons of water, and I estimate it to contain 55 tons of salt. The 2,000 odd acres of gathering ground is moorland on Millstone Grit for the most part with a little of the Yoredale Rocks. Needless to say it is a saltless area, yet the 55 tons of salt is renewed probably more than once a year in the course of filling up to replace the Corporate demands on the reservoir. The chlorides come down in the rain-water. We may take it then, as a working hypothesis, that rain has brought down the salt which has been derived from the sea, and that to the sea it quickly returns.

---

* Proc. Yorks. Geol. and Polytec. Soc., Vol. XIV. Pt. I., p. 17.

It will be convenient to speak of this salt as " sea-salt," in (
distinction to such as may be derived from the earth by solvent
dation, which will be referred to as " earth-salt," and the propor
amount of the sodium compound will be indicated in this pa
the chlorine figure.

CIRCULATION OF SEA-SALT.

The conditions obtaining in the Widdop area are of mor
ordinary scientific interest, and I have made an experimen
vestigation of them during the winter 1900-1, which has been
municated to the Chemical Society of London.* From Nov
12th to February 18th weekly chlorine determinations were
of the reservoir water, of the water in a rain-gauge kept ck
and of rain-gauge water on four other gathering grounds some
to the east of Widdop, viz., Walshaw Dean, Midgley, Warle
Ovenden Moors, besides daily tests of rain-gauge water from
Vue, in the town of Halifax. The area under investigatio
its position in England and the relative positions of the rain-g
are given in Plate LXVI. The results of the weekly tea
shown in the following table :—

CHLORINE, PARTS PER 100,000 IN RAIN-GAUGE WATEE

Height above sea level.	Widdop Reservoir.	Widdop. 1,050 ft.	Walshaw Dean. 1,380 ft.	Midgley Moor. 1,350 ft.	Warley Moor. 1,325 ft.	Ov : 1,
November 12	1·2	0·8	0·3	1·1	1·2	
,, 19	—	1·1	0·7	0·7	0·45	
,, 26	1·0	1·0	0·65	1·5	0·6	
December 3	1·0	0·5	1·05	0·7	0·9	
,, 10	—	0·2	0·6	0·6	0·5	
,, 17	1·1	0·2	0·8	0·7	0·8	
,, 24	1·15	3·3	2·55	3·1	2·1	
,, 31	1·1	1·1	0·9	1·6	1·1	
January 7	1·2	1·2	—	0·55	2·1	
,, 14	1·3	6·5	2·75	2·55	2·4	
,, 21	1·25	0·85	1·1	1·0	1·1	
,, 28	1·25	1·75	1·6	1·8	1·9	
February 4	(?) 1·4	0·8	1·0	2·7	1·2	
,, 11	1·2	1·75	1·6	2·45	2·2	
,, 18	1·3	—	—	—	—	
Average ...	1·188	1·50	—	—	—	

* Trans. Chem. Soc., Vol. 79, p. 673.

Y

OGDEN

DEN

OR

HAL

ue

in
uge

ren-
An
iem
ent
non
ime
rth
and
The
the
ave
ght
ited
ing

rts,

It will be convenient to speak of this salt as " sea-salt," in contra-
distinction to such as may be derived from the earth by solvent denudation, which will be referred to as " earth-salt," and the proportionate
amount of the sodium compound will be indicated in this paper by
the chlorine figure.

CIRCULATION OF SEA-SALT.

The conditions obtaining in the Widdop area are of more than
ordinary scientific interest, and I have made an experimental investigation of them during the winter 1900-1, which has been communicated to the Chemical Society of London.* From November
12th to February 18th weekly chlorine determinations were made
of the reservoir water, of the water in a rain-gauge kept close by,
and of rain-gauge water on four other gathering grounds some miles
to the east of Widdop, viz., Walshaw Dean, Midgley, Warley, and
Ovenden Moors, besides daily tests of rain-gauge water from Belle
Vue, in the town of Halifax. The area under investigation, with
its position in England and the relative positions of the rain-gauges,
are given in Plate LXVI. The results of the weekly tests are
shown in the following table :—

CHLORINE, PARTS PER 100,000 IN RAIN-GAUGE WATER.

Height above sea level.	Widdop Reservoir.	Widdop. 1,050 ft.	Walshaw Dean. 1,380 ft.	Midgley Moor. 1,350 ft.	Warley Moor. 1,325 ft.	Ovenden Moor. 1,375 ft.
November 12	1·2	0·8	0·3	1·1	1·2	1·3
,, 19	—	1·1	0·7	0·7	0·45	0·5
,, 26	1·0	1·0	0·65	1·5	0·6	0·8
December 3	1·0	0·5	1·05	0·7	0·9	0·7
,, 10	—	0·2	0·6	0·6	0·5	0·5
,, 17	1·1	0·2	0·8	0·7	0·8	0·7
,, 24	1·15	3·3	2·55	3·1	2·1	2·45
,, 31	1·1	1·1	0·9	1·6	1·1	0·9
January 7	1·2	1·2	—	0·55	2·1	1·8
,, 14	1·3	6·5	2·75	2·55	2·4	3·1
,, 21	1·25	0·85	1·1	1·0	1·1	0·75
,, 28	1·25	1·75	1·6	1·8	1·9	1·4
February 4	(?) 1·4	0·8	1·0	2·7	1·2	1·4
,, 11	1·2	1·75	1·6	2·45	2·2	1·8
,, 18	1·3	—	—	—	—	—
Average ...	1·188	1·50	—	—	—	—

* Trans. Chem. Soc., Vol. 79, p. 673.

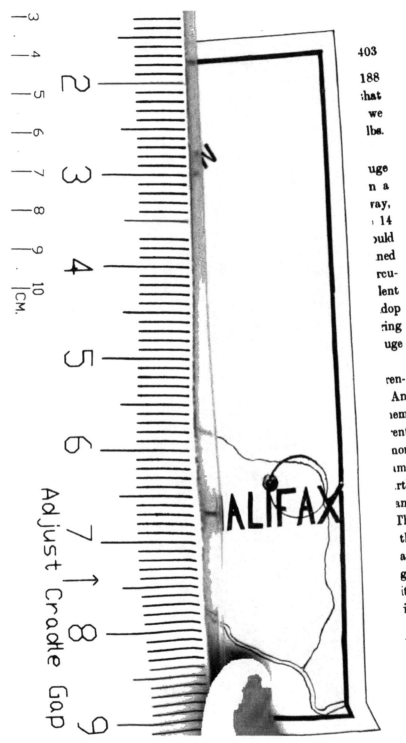

It will be o
distinction t
dation, whic
amount of t
the chlorine

The conc
ordinary scie
vestigation of
municated to
12th to Febru
of the reservo
and of rain-ga
to the east of
Ovenden Moor
Vue, in the to
its position in
are given in
shown in the fo

CHLORINE.

Height above sea level.	R
November 12	
,, 19	
,, 26	
December 3	
,, 10	
,, 17	
,, 24	
,, 31	
January 7	
,, 14	1
,, 21	1
,, 28	1
ary 4	(?) 1
11	1
18	1

188
hat
we
lbs.

uge
n a
ray,
14
uld
ned
rcu-
lent
dop
ring
uge

ren-
An
em
ent
non
me
rth
and
The
the
ave
ght
ited
ing

—

rts,

ordii
vesti
mun
12th
of tl
and
to tl
Ovei
Vue,
its p
are
show

Hei	
s	
Noi	
Dec	

The reservoir water was fairly constant with an average of 1·188 parts of chlorine per 100,000. This figure will be greater than that of the annual average rainfall on account of evaporation, and if we deduct 10 per cent. it would still be equivalent to a fall of 172 lbs. of common salt per acre with the 1899 rainfall of 43·17 in.

The rain-gauge figures fluctuated wildly. Widdop rain-gauge water reached 3·3 per 100,000 during a December week when a violent storm was blowing from the Irish Sea about 40 miles away, and on another occasion it got up to 6·5. The average for the 14 weeks of observation was 1·50. That the winter average should be in excess of the yearly average is in keeping with the results obtained by other observers, and gives some idea of the activity of salt circulation during the stormy part of the year when gales are prevalent from the sea. The abnormally high figures obtained at Widdop synchronised with similarly high figures for the other gathering grounds ; this is well seen in the plotted curves for the rain-gauge results. (Plate LXVII., fig. 2.)

Observations on chlorine in rain-water have been made at Cirencester since 1870 by Professors Church, Prevost, and Kinch.[*] An abnormal amount of chlorine has almost always been traced by them to salt spray from the Bristol Channel, 35 miles away. On different occasions immediately after a storm from the S.W. crystals of common salt have been found on the College windows facing west. The same kind of evidence has been furnished to me by Mr. J. H. Howarth respecting Malham ; concerning Leeds by Mr. J. E. Bedford, and Mr. J. Denison calls my attention to the following references : " The winds from the South-West have sometimes blown so strong that the pieces of cloth on the tenters in several parts of Halifax parish have been charged with a considerable number of saline particles brought from the sea,"[†] and further, " Sea-salts have been found deposited on the windows in West Park and York Place, Harrogate, thus giving practical evidence of the presence of the sea-breeze."[‡]

---

[*] Trans. Chem. Soc., Vol. 77, p. 1271.

[†] Watson's History of Halifax, 1775, p. 5.

[‡] The Use and Abuse of Harrogate Mineral Waters, by Arthur Roberts, p. 32.

The following is a synopsis of results up to date, the chlorides in the rainfall being calculated into annual downfall of salt per unit area :—

Place.	Distance from the Sea.	Deposit of NaCl per acre per year.	Observers.
Cirencester ...	35 miles	36 lbs. (average of 26 years)	Church, Prevost, and Kinch[*]
Rothamsted ...	—	24·59 lbs. (average 6 years)	Lawes, Gilbert, and Warrington[†]
Perugia	75 miles	37·8 lbs.	Bellucci[†]
George Town, Demerara	—	186 lbs.	Harrison[†]
Widdop... ...	40 miles	172 lbs.	Ackroyd

As to the particular nature of this salt circulation, we may suppose (1) that salt spray from the sea is carried to the land to distances varying with the force of the wind ; (2) that when the salt is dried by evaporation it is carried further inland with the dust ; and (3) that rains dissolve it and bring it down to the rivers by which it is carried back to the sea.

Of the carrying of the salt spray, abundant evidence, as I have shown, has been obtained in times of storm, and of the presence of sodium in dust at all times the spectroscope yields never-failing proof, while in the examination of rain-water I have found the highest chlorine contents in light rainfalls.

The full extent inland of salt circulation has yet to be determined. It may not be entirely limited by mountain chains barring the direction of prevalent sea-winds, as finely-divided sea spray may possibly be carried nearly as far as the dust of Krakatoa. This much, however, we know, that the farther one gets inland and the less the chlorine figure becomes for rain-water. Angus Smith has published some averages for rain[‡] in terms of hydrochloric acid per million. I have converted his figures to parts of chlorine per 100,000 of rain-water, and arranged them as follows in proof of this diminishing distribution :

Ireland, Valentia	..	..	..	..	4·72
Liverpool	..	..	..	..	3·49
English Towns ..	..	..	..	..	0·84
Darmstadt	..	..	..	..	0·094

* Trans. Chem. Soc., Vol. 77, p. 1273.   † Ibid.

‡ Air and Rain, Longmans, London, 1872, p. 281.

FIG. 2. CHLORINE CURVES.

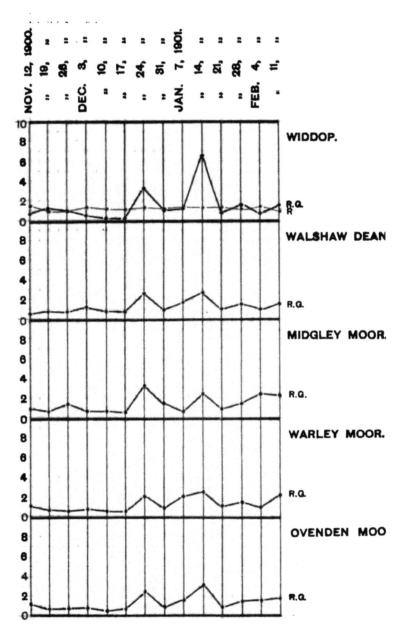

R.G.—Rain Gauge Water.

R —Reservoir Water.

THE CONCENTRATION OF SEA-SALT.

When rain-water reaches the earth its burden of salt necessarily becomes concentrated by evaporation. A flat sheet of water will in twelve months lower its level 20 inches,[*] so that it follows if the original depth were 40 inches, the result of a year's evaporation would be to double the amount of salt in a given weight or volume of the residual water ; much quicker must evaporation be from land surfaces where capillary action may come into play and constantly draw moisture upwards where it is exposed on a maximum of earth-surface to the evaporative influence of the sun, or of a higher atmospheric temperature.

The final result appears so apparent as to need no proof. I may, however, refer to peculiar cases of concentration I have come across in the course of this study. Well-waters in the Millstone Grit show evidence of concentration in higher chlorine figures, and also that the salt is cyclic *sea-salt* from the fact that *it does not increase at the same rate as the other solids in solution*, whereas, if it were derived from the soil, it ought to increase either at the same or at a greater rate, because of its greater solubility than that of the other bodies ; this is illustrated by the following data :—

	Parts per 100,000.		Ratio. Total Solids to Cl.
	Total Solids.	Cl.	
Upland surface water. Average of 19 Analyses extending over 2½ years, Sept., 1898, to Jan., 1901	9·68	1·33	100 : 13·7
Well No. 1 (pure)	38·	3·85	100 : 10·1
Well No. 2 (pure)	28·	2·00	100 : 7·1
Well No. 3 (pure)	38·	2·15	100 : 5·7

THE CASPIAN AND DEAD SEAS.

This concentration of chlorides is an important geological factor. To return to our Pennine reservoir. If there were no out-

---

[*] A four years' mean evaporation from the Torquay watershed, as kindly supplied to me by Mr. W. Ingham, C.E., is 20·88 in., 1897-1900.

let and a rate of evaporation equal to the inflow of water, it would, in 50,000 years, contain two and three-quarter millions of tons of salt, or if we halve the rate of accumulation it would still be 1,375,000 tons of salt, collected from a gathering ground of 2,223 acres into a reservoir of 93 acres surface, and of a depth of 65 feet. The reservoir at this lesser estimate would contain 32 per cent. of salt, as compared with the Dead Sea with its 24 per cent.,* and the accumulation, derived nearly entirely from rain-borne sea-salt, and some of it already precipitated, would have been amassed during a period which is comparatively trifling, as geological time is reckoned, being probably less than a seventh of the Pleistocene Age.

We have here then a process at work which must be taken into account in framing theories of salt-lake and salt-deposit formations; hitherto it has been entirely overlooked, and only contributions of earth-salt derived by solvent denudation have been drawn upon for the purpose, with, in many cases, the added supposition that the first step in their formation was the disconnection of an arm of the sea, a supposition which is not always required. Thus, with regard to the salt deposits of Northern Africa, Prof. Zittel† has shown on what appears to be overwhelming evidence, that the popular idea of the Sahara having been the basin of a sea in Pleistocene times is without foundation, and as the rainfall was heavier and the climate damper in those days than now, rain-borne sea-salt must have played an important part in the formation of the salt hills of the Sahara.

The usual theory of the cause of the saltness of the Caspian Sea appears also somewhat anomalous. The Mediterranean, Black Sea, Sea of Azof, and Caspian Sea, decrease in saltness in this order, on which Ramsay has remarked :‡ "It will be seen that the Black Sea is fresher than the Mediterranean. . . . The Caspian is still fresher, and its fauna and fossils in recent deposits in the neighbourhood prove it to have once had connection with the Black Sea, from which

---

* This figure, often quoted, refers to the "saline residue" in which the quantity of common salt, as will presently be shown, is governed by the amount of magnesium salts present.

† Nature, Vol. XXIX., pp. 121 and 122.

‡ Nature, Vol. VII., p. 313.

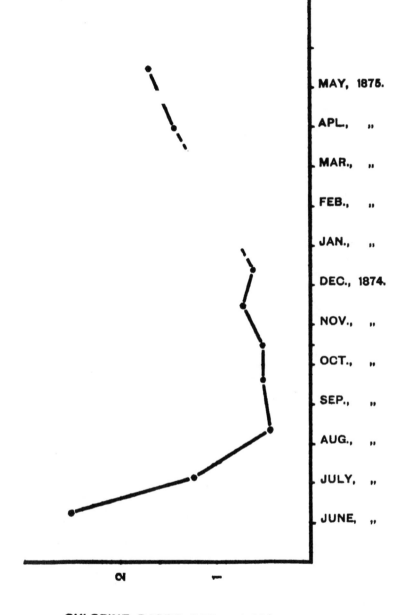

MAY, 1875.

APL., „

MAR., „

FEB., „

JAN., „

DEC., 1874.

NOV., „

OCT., „

SEP., „

AUG., „

JULY, „

JUNE, „

Geol. and Polytec. Soc., Vol. XIV., Plate LXVIII.

CHLORINE PARTS PER 100,000.

has been separated by changes in physical geography; it was then salter than at present, but is now growing salter again every year, and the fauna now inhabiting its waters have likewise considerable affinities with North Sea types." Now I should think the whole physical facts of the case are comprised in the simple statements that the Caspian is an inland sea of large volume and variable composition, fed by inland rivers rising in and flowing through regions where the sea-salt in the rainfall, not reckoning that derived from the lake itself, must have reached a minimum; where the chief source of saltness must be the earth-salt from solvent denudation, and how small that may be expected to be will be presently shown in the case of the Aire at Malham Cove; where a large portion of the evaporated water must come down again to the sea itself, so large is its area; from all of which statements it follows that the Caspian must increase in saltness at so phenomenally slow a rate that one is not surprised to find it least salt of all the waters which have been mentioned. As an inland lake without outlet, one would think that it can never have been salter than it is at present, and it will be well to revise the biological data upon which such an idea is based.

Let us next consider the Dead Sea from this point of view. Little further removed from the Mediterranean than Widdop is from the Irish Sea, with a rainfall in Palestine higher than that of the Peninne Hills, and an intensity of meteorological conditions in the past of which we can at the present day form but a poor conception,[*] it is reasonable to suppose that the salts in the Dead Sea have been for the most part derived from the Mediterranean—carried by winds and brought down by rains—and, barring the contributions of Jebel Usdum, it will probably be found that over 90 per cent. of the salts come from this source. On such a theory we look for some measure of likeness between the dissolved matters of the two waters, but an initial obstacle to comparison presents itself in the much greater concentration in one case than in the other, and in the variability of the Dead Sea itself from the North, where the Jordan enters, to the South, where the salt hill of Jebel Usdum overlooks it. In

_____

[*] The Land of Israel, Tristram, p. 320.

instituting a comparison I have, therefore, confined myself to the
ratio of the related ions, chlorine and bromine.   Usiglio's figures
for the Mediterranean at Cette give

Cl.   Br.

100 : 2·1

which are practically the same as Von Vibra's for the Atlantic at
the Equator—

Cl.   Br.

100 : 2·09

The analyses of Dead Sea surface-water made by Terreil give
the following ratios :—

	Cl.	Br.
Near Rasdale	100 :	0·95
Lagune, North of Sodom	100 :	2·7
Near Island	100 :	3·6
Average	100 :	2·4

The closeness of this average to the Mediterranean and Atlantic
Ocean figures is somewhat remarkable, and although it will not be
taken as absolute proof that the Dead Sea has obtained its salt from
the Mediterranean, it at least destroys the contention that the dissolved constituents of the two seas are so dissimilar that one cannot
have been derived from the other.

THE GEOLOGICAL AGE OF THE EARTH AS MEASURED BY THE RATE OF SOLVENT DENUDATION.

I now pass to the bearing of the foregoing facts and others I
shall adduce on Professor Joly's way of calculating the age of the
earth since the first ocean was formed.*   The method may be thus
briefly put—

$$\frac{NaCl + NaNo_2 + Na_2SO_4 \text{ in the seas.}}{\text{Annual addition of } NaCl + NaNO_3 + Na_2SO_4 \text{ by rivers.}} = \text{Age of the Earth.}$$

or more particularly, the sodium contents of the seas, divided by the
sum of the annual sodium increments furnished by the affluent rivers.

* Trans. Roy. Dublin Soc., VII. (Series II.), 23-66 ; and B.A. Report,
1900, pp. 369-379.

equals the geological age of the Earth. There can be no mathematical precision in any method for getting at the age of the Earth, and this latest method of attacking the problem appears no more precise than the rest.

If we take the numerator, for the present, to be approximately correct. the degree of accuracy attainable will depend upon the reliability of the denominator, and to this question my attention will be solely directed.

It is assumed, in the first place, that the solvent denudation which yields the annual load of combined sodium to the seas has been uniform in its action from the first. Now, as a chemist, I have tried to picture to myself the condition of things immediately before and after the " consistentior status," when the cooling globe allowed of the condensation of aqueous vapour to form the first ocean.

Oxides of the alkalies and alkaline earths, silicated and otherwise, would exist among the first compounds. Hydrochloric acid would admittedly be another, and as the temperature gradually lowered their incompatibility would result in the production of silica, aqueous vapour, and salts like sodium chloride.

There is no more remarkable fact than the thoroughness with which hot water takes up chlorides from a hot surface with which it is in contact or prevents their deposition. I need only instance the formation of boiler scale. The boiler water increases the amount of its chlorides in solution so regularly with the amount of evaporation that it has been made the basis of a technical method, now often used for finding the evaporative power of coal. The boiler scale contains only the minutest trace of chlorides.

. I imagine then that the seething Archæan Sea would dissolve all soluble chlorides from the ocean floor and rocks of those times, and as the deliquescent salts appeared on new land surface they would be washed away by every shower, so that to all intents and purposes " the sea was salt from the first." My ideas here run somewhat parallel to those of Professor W. J. Sollas, who, in his able address to the Geological Section of the B.A. at Bradford, observes : " The ocean when first formed would consist of highly-heated water, and this, as is well known, is an energetic chemical

reagent when brought into contact with silicates like those which
formed the primitive crust. As a result of its action saline solutions
and chemical deposits would be formed ; the latter, however, would
probably be of no great thickness, for the time occupied by the ocean
in cooling to a temperature not far removed from the present would
probably be included within a few hundreds of years."*

This view of things is opposed to uniformity of rate of solvent
denudation, and leads to the direct inference that in the later age
of the earth such changes are proceeding at a very much slower rate.
I see no objection to a varying rate in the facts adduced by Dr. Joly
as confirmation of his hypothesis of uniformity from the first. He
points out that the sum of the soda now in the ocean and in the
sedimentaries would nearly suffice to effect the full restoration of this
constituent to the original igneous rock. Such facts are equally
in keeping with an unequal distribution of the work of denudation
over the period of the Earth's age, that it was rapid at first and afterwards slower and slower with the flow of time, and the slower the
rate at which one can prove it to be now proceeding, the more
prodigious must have been the initial rate to yield the final colossal
sum of oceanic contents.

ON THE RATIO OF SEA-SALT TO EARTH-SALT IN RIVER WATER.

It will be apparent that if we have to get an idea of the rate at
which solvent denudation is now proceeding it will be necessary
to find what proportion of earth-salt rivers are carrying to the sea
along with the cyclic sea-salt. I have to offer the following attempt
at a solution of the problem for the source of the Aire.

A gravimetric analysis of Craven limestone gave 0·01 per cent
of chlorine. The water at Malham Cove is of 10° of hardness, and
contains a total of ·7 parts of chlorine per gallon. Therefore, every
gallon of water at the Cove contains

					Grains.
Calcium Carbonate	..	..	..	..	10·000
Chlorine in earth-salt		..	..	..	0·001
Chlorine in sea-salt		..	..	..	0·699

* B.A. Report, 1900, p. 716.

IG. 4. COMPOUND SOLUBILITY CURVE.

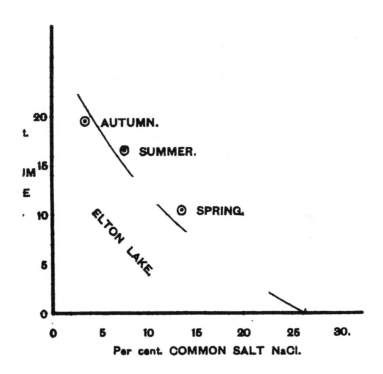

r of the load of salt in this special case carried to the sea fully 99·8 per cent is sea-salt, and only less than 0·2 per cent earth-salt, or that art which is to be taken into account in ascertaining the rate of olvent denudation now in progress, and even this is cyclic in a sense, s it comes from a Carboniferous and goes back to a present-day sea.

I draw practically the same conclusion from other and more general considerations. Thus, the solids dissolved in average river water yield the atomic ratio of sodium to chlorine of 1 : 0·345, as calculated from Sir John Murray's data,[*] while in the Earth's crust, using Professor F. W. Clarke's data, we have a ratio of 1 : 0·0024 ;[†] n other words, the chlorine in river water, which we take as a measure of the salt going to the sea, exists in 143 times greater proportion than we find it in the crust of the Earth, or to put it in yet another orm probably only 0·7 per cent. of earth-salt, and 99·3 per cent. of salt from some other source, say cyclic salt, is being carried by rivers to the sea.

How do these results square with other facts ? Lawes, Gilbert, and Warrington[‡] have shown that the amount of chlorides in drainage from drain-gauges in unmanured land and uncropped soil is most exactly equal to that contained in the rain. In other words ley were unable to estimate the amount of earth-salt, it was so small. gain in chalk, granite, sandstone, and other rocks. the sodium chloride is so inconsiderable a trifle as usually to find no place in recorded analyses. It is improbable then, that after extended investigations. an average of 1 per cent. of the sodium carried to the sea as chlorides, &c., will ever be found as the result of solvent denudation, and consequently that Prof. Joly's 90 millions of years estimate of the age of the earth based on uniformity of action, would come 8.000 millions of years, a quantity of time which I imagine ill be too excessive even for the exorbitant demands of the biologists.

SEASONAL VARIATION OF SALT CIRCULATION.

In all such problems again seasonal variation of fluviatile salt d must also be taken into account. It has a tendency to average

---

* The Scottish Geographical Magazine, Feb., 1887, p. 12.
† Bull, U.S. Geol. Survey, No. 148, p. 13.
‡ Trans. Chem. Soc., Vol. LI., p. 94.

itself where large volumes of water are concerned ; thus in the hundreds of samples analysed from the source of the Aire during our investigations of the underground waters, there has been no noticeable variation from normal ; these, however, have only been summer experiments. There must be variations due to periods of excessive evaporation and of storm. The Nile furnishes a striking example. Wanklyn.* in 1874 and 1875, obtained the following chlorine figures for this river :—

		Parts per 100,000.
1874.—June 8th	.. ..	2·5
July 9th	.. ..	1·3
August 12th	.. ..	0·4
September 20th	.. ..	0·5
October 12th	.. ..	0·5
November 12th	.. ..	0·7
December 12th	.. ..	0·6
1875.—April	.. ..	1·4
May 13th	.. ..	1·7

The annual fluctuation is better grasped from the plotted curve. (Plate LXVIII., fig. 3.)

Similar observations are wanted for all the rivers of the world.

CONCLUSION.

To sum up, then, salt circulation in Nature forms an important and, up to now, inadequately realised phenomenon, and to it we undoubtedy owe much of the saltness of present day salt-lakes. In the solvent denudation method of arriving at the age of the Earth. too little allowance has been made for the mass of sodium chloride carried by the rivers, which has been brought from the ocean by atmospheric transportation, and consequently too high a figure has been adopted for the variable proportion of sodium derived from the Earth. But although little satisfaction may be received from the numerical results obtained by this in common with other ways of trying to solve the mystery of the Earth's age, they are at least

* Water Analysis, 6th Edition, p. 151.

lcome in this respect that we are led to garner and make use of an
undance of facts which lie to hand and only need the trouble of
ping.

ADDENDA.

Since this paper was read, Mr. John Scarborough, of Halifax,
as been good enough to communicate to me a reminiscence of his
oyhood connected with the subject. He remembers a storm
which brought salt with it in such quantity that it formed an incrustation on the tiles and window panes ; he saw and tasted it.
He fixes the date approximately by the roofing of St. Marie's Roman
Catholic Church, which was early in 1839.

THE GREAT SALT STORM OF 1839.

I determined to look up contemporary records for further par-
iculars, and my investigations were considerably lightened by Mr.
oseph Whiteley, the Borough Librarian, placing in my hands a
mall volume entitled, " Narrative of the Dreadful Disasters occasioned
y the Hurricane which visited Liverpool and various parts of the
ingdom on the nights of Sunday and Monday, Jan. 6th and 7th,
839."

All Sunday, Jan. 6th, the wind was blowing strongly from the
E. ; the barometer fell considerably, and the wind shifted suddenly
o the S.W., and, increasing in rapidity, became a perfect hurricane
on after midnight. Vessels were driven ashore, trees up-rooted,
ouses unroofed, chimneys and thick walls were blown down. The
orm still raged until Monday afternoon, when it began to abate,
nd about 10 o'clock there was a perfect lull, followed by furious
usts from the N.W. The following extracts are references to the
renching with brine to which the country was subject, and affording
undant evidence that during this storm salt was driven from the
ish Sea and Atlantic Ocean right across the island :—

ST. HELENS.—" Such was the force of the tempest here that
e salt spray from Liverpool covered all the shrubs and hedges in
e whole neighbourhood, and the plants, windows, &c., were white
d covered as with hoar frost."

MANCHESTER.—" One very remarkable fact attending the recent storm has been the extent to which objects exposed to the violence of the wind have been covered with saline incrustations, no doubt produced by spray brought from the sea by the violence of the storm. Windows, the branches of trees, and many other objects have been so completely incrusted with salt as in many cases to appear as if covered with hoar frost, so as to attract the notice of the most careless observer. We have now lying before us about a quarter of an ounce of salt which was collected in a few minutes by G. L. Ridehalgh, Esq., from trees in the neighbourhood of Flixton, and a number of branches exhibiting saline incrustations in the most palpable manner have been brought to us from different quarters. We understand that this appearance is not confined to this neighbourhood, but prevails in Saddleworth and other places to the eastward."—*Manchester Guardian.*

ROCHDALE.—" The hedges and trees in the neighbourhood, and as far inland as Saddleworth, were literally pickled with brine, or salt spray, brought from the sea by the strength of the wind."

CHESTER.—" We understand that the hedges, trees, and fields in some districts are covered with a substance having the appearance of a white fur, which is perfectly salt to the palate."—*Chester Courant.*

ECCLESTON.—" The hedges were covered apparently with hoar frost . . . salt and brackish to the taste."

BURY.—" The atmosphere early in the morning was impregnated with saline particles, which it was found on the appearance of daylight had been deposited on the trunks and branches of trees in the country, and on some windows near the town, forming a white incrustation, the nature of which was very evident to the taste. The minute particles must have been borne on the wings of the winds from the bosom of the ocean to these comparatively inland districts."

LONGTON. —" Monday, twelve o'clock at noon, wind W. The trees and hedges appear as covered with hoar frost by reason of particles of salt. Two o'clock : The wind has increased with redoubled violence and the air is darkened with particles of saline matter as of a mist or fog."

)DERSFIELD.—" It is a singular fact that trees about Hudders-
ing the storm on Monday appeared as if covered with hoar
t as it did not freeze, it led to an examination, when it turned
this was a briny deposit which the wind had probably brought
: Irish Sea."

TON.—" Not within the memory of the oldest person has this
lford) been visited with such a tremendous gale as set in
west on Tuesday morning, the 8th inst (Monday, the 7th ?),
o'clock, and continued unabated until 11 at night. The
narkable feature in the phenomenon here was that before
every tree and hedge in the bleak situations were encrusted
:e a hoar frost) with a powerful alkali which an eminent
pronounced to be muriate of soda It was easy to collect
iwing the branches over with a piece of dark woollen cloth,
ion became as white as if chalked. Several times something
observed within seven or eight miles of the German Ocean*
e wind has blown from the east, and it was supposed the
iorbed it from the vapours of the sea ; but the wind having
vn from the west, if such was the case, it must have been
l completely across the island from the Irish Sea. It
. that the greater the elevation the greater was the deposit
ete, which was clearly confined to the bleak side, particularly

instructive comment on this newspaper observation is afforded by
)f the rain-gauge water of Ingleby Greenhow Vicarage, in Cleveland,
above sea-level and some 20 miles from the German Ocean. The
rere kindly collected for me by the Rev. J. Hawell, who, knowing
t in view, also supplied me with the meteorological data in the last

m the Rain Fell.	Chlorine, Parts per 100,000.	Equivalent in Grains of Common Salt per Gallon.	
1900.			
2, 3, 4, 5, 6, and 7	1·00	1·15	
9, 12, 14, and 15...	0·65	0·75	
, 18, 20, and 21 ...	1·40	1·61	
1901.			
... ...	8·60	9·92	Violent storm from the east. Rainfall, 0·34 in.
... ...	4·00	4·61	Strong wind blowing from east coast. Rainfall, 0·86 in.

observable on well-painted gates and on the upper windows of houses, which were rendered quite dim by the incrustation. It was also observed to come through the crevices of a barn-door like smoke, and the reflection of the light on it caused it to glisten very brilliantly."

COUNTY OF DURHAM.—Canon Tristram, whose attention I had called to this subject, kindly wrote on the 11th of March, 1902:— " It may interest you to know, with respect to the storm of 6th and 7th January, 1839, which I well remember, a fact which came under my own observation. At the Castle, Castle Eden, which stands on a bluff not far from the east coast of the County of Durham, and overlooking the sea, on the morning of the 7th January all the windows of the Castle facing west were covered with a saline incrustation like hoar frost, while those on the east face had not a trace of salt on them."

FURTHER EVIDENCE OF SALT CIRCULATION IN CALMER TIMES.

It is probable, as I have attempted to show by chemical evidence, that winds blowing from the sea are always more or less salt-laden. Armand Gautier has quite recently made a determination of the amount of salt in sea air (Bull. Soc. Chim., 1899 (iii), 21, 391–392). A known volume of air at the Rochedouvres Lighthouse was aspirated through a long plug of asbestos wool, which was afterwards washed with hot water in which the soluble chlorides were then determined. The air was collected both during day and night in *fine dry* weather with a W.N.W. wind blowing from the open sea. The aspirator was nine metres ($29\frac{1}{2}$ ft.) above the sea-level. For a mean temperature of 15° C. the chlorides found worked out to 22 milligrams per cubic metre of air (equal to 0·259, or say $\frac{1}{4}$ of a grain per cubic yard), which Gautier regards as the maximum quantity of salt which air can retain in suspension.

ON THE ORIGIN OF THE SALTNESS OF SALT LAKES.

These various facts enable one to realise that with extreme meteorological conditions, such as have admittedly obtained in the past history of the Jordan watershed, it may have been no uncommon thing for Palestine to be drenched with sea-salt by the prevalent westerly gales, in which case a more important contribution towards the saltness of the Dead Sea would be made than could be furnished by its widely-distributed limestone rocks. On this subject I hope to get definite facts from the analyses of Palestine limestones.   Through the kind offices of one of our Vice-Presidents (Mr. Walter Morrison) a number of samples have been submitted to me for this purpose by the Palestine Exploration Committee. In opposition to this view it has been pointed out that wide differences exist in the composition of the waters of salt lakes, and the opinion has been expressed by Professor Joly* that " the whole facts of the case entirely negative the wide deductions," which I have founded on my calculations.   He further says : " The very variable composition of salt lakes must be regarded surely as an insuperable objection. . . . ! The Dead Sea, for instance. shows a very large excess of magnesium salts over sodium salts, the chlorides constituting 15·9 per cent. and 3·6 per cent. respectively of the total solids.   There is even a large excess of calcium over sodium in its waters.   In the Great Salt Lake the proportions are just the other way.   The percentages are nearly marine, 11·9 per cent. of sodium chloride and a very little magnesium chloride, but 1·1 per cent.   There is relatively very little calcium.   . . .   Thus, the lake which is most favourably situated for the rain supply of sea-salts is just that one which most completely departs in its chemical composition from that of the ocean. Again, we find a lake, such as the Elton Lake of the Kirghis Steppe, 200 miles from the Caspian. possessing a chemical composition approximating to that of the Dead Sea—19·7 per cent. $MgCl_2$, 5·3 per cent. of $MgSO_4$. and 3·8 per cent. NaCl.   Calcium is. however, in its case absent or inappreciable." (*Chemical News*. June 28th, 1901, p. 301.)

These seeming discrepancies disappear in the light of a principle which I enunciated recently (*Geological Magazine*, Decade IV, Vol.

---

* The controversy on these matters between Prof. Joly and myself will be found in the Chemical News, June 28th and August 2nd, and Geological Magazine for August, October, November, and December, 1901.

K

VIII. No. 448, p. 446, October 1901). It can be best shown here by reference to and explanation of the Elton Lake. The waters of this lake vary in composition, becoming more concentrated as the year advances. The seasonal change is exhibited in the following analyses :—

	In Spring (Göbel).	In Summer (Erdmann).	In Autumn (Rose).
Sodium chloride, NaCl	13·1	7·4	3·8
Magnesium chloride, MgCl₂ ...	10·5	16·3	19·7
Potassium chloride, KCl	0·2	—	0·2
Calcium chloride, CaCl₂	— —		—
Potassium sulphate, K₂SO₄ ...	—	0·04	—
Magnesium sulphate, MgSO₄ ...	1·6	2·20	5·3
Water	74·4	73·50	70·8
	99·8	99·44	99·8

The two main bodies in solution are magnesium chloride and sodium chloride, and as concentration proceeds on the approach of autumn the former increases and the latter decreases in amount. Regarding the Elton Lake as a saturated solution we may plot these results along with the fact that a saturated solution of common salt with no magnesium chloride in it holds about 26·5 per cent. of sodium chloride at normal temperature. We get the compound solubility curve as in Plate LXIX., fig. 4.

It follows that solutions containing less than 10 per cent. of magnesium chloride when saturated with common salt will give analytical data which, upon co-ordination, will fall approximately on the curve. This I have proved.

There is nothing anomalous therefore in the differences one observes between different lakes any more than there is in the seasonal differences existing in the Elton Lake itself. It is very largely a question of the amount of one body in solution being conditioned by the amount of another that there may be, to which of course must be added the geological nature of the gathering-ground supplying any particular lake. The Great Salt Lake, with a minimum quantity of magnesium chloride, may have a maximum quantity of common salt in solution ; and the Dead Sea, on the other hand, with magnesium salts in excess from the excessive evaporation, may be continuously precipitating its salt, giving us in the southern parts " quite a paste " (Tristram), while on the floor of the lake itself salt is being deposited (Hull : Geol. and Geog. of Arabia Petræa, p. 122). Herein I take it lies the explanation of the origin of salt hills, like Jebel Usdum, which have always been a mystery to geologists.

THE LOST IODINE.

A final objection to the theory that the Dead Sea owes its salt-ness largely to transported sea-salt is that of the absence of iodine in its waters. How far this objection is a legitimate one is a matter for serious consideration. Its absence, seeing that it occurs in such minute quantity in the ocean—a few parts per million—is not so wonderful as that of the loss of CO_2 which has entered the Dead Sea in combination with lime from the limestone hills of Palestine, and yet gives practically no evidence of its presence in its waters. It may be that the iodine has similarly vanished in some occult chemical change which remains to be discovered. Another possible view is that the *iodine has not been found* and not that *it is not there*. The difficulties of detection are many, arising not only from the minute quantity of it, but also from its state of combination. Sonstadt regarded it as existing as iodate in sea-water, a form that would elude the ordinary tests for iodides. Gautier also has come to the con-clusion that the iodine in sea-water is not in the form of metallic iodides, but that it exists in organic compounds, four-fifths of it being soluble, and the remaining fifth forming part of the substance of the infusoria inhabiting the superficial layers of the ocean (Comp. Rend., 1899, 128, 1,069—1,075). Finally it exists there with a prodigious excess of other haloid compounds, which increase the difficulties of detection and separation. These considerations lead one to think that the presence or absence of this element in the Dead Sea or its deposits is still a question for investigation, in which the modern methods of chemical science could possibly be now effectively employed. In the meantime any arguments founded on its supposed absence have very little value.

THE AGE OF THE EARTH.

From the ground of solid fact we come to an atmosphere of speculation in returning to a final consideration of the validity, or otherwise, of the method of finding the Earth's age, since seas were formed. by the expression :

$$\frac{\text{Sodium in the seas}}{\text{Sodium annually added by the rivers}} = \text{Age of the Earth.}$$

My view of the matter is directed by one striking fact, which is that all the sodium in the sea exists, to all intents and purposes, in the 39.782 billions of tons of common salt which has been calculated to be there in solution. Now if there has been uniformity from the first, or " constancy in the nature and rate of solvent actions going on over the land surfaces " (Joly, Trans. Roy. Soc. Dublin, Ser. II., Vol. VII., page 24), it necessarily follows that all this sodium chloride in the ocean has gone there as such, or, at any rate, that all the sodium going there has finally taken the form of chloride. It is on such considerations that my criticisms of this method have, from the first, been based, and, to be strictly consistent. if the numerator consists of common salt only, then the denominator ought to be made up of the same material. Let us critically examine, however. this denominator, and we shall really find that much. if not all, of it consists of cyclic sodium as I have already attempted to prove by other considerations. The sodium in it comprises 10,303 tons as sulphate, 7,252 tons as nitrate, and 6,549 tons as chloride. Of this last not less than 43 per cent. has come in the prodigious quantity of calcium and magnesium carbonates. and is represented in analyses of these bodies as 0·01 per cent. of chlorine. Then an unknown proportion of " fossil sea-salt from other sources " is included in this chloride, as shown by the researches of Sterry Hunt. Osmond Fisher,[*] and A. R. Hunt.[†] Probably all the rest is cyclic sea-salt. As regards the nitrate, it is not improbable again that all of it owes its origin to cyclic sea salt. The chemical argument on this point is interesting. and may be thus given :—In the first stage of nitrification we have

[*] Geol. Mag., March, 1900, pp. 129 and 130.
[†] Geol. Mag., March, 1901, pp. 1-3.

the genesis of ammonia and carbonic acid ; next the formation of ammonium carbonate, which is changed to ammonium nitrite. The nitric organism transforms the nitrite to nitrate, and finally common salt yields its sodium in the change of ammonium nitrate to sodium nitrate. That the common salt is of marine origin is proved by the composition of the *caliche* of the South American nitrate industry, for the mother liquor left after the nitrate crystals have been removed contains iodine in the form of iodate, the state of combination in which it exists in sea-water. Although unaware of this chemical fact, geologists have looked to salt of marine origin as the source of the sodium in sodium nitrate. There remains now only the sodium in the sodium sulphate. This is of too uncertain origin for one to say how much or how little is available for the purpose under consideration, but sufficient has been said to show that the quantities of which the denominator is made up are not definite enough for calculating the age of the Earth. Nor does the uncertainty end here, for while, as Prof. Sollas mentions, the outflow of the rivers yielding the data relied upon forms only some $7\frac{1}{2}$ per cent. of the total quantity of water being annually delivered into the ocean, there is, besides, the fact that in some of the largest of these rivers having their origin in the torrid zone, the saline contents, as in the case of the Nile, vary some 400 per cent. in the course of the year.

ON THE CHARACTERS OF THE CARBONIFEROUS ROCKS OF THE PENNINE SYSTEM.

BY WHEELTON HIND, M.D., B.S., F.R.C.S.

It is a well-known and undisputed fact that the Carboniferous succession of the Midlands differs very considerably from that which obtains in North Yorkshire, Northumberland, and Scotland, and the correlation of the Carboniferous sequence in various parts of Great Britain and Ireland has been a matter of difficulty and dispute. A study of the literature of the subject, voluminous and scattered though it be, seems to show one important fact, and that is that very few, if any, of the writers had studied the succession in more than one or two localities, or had given any attention to the evidence afforded by paleontology. To Yorkshire geologists, the sequence of the Carboniferous rocks in their county should be of the highest interest, not only on account of the large number of sections exposed in the romantic dales, for which the county is so justly famous, but because the change from the northern to the southern type of stratigraphical succession takes place in the county, and because, I am convinced, an accurate knowledge of the geology and paleontology of the Carboniferous rocks of the West and North Ridings will go far to settle the whole of the vexed question of correlation.

During the last five years I have published a series of papers on the correlation and sequence of the Carboniferous rocks of the Pennine axis and the south of Scotland, in which the following theses have been developed :—

(a) That the differences in the northern and southern types of the Carboniferous sequence is due to conditions brought about by the proximity of continental land to the north and north-east : that the main difference between the types is due to the very much greater amount of detrital material deposited as sediment in the area occupied by the northern type ; that this area received the muds and sands brought down by a large river and deposited out at sea, thus hindering the deposit of the non-detrital but organic limestone, which went on uninterruptedly for a long period in the area occupied by the southern type of rocks.

) That the distribution of fossils shows most emphatically that the fauna of the Carboniferous Limestone of the southern type of rocks is identical with the fauna of the Great Scar Limestone and Yoredale series of the northern type.   Hence the Yoredale series are the homotaxial equivalents of the upper part of the Carboniferous Limestone " massif " of the southern type, and do not in any sense overlie it.   The comparative thickness of the Great Scar Limestone plus the Yoredale series and the mass of limestone further south further corroborates this view.

) That the detrital deposits of shales, dark limestones, and quartz-grits which overlie the massif of limestone in the southern area are not the equivalents of the Yoredale series ; that this deposit is extremely local and lenticular, and that the boundaries of this lenticle can be fairly accurately mapped by measured sections, and that the fauna contained in these beds is entirely different from that found in the limestone massif below, and also from that of the Yoredale series of the northern type.

) That many Genera and species of Carboniferous fossils which occur low down in the Carboniferous rocks of Scotland appear for the first time at higher and higher horizons as the beds pass south, demonstrating a migration southward, and indicating a passage south of similar conditions of environment.

The evidence for these theses is given at length in a paper by Howe and myself, published in the Q. J. Geol. Soc., Vol. LVII, and much of it need not be repeated here, but I propose to briefly w the chief stratigraphical and paleontological facts on which views are based.

The southern type of Carboniferous rocks is well seen in the inal mass of limestone which occupies parts of North Staffordshire and Derbyshire forming the tectonic centre of the great Pennine anticlinal. This mass, from 2,000 to 3,000 ft. thick, whose base has never been seen, is practically one mass of limestone, divided by very thin partings of shale into beds of stone, which are occasionally very thick and with very obscure bedding planes. In the quarry opposite the High Tor at Matlock and at Coombsdale the officers of the Geological Survey describe a thin coal and its underclay in the limestone. I have been informed that a piece of *Stigmaria* was obtained in the former locality.

The upper part of the Limestone is extremely variable, being thin bedded, thick, and crinoidal, massive and highly fossiliferous, or even cherty at apparently the same horizon in various places. But at or near the top is a bed containing many rolled water-worn shells and pebbles (?) of limestone, which is fairly constant over the whole area. At times the top of the limestone series is abrupt and sudden, and shales come on at once; at others there exists a well-defined series of passage beds of thin shelly limestones and shales, possibly a phase of the rolled shell bed, consisting of more comminuted but less rolled shell material. In places the upper 30 or 40 ft. of this Limestone Massif must have been a shell bank, for the limestone is made up of fossils, many of the shells being perfect, but others being slightly eroded; but, evidently, from the fact that all kinds of fossils are present massed and jammed together, not exactly in the place where all lived, but while there is no doubt that some species were able to and did live on the bank, the shells of others were washed there by currents.

At Castleton and Waterhouses the bed of rolled shells consists chiefly of fragments of *Productus giganteus*, but this shell occurs again in the next 15 or 20 ft. which forms the top of the limestone. This line is at once the upper limit of the massive limestone and of *P. giganteus*, for in this district this shell does not appear again at any higher horizon.

The massif of limestone is succeeded by a series of black shales and thin limestones, for which I have adopted the name PENDLE-SIDE SERIES, with a peculiar fauna of *Posidonomya Becheri*, *P. membranacea*, *Posidoniella lævis*, *P. minor*, *Chænocardiola Footii*, *Pterino-*

ecten papyraceus, Glyphioceras reticulatum, G. spirale, G. bilingue, and other species, none of which occur below.   Only a few brachiopoda of the Carboniferous Limestone pass up into these beds, but the peculiar fauna enumerated above recurs at various horizons in the Millstone Grit series and lower coal measures.   The limestones become thinner and thinner, and towards the top of the series sandstones and quartzose gannisters appear in the shales, and become more and more pronounced until the Grits come on.   So much for the general sequence in Derbyshire.

The sections from Pendle Hill to Clitheroe show a sequence perfectly parallel to that which obtains in Derbyshire.   The shelly detrital beds and crinoidal limestones being specially well marked. Very fine and almost complete sections are to be seen in the Angram and Pendleton Brooks, and the cloughs on the west flank of Pendle Hill.   These beds yield a typical fauna, as may be noted by an examination of Appendix B (page 461).

The sections in the Massif of limestone near Clitheroe show white massive limestone full of shell fragments or made up of crinoids at the top, while lower down, as at Chatburn, a series of dark blue limestones with their shale partings are in evidence.   Further west liers of the Limestone Massif, consisting of the upper and fossiliferous beds, protrude through the Pendleside shales at Withgill, Ashnot, Doe Barn, Whitewell, Chipping, Sykes, and Slaidburn, at which places good specimens of the typical fossils of this horizon can be obtained. The beds at Sykes were largely altered by vein stuff, but contain *Amplexus coralloides* and *Lithostrotion* sp.   The Pendleside series are to be seen in the brook courses and a few quarries round these liers at Black Hall and Cold Coates, S.W., and Thornley Hall, south of Chipping; the river Hodder; quarries below the Longridge Fell escarpment; below Ashnot Barn; Holden near Bolton-by-Bolland; streams at West Bradford and Grindleton; the river Ribble near Hinckley Hall; and the typical fauna of the series has been obtained at all these localities.

A most interesting set of beds occur at or about the top of the limestone Massif, along a line extending from Thornton to Barnoldswick.   This is an anticlinal hill, a N.E. and S.W. axis, the limestone beds forming the limbs of the anticlinal being very thin and separated by shales, which, toward the top, become fairly well developed. This portion of the limestone is a southerly continuation of the anticlinal of Haw Bank and its continuation to Bolton Abbey, and its secondary folds round Draughton.   Here, the upper part of the Limestone Massif is split up by shale beds, which attain some little thickness. Sections at Rain Hall and Gill rock quarries, Barnoldswick, show this condition of deposit very clearly.   Further south-west a similar set of beds are seen in the Lancashire and Yorkshire Railway, one mile north of Rimmington Station, and still further S.W. at the large quarry two miles south of Chipping.

I consider the beds along this line to be a local phase of the close of the limestone deposit, probably due to a current which brought detrital sediment and deposited it in a definite and strictly limited area.   The fauna of the shales intercalated with these limestones is peculiar and interesting.   At Draughton, in a small quarry near the railway, the shales yielded a fine specie of *Ctenacanthus tenuistriatus.*

The limestones at Draughton yielded *Cardiomorpha ovata,* and the Rev. A. Crofton tells me he has obtained species of *Spirifer, Productus, Phillipsia,* and Corals here.

At Thornton Quarries I obtained—

> *Cladochonus* sp.   Very frequent.
> *Palæchinus sphæricus.*   Very frequent.
> Crinoid-stems.
> *Conocardium Hibernicum.*

The following specimens were found at Rain Hall Quarries :—

> *Cladochonus* sp.
> *Palæchinus sphæricus.*
> Crinoid-stems, three species.
> *Productus semireticulatus.*
> *Orthotetes crenistria.*
> *Athyris planosulcata.*
> *Conocardium aliforme.*
> *Syringopora geniculata.*
> *Zaphrentis Enniskilleni.*

The Chipping Quarry also yielded—

Palæchinus sphæricus.
Many Crinoid-stems.
Chonetes papilionacea.
Productus semireticulatus.
Productus longispinus.
Spirifer trigonalis.
Orthis Michelini.
Cælonautilus cariniferus.

This series of fossils is altogether distinct from the fauna of the Pendleside series, but on the other hand many of the fossils obtained at the top of the Massif of Limestone are absent in these localities, a fact probably due to the conditions under which deposition took place. *Palæchinus sphæricus* and *Cladochonus*, so common in these calcareous muds, are rare in the limestone itself. A very similar local deposit with a similar fauna occurs in the neighbourhood of Bradbourne, Derbyshire, also situated at the top of the Limestone Massif.

I regard the appearance of this set of shales in the massive limestone, so far south of and altogether independent of the Craven faults, as important. The presence of this muddy deposit indicates that the Yoredale phase was not restricted by these faults, and consequently that the faults could not possibly have had any causal influence on the different character of the deposits of the northern and southern types. The extent to which muddy sediment was carried out to sea would vary very largely, being affected amongst other causes by currents, the formation of bars, and the size and flow of the river bringing down sediments. Consequently, the area, where a pure and continuous deposit of calcareous ooze was going on, would be liable to temporary invasion, resulting in a series of shales becoming intercalated in a limestone deposit, the general facies of the fauna remaining unchanged.

Further north, the Limestone Massif is thrown up in the Craven district, where it forms a long anticlinal, which has been peculiarly scarped into rounded domes, from Greenhow Hill, on the east, to

Old Cumnor, on the west. This limestone is the usual massive, white stone, crammed with shells, like that seen at Clitheroe, and the great coral localities of Derbyshire and Staffordshire, and moreover contains an identical fauna. It is known to be overlaid by shales and a black limestone below Thorpe and Rylstone Fells, which contain the typical Pendleside fauna. But a most instructive and interesting section is to be seen on the south side of Dibbs Bridge, about four miles east of Grassington, on the Pateley Bridge road, which is identical with the section at the foot of the Winnats, at Castleton, containing a bed which Messrs. Barnes and Holroyd have described as a beach bed. Trans. Manch. Geol. Soc., Vol. XXV., page 119).

The section shows—

Post of limestone with shell fragments and *Productus giganteus*. 15 feet.

Post of shelly limestone not much rolled. About 10 feet.

*Bed of rolled shells, lenticular shell fragments, and rolled pieces of limestone. 4 feet.

Post of limestone with many specimens of *Productus giganteus*. 6 feet.

Hard blue limestone. 15 feet.

The bed marked * is quite undistinguishable from the similar bed at Castleton. I saw several samples of this rock some four miles further west, on the road between Threshfield and Linton, and Threshfield and Grassington, in the stones piled for road-making. In this I got the peculiar dark masses of foraminiferal limestone imbedded in white limestone, which Messrs. Barnes and Holroyd have recently described. The exact similarity of the upper beds of the limestone of Craven and Derbyshire is very striking, and at any rate points to similarity of conditions of deposit. I myself believe that the two sets of beds are on the same horizon, and present a well-marked stratigraphical line. In the Craven district the Pendleside series quickly thins out, so that from Appletreewick to Greenhow Hill, and for some distance from Grassington northward, the grits immediately overlie the Limestone Massif, and nowhere north of a line passing from Settle, *viâ* Malham, to Hebden (Wharfedale) has the Pendleside fauna been yet obtained, but it occurs in the shales in the

near Linton stepping stones. and in the beds seen in the
between Burnsall village and the grits of Thorpe Fell.   North
Wharfe I got *Posidoniella lævis* in shales near the bridge over
. Gill and the bank opposite St. Michael's Church, Linton.
es in the river Wharfe, near the stepping-stones at Linton,
Posidonomya membranacea in abundance, with remains of
tes.   It has generally been taught that the knolls of lime-
ı Craven are the representatives of the Pendleside Limestone,
 of the fact that the stratigraphical succession of the Thorpe,
and Pendle-Clitheroe districts are identical. that the limestones
eroe and Cracoe are exactly similar in composition, and both
gether different from the Pendleside Limestone, and notwith-
g the fact that the faunas of the Clitheroe and Cracoe Lime-
ıre identical and differ entirely from that of the Pendleside
ıne, and that the latter fauna is abundantly present in the
ıries which immediately succeeds the limestone of the knolls.
ı the knolls themselves on different horizons, but they belong
same anticlinal fold cut into domes by small water courses
ı sub-aerial action of water containing CO_2 in solution, and
ıe not always weathered equally, by which I mean that
t posts of limestone are to be seen on the different hills.
ıween Butterhaw and Hill Skelterton is a fairly large swallow
 the upper part of which is seen a section of shales enclosing
limestone, dipping at an angle that I think quite sufficient to
hem over the white limestone, but if not, the true position
,ve been altered by the undermining which has occurred in
ıllow.   The black limestone is crammed with *Posidonomya*
. *Posidoniella lævis,* and *Glyphioceras reticulatum.*
ı presence of the Pendleside Limestone is thus demonstrated
he white limestone of the knolls, and the closeness of the bed
white limestone of the swallow also shows the rapid thinning
 the mass of shales, which, at Pendle Hill, separate the Pendle-
nestone from the Clitheroe Limestone.   This thinning away
ʾendleside series is borne out by the presence on Simondseat
ow holes in the Grit, which shows that the limestone is no
ıstance below the surface.   North of Grassington and almost

as far north as Kettlewell the Grits repose on the thick limestone, showing that the Pendleside series has quite disappeared.

Mr. Tiddeman quotes a conglomerate bed which he thinks has been formed by masses of white limestone rolling down from cliffs or reefs of limestone and becoming embedded in shales deposited subsequently round these hypothetical structures. There is a crushed bed of limestone which might be called a conglomerate in a stream section east of Keal Hill, but unfortunately for this view the masses of limestone are not white, nor do they contain the fossils of the so-called reefs and chemical analysis shows that the latter limestone contains 97·5 per cent. of $CaCO_3$ and ·6 per cent. of silica, the limestones in the conglomerate (?) containing 36·7 per cent. of $CaCO_3$ and 54·0 per cent. of silica.

The country between the Midland Railway and the boundary of sheet 60 of the one inch ordnance map is largely mapped as shales, through which some few inliers of limestone appear. An examination of the area, the numerous quarries, the contour of the ground, the absence of trees, boggy ground and streams point to a far larger area of limestone than is mapped.

Commencing on the west, the side of the fell at Tosside shows at the Knotts a massive white limestone cropping out below the grit as a lenticular patch, but, owing to absence of stream sections, its extent cannot be well traced downwards. This limestone is lithologically quite different from the Pendleside type, and contains a characteristic Carboniferous Limestone fauna.

CORALS.

Lithostrotion and *Zaphrentis.*
Crinoids.
Chonetes papilionacea.
Productus margaritaceus.
Productus giganteus.
Glyphioceras crenistria, &c., &c.

About three-quarters of a mile north of this, in an exposure behind the farm at Brockthorns, is another very interesting section, the beds being almost horizontal. This shows at the top :—

Dark soft shales, 3 ft.

Limestone composed of rolled shells, corals, crinoids,
and fragments of limestone, comparable to the
Beach Bed of Castleton, 3 ft.

Shales, 2 ft.

Shaley limestone, 1 ft.

Shales, 1 ft.

Thick limestone, 6 ft. to sole of quarry.

At the north end of the quarry the middle beds are disturbed
folded, but the upper and lower limestone are not affected.

The fauna here is extensive. Some of the calcareous shales
rammed with crinoid joints and fragments of Palæchinus. The
y limestone contains :—:

Glyphioceras crenistria.
 ,, *sphæricum.*
Rhynchonella pleurodon.
Productus giganteus.
 ,, *longispinus.*
 ,, *striatus.*
 ,, *semireticulatus.*
Chonetes papilionacea.
Spirifera trigonalis.
 ,, *lineata.*
 ,, *glabra.*
Dielasma sacculus.
Strophomena analoga.
Lithostrotion Martini.
Crinoids sp. 3.

A very noticeable feature in a rolling hill passes south from
quarry in the direction of the Knotts, which we cannot but think
ates a continuance of the limestone.

About a mile and a half north-west a narrow lenticle of lime-
e is mapped. In this occur an interesting series of quarry
ons showing a similar series to the Brockthorns quarry.

The quarry at the dip mark, north of the word Bollards, shows the following sequence :—

Crinoid limestone, 3 in.
Black shale, 4 in.
Finely stratified limestone, 9 in.
Gypsum, 1 in.
Black shale, 3 in.
Platy limestone, 3 in.
Crinoidal limestone, 2 in.
Black shales, 2 ft. 3 in. with *Productus longispinus.*
Crinoidal limestone, 5 in.
Earthy limestone, 2 ft.
Solid limestone, 15 ft. Base not seen.

The following fossils were obtained :—

Productus mesolobus.
..   *plicatilis.*
..   *punctatus.*
..   *semireticulatus.*
Spirifera trigonalis.

At Pythorns, a little more than a ¼ mile further east, the conglomerate bed of Brockthorns is again seen crammed with rolled fossil debris about 3 ft. thick. This is covered in by a few inches of shales, while below is fairly massive but well-bedded limestone, with bands of apparent brecciation.

The following fossils were obtained :—

Zaphrentis.
Crinoids.
Phillipsia sp.
Productus giganteus.
..   *punctatus.*
Spirifera lineata.
..   *glabra.*
Orthis resupinata.
..   *Michelini.*
Spines of *Palæchinus.*
Euomphalus Dionysii?

wo miles still further east is the quarry at Teenly Rock, which
fine well-bedded limestone, the beds being nearly horizontal.
ds the base of the quarry is a bed of oolitic limestone. Fossils
it plentiful here, but a typical Carboniferous Limestone fauna
I :—

> *Productus giganteus.*
>> „ *semireticulatus.*
>> „ *plicatilis.*
>> „ *punctatus.*
> *Spirifera glabra.*
>> „ *lineata.*
> *Athyris planosulcata.*
> *Ortholetes crenistria.*
> *Orthis resupinata.*
> *Conocardium aliforme.*
> Crinoidal fragments.

etween Teenley and Pythorns thin limestone and shales are
n the stream at Becks Brow Bottom, and a sulphuretted
gen spring, indicating a roll and temporary disappearance of
mestone from the surface.

ast of the Ribble is a patch of country bounded on the north
1st by the Hellifield and Skipton line. Numerous dip marks
uarries are shown, nearly all of which show the upper beds
massif of limestone. The country has a rolling domed contour
we have found to be characteristic of limestone rather than
areas, and is moreover covered by stone walls, and is treeless
raterless. Massive bedded limestone is seen in a quarry just
d Bell Busk station, and another quarry rather less than ¼
rest of Bell Busk viaduct.   This shows 18 ft. or more of thick
rely bedded limestone with no shales, and yielded the following
:—

> *Chonetes papilionacea.*
> *Spirifera lineata.*
> *Syringopora geniculata.*
> *Cyathophyllum Stutchburyi.*
> Crinoids.

The Cold Coniston limestone mapped as isolated beds, regarded in connection with the beds exposed in the hill 200 yards west, are evidently more extensive.

This limestone occurs in a series of domes or knolls, one of which is bisected by the Skipton-Settle road at Fogga.

The upper part of this limestone is massive, not well bedded, whitish in colour, but the lower beds are more regularly stratified. A fairly extensive fauna, typical of the Carboniferous Limestone, occurs here.

Two hundred yards west is another quarry on the south side of the road in the side of a large well-rounded rolling hill. The beds are well marked, and dip at 30° N.N.E., and thus would, if produced, pass below the limestones of Fogga.

About 30 ft. of beds are exposed.

The following fossils were obtained :—

> *Productus semireticulatus.*
> *Spirifera glabra.*
> *Rhynchonella pleurodon.*

LARGE CORALS.

> *Cyathophyllum Stutchburyi.*
> *Zaphrentis cylindrica.*

Three-quarters of a mile further west, and a little north, another exposure is found in a wood at Old Rock plantation, on the 1 inch map.

Here several feet of well-bedded limestones covered by black shales are seen dipping N.W. at 15°.

On the rolling ground south of the road are the following sections. A little more than ¼ mile S.S.W., at the word quarry on the 1 inch map is the following section :—

> Hard blue limestone. 1 ft.
> Hard nodular shale with crinoids, Zaphrentis, and fragments of shells. 12 ft.
> Well-bedded limestone.

Half a mile south-east, between the woods Camp and Hall Field, are two exposures in the same beds, showing massive well-bedded grey limestone overlaid by calcareous shales.

The limestone yields *Productus giganteus*, and a very large coral, *Cyathophyllum Stutchburyi*, and Crinoid débris.

The dip varying from 10° to 20° almost due north.

About three-quarters of a mile west of the above, five exposures are seen in the high banks of the northern tributary of the Swinden Beck. Both exposures have been quarried.

The most northern quarry shows—

Shale, becoming calcareous at base. 6 ft.

Hard compact limestone. 2 ft. 3 in.

Earthy shale. 9 in.

Hard limestone, base not seen. 6 ft.

Talus. 12 ft.

Dip a little W. of N. 10° to 20°.

The middle quarry continues the section downwards, and shows several feet of hard massive limestone with the large corals, *Cyathohyllum Stutchburyi* and *Zaphrentis cylindrica*.

The lowest quarry still continues the section downwards, and shows a further 18 ft. of well-bedded thick limestones and no shales.

Unfortunately, a gap in the sequence occurs here, but almost a hundred yards south the stream shows a succession of black shales, the exact position of which is doubtful, but a series of dark limestones of Pendleside type occurs in them just east of the railway in Swinden Gill.

Further north massive limestone is seen between the Craven Faults, and is well exposed in the scars of Malham, Settle, and Giggleswick. The upper beds at Malham and Settle yield the usual rich fauna of the Carboniferous Limestone. No trace of any interstratified shales or sandstones are to be seen.

The shale series which underlies the Millstone Grit rocks west of Malham is shown by its fossils to belong to the Pendleside series (Hind and Howe, of supra. cit., pp. 359, 360), and possibly some of the shales immediately below the grits of Black Hill may belong to this series. But there is a thin limestone which is seen in a stream a quarter of a mile west of Black Hill, which contains *Productus latissimus*, and is therefore shown to belong to the Yoredale series. At Glattering Sykes, in Outside Grizedale, is a spring which washes out numerous fossils. This water has evidently pierced a limestone by a swallow, and is thrown out by a fossiliferous bed of shale which underlies it. The passage of the subterranean stream over the fossil bed washes out numerous fossils, amongst which are—

Orthis Michelini.
Spirifera trigonalis.
Athyris planosulcata.
Ortholetes crenistria.
Zaphrentis sp.
Platycrinus sp.
Scaphiocrinus sp.

The fauna in this bed is, therefore, totally different from that characteristic of the Pendleside series, and demonstrates that the beds have some relation to the Yoredale series of Wensleydale.

An interesting section is seen in the brook east of Scaliber Force, commencing where the road to Kirkby Malham crosses the stream by a culvert :—

Sandstone with a curiously mammillated under surface. 3 feet.
Sandy shales, 9 feet.
Hard quartzose sandstone, 3 feet.
Black shales with bullions, 20 to 30 feet.

A fault passes across the shales, and where the shales are contorted several masses of limestone occur with large corals, *Productus giganteus, Spirifera trigonalis, Athyris planosulcata*, and crinoids. Below the disturbed shales, the stream section shows more shales dipping regularly and not contorted. It is very doubtful indeed if these fossils are *in situ*, but the masses of limestone seem to me to have been dragged in along the fault and crumpled up, as the seam of limestone is not apparent anywhere else in the stream section.

Negative evidence is important at this spot. Nowhere is there to be seen between the Millstone Grit and the top of the Carboniferous Limestone any beds corresponding to the Yoredale series of Wensleydale.

The western boundary of the Yoredale phase of Rocks is fairly well marked, the great splitting up of the main mass of limestone does not seem to take place west of a north and south line passing through

ɩby Lonsdale, though one well-marked bed of yellowish-white
stone is to be seen in shales north of Whittington village. Unfor-
ɩtely no fossils were found either in the stream section or the small
ɩry ; but between this bed and the Massif of Limestone, in a small
ɩm, a quarter of a mile south of Sellet Hall, a calcareous shale with
ɩules yielded :—

Athyris planosulcata.
Chonetes Laguessiana.
Productus longispinus.
Pr. punctatus.
Spirifera glabra.
Sp. trigonalis.
Sp. pinguis.
Rhynchonella pleurodon.
Edmondia unioniformis.
Sanguinolites striatolamellosus.
Monticulipora sp.
Fenestella sp.
Crinoid-stems.

A fauna with a Carboniferous Limestone facies.
The study of the belt of Carboniferous rocks deposited round
older rocks of the Lake District is of interest and importance.
the eastern side the base of the Carboniferous rocks is seen in
neighbourhood of Shap, where the basement beds have the
racter of a conglomerate. Around Carnforth and as far east as
kby Lonsdale, there is no evidence that the Carboniferous Lime-
ɩe is sub-divided by intercolations of shale and sandstone, and the
ɩstone here is at least 500 feet thick ; but east of Shap the lime-
ɩe is split into well-marked beds, of which the lowest is the Shap
ɩestone, and the next in series is the Knipe Scar Limestone. The
ɩr yielded me—

Lithostrotion junceum.
Cyathophyllum regium.
Athyris planosulcata.
A. globulina.
A. expansa.

Chonetes papilionacea.
Ch. Buchiana.
Dielasma hastatu.
Productus cora.
Pr. giganteus.
Spirifera trigonalis.
Sp. lineata.
Syringothyris cuspidata.
Orthotetes crenistria.
Edmondia sulcata.
Solenopsis minor.
Euomphalus pentangulatus.
E. cirrus.
Loxonema sp. (cast).
Naticopsis plicistria.
Orthoceras Breynii.
Cephalopod fragments.

The lowest or Shap Limestone yielded large masses of *Chætetes tumidus, Cyathophyllum regium, Syringopora ramulosa, Chonetes papilionacea, Productus giganteus, Spirifera glabra, Sp. pinguis* near Askham and Rossgill, and plant remains near Shap.

The list of fossils given on pp. 85-88, Mem. Geol. Surv., Geol. Country round Kendal, &c., is fairly long and accurate; the nomenclature of the lamellibranchs and cephalopods, however, needs revision, but it will be sufficient here to point out that *Edmondia sulcata* and *Prolecanites compressus,* so characteristic of the lower limestones of Scarlet and Ballasalla, Isle of Man, also occur here.

Very valuable contributions to our knowledge of the Carboniferous Rocks on the south and east of the Lake District, have been made by Mr. J. D. Kendall, F.G.S. His papers on the hematite deposits of Furness and West Cumberland, and the Carboniferous Rocks of Cumberland and Furness, are published in the Trans. N. Eng. Inst. M. Mech. Engineers, Vols. XXVIII., XXXI., and XXXIV. In the latter, especially, he shows that the Carboniferous Limestone in Furness, i.e., to the south of the Lake District, is almost undivided by shales, but that as it passes north shales and sandstones set in and split up the mass.

ı West Cumberland he gives the following section as obtaining
Ullock to Egremont :—

1st or Langhorn Limestone, 30 to 60 feet.

Shales, 10 to 14 feet.

2nd Limestone. 14 to 24 feet.

Sandstones and shales, 40 to 60 feet.

3rd Limestones, 10 to 16 feet.

Shale, 2 to 6 feet.

4th or Clint's Limestone, 235 to 310 feet.

Shale or sandstone, 14 to 24 feet.

5th Limestone, 50 to 70 feet.

Shale, sandstone, or thin limestones, 14 to 24 feet.

6th Limestone, 54 to 70 feet, or 105 at Lamplugh.

Shale, thin and variable.

7th Limestone, 40 to 182 feet.

Shale.

Skiddaw Slate.

hows a proportion, taking the greatest thicknesses, of 757 feet
estone and 130 feet of detrital rocks.

)uring the last few weeks I have examined several quarries and
as at Bigrigg, Yeathouse, Rowrah, and Lamplugh, and although
are not very common I obtained *Productus giganteus* at all
places. The Yeathouse Quarry gives the following section,
ıds dipping west at about 20°.

Limestone massive, beds much mottled with

Productus giganteus, 50 feet.

Bluish Limestone with peculiar black inclusions, with

Productus giganteus, 30 feet.

„   *semireticulatus.*

Edmondia sulcata.

„   *Lyelli.*

Solenopsis minor.

Solenomya primæva.

Pinna flabelliformis.

Loxonema sp.

Grey Limestone, with *Productus giganteus* and *Productus
cora.* 40 feet.

Shale, 2 feet.

Thin Limestone, 1 foot 6 inches.

Shale, 4 feet.

Sandstone, 2 feet.

Sandy Shales, 6 feet.

Shales, 6 feet.

Limestone, 75 feet.

In his Furness paper (Op. supra cit., Vol. XXXI., page 215), Mr. Kendall estimates, from borings and sections, the thickness of the limestone in the Furness district as 946 feet, and this mass contains only about 34 feet of interbedded detrital rocks.

Above the mass of the limestone of Furness shales with limestones are to be found, and it will be a matter of importance to ascertain whether they contain the Pendleside fauna or not. Personally I have only examined the shales in the stream section north of Borwick Hall. A series of sandy and muddy shales with sandstones were seen, but the only fossils obtained were fragments of plant remains and scales of *Megalicthys Hibberti.*

Unfortunately no paleontological evidence is mentioned in Mr. Kendall's paper, but the important fact remains that during the deposition of the limestone, currents bearing detrital mud and sand did not reach Furness to any extent, but they did reach the district of West Cumberland, and caused the division of the limestone into distinct beds.

The plates of Mr. Kendall's paper, Vol. XXXIV., Plates XIII. and XIV., of supra cit., are very instructive, the gradual change of the series being shown by parallel columns of actual sections, and finally compared with the section of the Carboniferous series at Weardale and Allen Head. The change of the Yoredale phase into that of the southern type of Carboniferous rocks is, in the West Cumberland and Furness districts, therefore demonstrated to be very gradual, and a comparison of figures representing the proportion of shale to limestone in the Furness, Egremont, and Alston Moor districts is very instructive.

The numerators represent the thickness of shale, the denominators the thickness of limestone :—

Furness.	Egremont.	Weardale and Allen Head.
34	130	1,602
946	757	480

A comparison of these amounts with other sections (page 443) hich show the gradual change in the Carboniferous series as it passes rth from Wensleydale to Scotland, demonstrates the same gradual ange, and proves mathematically that the detrital rocks replace the ganic limestones to a greater and greater extent as the beds go north.

The Cumberland-Furness area is not, however, affected by the aven faults, which is most strong evidence against the view put rward by Mr. Tiddeman to account for the apparent sudden onset the Yoredale phase of deposit in Yorkshire.

A theory has been advanced by him that the change from the rthern to the southern type of the Carboniferous succession was rgely due to the fact that the Craven faults were contemporaneous ith the deposit, and in some way caused the damming back of the trital sediments. It is difficult to see how this could have been. he Yoredale type of rocks is not seen between the Craven faults, d on the western side of the Craven area Mr. Dakyns says : " The oredale type of beds can hardly be said to exist south of Kettle- ell " (Proc. York. Geol. and Poly. Soc., 1890. Vol. XV., p. 361). . also seems to me to be the case that the peculiar shaley beds found tending from Thornton and Barnoldswick west towards Chipping e an indication of the Yoredale phase. Then we know that the ansition from the northern to the southern type is very gradual, d that the Yoredale series becomes developed gradually, and finally, lthough Phillips and other writers never admitted it. the whole arboniferous series further north takes on the Yoredale phase ; ı fact, the Yoredale series has no true base, but the base is at a lower d lower horizon as the series passes north. If Tiddeman's view ere correct. there ought to be a north and south fault in Upper Vharfedale to account for the change which takes place from east) west from Kettlewell to Fountains Fell. The change of type is here, but no fault and no barrier.

I am unable to accept the theory that the Craven faults were ontemporaneous with the deposition of the beds in which the faults occur. A glance at the map will show that the southern limb of the fault faults limestone unconformably against limestone, shales, grits, and coal measures, and this would surely not have been the case if the faults had been formed as the material was deposited.

In Wensleydale the Carboniferous succession is entirely different from that which obtains further south. At the base there is a mass of limestone about 500 ft. thick, with a few feet of basement conglomerate, resting on the upturned edges of the older Palæozoic rocks. This is succeeded by about 1,000 ft. of alternating shales, sandstones, and limestones, the latter being about six in number, and forming well-marked features along the escarpments of the valleys, and often giving rise to waterfalls in the tributary streams. As the beds pass north the lower undivided mass becomes split up into beds by the intercalation of shales and sandstones, and becomes the Melmerby Scar series. Coal seams are also developed at several horizons. The tendency as the beds pass north is that detrital sediment increases and limestones thin out.

The most southerly point at which any great development of these conditions occurs is in the flanks of Fountain Fell, in which the number of mappable limestones falls short of those seen in Wensleydale. On Ingleborough only four distinct beds of limestone are to be seen. The limestones of the Yoredale series are all fossiliferous, and *Productus giganteus* is found in all of them in Wensleydale. The intervening shales are often very fossiliferous, and contain a fauna very similar to that of the Carboniferous Limestone.

The rivers in all the great dales have cut through the Yoredale series, and splendid sections of the series are to be studied, while the collecting of fossils can be carried on in the intervening beds.

The top bed of Yoredale Limestone in Wensleydale is well seen near Leyburn, and yields a plentiful store of fish remains and other fossils. The fish all seem to belong to species which are obtained in the upper beds of the Carboniferous Limestone massif elsewhere. *Psammodus, Psephodus,* &c., &c.

The following columns show in tabular form the changes which the beds of limestone undergo as they pass north :—

I. INGLEBOROUGH.	II. WENSLEY AND SWALE DALES.	III. WEARDALE AND ALSTON MOOR.	IV. NORTH NORTHUMBERLAND.	V. WEST OF SCOTLAND.	VI. EAST OF SCOTLAND. FIFE.
Main Limestone, 60 fet. Middle Limestone, 20 feet. Simonstone Limestone, 30 fet. Hardraw Scar Limestone, 40 feet. Great Scar Limestone, 600 feet. Basement beds. Total, 1,500 feet, of which 750 feet are limstone.	Red beds or Crow Limestone. Main Limestone, 60 feet. Underset Limestone, 20 fet. 3 yards Limestone, 9 6 fet. 5 yards Limestone, 9 feet. 5 yards Limestone, 15 feet. Middle Limestone, 30 feet. Same Limestone, 20 to 30 feet. Hardraw Scar Limestone, 50 fet. Gayle Limestone. Great Scar Limestone, 500 eft. Basement beds, 10 ft. Total, about 1,600 feet, of which 750 feet are limestone.	Fell Top Limestone, 4½ feet. Little Limestone, 6 feet. Great or Main Limestone, 63 fet. 4 fathom Limestone. 24 feet. 3 yards Limestone, 9 feet. 5 yards Limestone, 15 feet. Scar Limestone, 30 feet. Cockleshell Limestone, 14 feet. Tyne Bottom Limestone, 24 eft. Jew Limestone, 24 f et. Little Limestone, 15 feet. Smiddy Limestone, 31 feet. Robinson's Limestone, 21 feet. Melmerby Scar Limestone, 142 feet. Measures, 519 feet. Total, 2,082 feet, of which 480 feet are limestone.	Upper Fell Top, 20 feet. Lwr Fell Top, 6 feet. Limestone, 14 feet. Dryburn Limestone, or 10 yards or Khw Nook Limestone, 30 feet. Denwick: Low Dene, or 8 yards, 28 feet. Acre or 6 yards, 22 feet. Thin Limestone, 2 feet. Thin Limestone, 8 fet. Kelwell and Main or 9 yards, 27 fet. Several Thin Limestones. Oxford, or 5 yards limestone, 16 f et. Thin Limestone, 5 feet. Woodend, Hobberlaw, or 4 fathom, 15 feet. Dun or Redesdale Limestone, 6 feet. Total, 692 feet. Scremerston Coal Series, 808 ft. 8 in. Total, 2,080 feet, of which 240 feet are calcareous.	Linnspout, or Castle Cary Limestone, 30 feet. Lower Posts, 26 feet. Arden or Calmy Limestone, 15 feet. Hig Hid or Index Limestone, 6 feet. Middle Sandstones, Shales with Coals and Ironstones, 600 feet. Kingshaw and Kerriland (Glen Limestone, 7 feet. Hosie Limestone, 4 feet. Neilsth, or Macdonald Howrat, 100 to 60 feet. Volcanic Series, 1,500 feet. Calciferous Sandstone Series, 1,500 feet, with about 40 feet of limestone. Total, 2,500 eft of stratified ocks, of which 244 feet are limestone.	Upper Group, 900 feet, with 10 to 28 fet limestone. Middle Group, ...tons, Shales, Fireclays, and Coals, 1,000 feet. Lower Group, Hosie and Hurlet Limestones, 250 feet, with 20 to 53 feet 1 limestone. All four Sandstone Series, 3,800 feet, with 50 feet lime limestones. Note.—Passing out in the Carluke district the limestones of the Upper and Lower Dl are thin ut. The three beds of the Upper Group only giving 12½ feet of limestone; the 15 beds of the Lower Group are 34 feet thk, so that out of a total thickness of 1,032 eft of strata only 46 eft are limestone.

I make no attempt to correlate the va
can be little doubt that the Main limeston
sented by the great Limestone of Teesdale a
by the Dryburn Limestone of North Nort

At present there is little or no paleont
definitely that any life zones existed in the a
any certainty, the exact bed of limestone '
definite band further south. The limestone
up by masses of shale as they pass north, a
and a correlation by the numerical position (
counted either from below or from above, i

Mr. W. Gunn has attempted a precis
stones of Dunbar and North Northumb
Yoredale limestones, but it seems to me tl
of his assumptions is almost *nil*. The pre
garded as of prime importance as a stratigra;
land and East of Scotland, but consider
in the thickness of the purely sedimentary
it is not impossible that an oil shale was dev
in the two fields. Oil shales are not unl
district at other horizons. The Oxford l
the representative of the Hardraw Scar Lir
and Dun Limestones are supposed to repre
stone, but there is absolutely no evidence

In discussing the Carboniferous sectic
says : " Opposite Pinkhead .   .   . there
crinital limestone, which seems most probab
Thus nearly all the lower limestones are dyi
as we proceed westward. and at Skateraw r
between the Oxford and the Eelwell have
Craig the lowest limestone is the Eelwell
has collected extensively both from the .
and from the second limestone (counting '
has come independently to the conclusion
the same, because they attain a similar a

This is important, but unfortunately the Four Laws or Wood-
ld Limestone at the Coomb, south of Redesdale, has yielded to Mr.
'unn and myself a fauna which contains all the special fossils (*lamelli-
ranchs* and *gasteropoda*) which have been found at Lowick. Now,
Lr. Gunn places the Four Laws Limestone much lower than the
Lcre. Who shall then decide whether the Cat Craig Limestones
re the equivalents of the Acre or the Four Laws Limestone, more
specially when it is known that the beds are undergoing rapid changes
s they pass north?

It seems to me more than possible that the initial mistake in the
lomenclature of the Yoredale Limestones has been to reckon them
rom below upwards. The Millstone Grit at the top might have
een taken as a base line with a good deal more reason than the
ontinually altering top of the Great Scar Limestone as it becomes
plit up by intercalations of shale to the north.

Perhaps a full and accurate paleontological survey would give
ome more certain ground for correlation, though I am bound to
ay at present I have obtained no direct evidence to enable one to
pproach the subject with any degree of accuracy. As the beds
ass north, owing to the muddy sediment and the probable shal-
wing of the sea, new forms of life occur plentifully, which were
ot met with in the massive limestones, but similar faunas occur
gain and again at different horizons.

Professor Lebour has given some excellent lists of fossils from
ertain of the Northumberland limestones in his Handbook to the
ieology and Natural History of Northumberland, from which it
rill be seen how very similar were the faunas obtained at different
orizons. I append lists of fossils which I have obtained in the Redes-
ale district from the Redesdale and Four Laws Limestone series.

Professor Lebour shows that *Productus giganteus* is found in
he Fell Top Limestone of Northumberland, but at present I have
o evidence that this fossil occurs above the Main Limestone in Wear-
lale or Wensleydale.

Last August my attention was called to a book on "The Laws
rhich regulate the Deposition of Lead Ores," by W. Wallace, published
n 1861. In it is a plate (Plate IV.) which gives his idea of the correlation of the Carboniferous deposits of England.

He considers the Main Limestone of Weardale and the north as the equivalent of the top of the Mountain Limestone of Derbyshire, and shows the shales of Derbyshire (the Pendleside series) as being altogether above the Yoredale series.

In company with Mr. J. Barker, of Frosterley, I examined the Carboniferous succession of Weardale where quarries extending for miles on both sides of the Wear are opened in the Main Limestone, and streams and the Wear show sections from the Millstone Grit to the Four-fathom Limestone. Further west, near Wearhead, the Lower Limestones and intervening beds are exposed down to the Scar Limestone, but unfortunately I had no opportunity to examine these beds thoroughly, and they are not quarried to any extent.

Below the Fell Top Limestones, in which I got no fossils, the sandstones and gannisters were full of plant remains.

The Little Limestone is 7 feet thick, and the following section yielded the fossils enumerated below : —

A Bed of Quartzose Sandstone above Little Limestone, Wolsingham.

Chonetes Laguessiana, Productus longispinus. P. muricatus, P. semireticulatus. Spirifer oralis. Ortholetes crenistria, Edmondia sulcata, Lithodomus lingualis, Bellerophon (cast), *Naticopsis* (large cast), *Phillipsia* sp. (common).

Pattinsons Sill Sandstone, 12 feet.

Shale, 24 feet. Full of round black concretions at base.

Athyris ambigua, Chonetes Laguessiana. Rhynchonella trilatera, Productus semireticulatus, Spirifera glabra, Sp. oralis, Sp. trigonalis, Spiriferina octoplicata. Cypricardella annæ, C. rectangularis, Bellerophon Urei, Orthoceras Morrisianum. Zaphrentis sp., Crinoid-stems.

Little Limestone, 7 ft.

Chaetetes radians, Syringopora geniculata. Cyathophyllum sp. *Productus longispinus.* Tooth of *Cochliodus.*

High and Low Coal Sills.

ut 30 feet
Jittle Lime-

Discina nitida, Productus punctatus, Rhynchonella pleurodon, Orthotetes crenistria, Aviculopecten sp., *Allorisma sulcata, Edmondia unioniformis. Nucula gibbosa, Nuculana attenuata. Protoschizodus axiniformis, Bellerophon decussatus* var. *striatus, Pleurotomaria atomaria, Orthoceras* sp., *Fenestella* sp., Crinoid ossicles.

ove Great
ie. A few
γ.

Athyris planosulcata, Chonetes Laguessiana, Orthis Michelini, Productus latissimus, Fenestella, and Crinoids.

reat Lime-
) to 70 feet.

CORALS :—*Lithostrotion basaltiforme, Cyathophyllum regium, Lonsdaleia floriformis, Clisiophyllum* sp., *Chaetetes radians, C. septosa, C. tumidus. Pyrgia* sp., *Syringopora ramulosa, Cladochonus* sp., *Athyris planosulcata, Dielasma hastata, Camarophoria crumena, Lingula mytiloides, Orthis Micheleni, Chonetes Buchiana, Productus aculeatus, P. cora, P. fimbriatus, P. giganteus, P. latissimus, P. longispinus, P. punctatus, P. sinuosus, P. semireticulatus, Spirifera crassa, Sp. lineata, Sp. trigonalis, Sp. distans, Orthotetes crenistria, O. crenistria* var. *senilis, Rhynchonella pugnus, R. reniformis, Aviculopecten carlatus, Pinna flabelliformis, Allorisma sulcata, A. monensis, A. variabilis, Edmondia sulcata, E. unioniformis. Cypricardella annæ, Sanguinolites plicatus, Solenomya costellata, S. primæva, Dentalium ingens, Euomphalus pentangulatus, E. catillus, Leveillia Puzo, Loxonema* sp., *Macrocheilina acuta, Naticopsis plicistria, N. ampliata, Pleurotomaria altavittata, P. carinata* Sow., *Bellerophon cornuarietis, Actinoceras giganteum,*

*Orthoceras cinctum, O. sulcatum, Temno-
cheilus pentagonus. Solenocheilus crassi-
venter (?). Phillipsia Eichwaldi var. mucronata. Phillipsia sp., Gyracanthus (spine),
Psammodus porosus.*

One post. known as the Frosterley marble, is made up of fine specimens of *Cliniophyllum*.

It may be remarked that the fauna of the Great or Main Lime-stone of Weardale is therefore identical with that of the Carboniferous Limestone, though containing somewhat fewer species.

The one important fact which the study of the various sections of the Carboniferous rocks given above demonstrates, is the rapid increase of detrital sedimentary deposits and the diminution of organic deposit (limestone) as the beds pass northwards.

It is found that practically no change in the nature of the deposit takes place between Derbyshire to Settle and the Craven district, but from this point northwards the change comes on rapidly. The nature of the change, the substitution of sediment obtained by the denudation of pre-existing rocks, for an organic deposit, due to living things, points without any equivocation to the cause of the change. This was the presence of continental land within a very short distance to the north of a line drawn across Scotland from the Firth of Tay to the Firth of Clyde.

I believe I am correct in stating that the most northerly deposits of Carboniferous age in Scotland are to be seen in the Pass of Brander and near Innimore of Ardtornish, on the mainland of Morvern. It is probable that the greater part of the Highlands of Scotland were dry land during Carboniferous times. The products of land erosion are deposited out at sea by the rivers which bring them down, and unless affected by strong local currents are deposited over a more or less pear shaped area, the deposit being thicker the nearer it is to the actual mouth of the river. Lighter materials are carried further out to sea, consequently shales would be laid down somewhat further from land than sandstones. The elevation of the land above the sea and its consequent effect on the rapidity of the flow of the denuding streams, will also have an important influence on the area and nature of the deposit.

The Yoredale series have the following general sequence :—

Sandstone,

Shale,

Limestone,

da capo, which denotes an area of clear sea unaffected by sedi-
t, invaded by detrital mud at first, and later by heavier detrital
l. Then a condition when detrital matter was no longer carried
ır south, and a return of a clear sea with conditions suitable for
environment of animals producing calcareous matter. This
ıge may have been brought about either by the formation of a
which prevented the carriage of sediment to the south, or an
lation of level sufficient to permit the repeated advance and
eat of conditions from north to south and *vice versa*.

South of Derbyshire we know that the Lower Carboniferous
s soon disappear. In South Staffordshire the Coal Measures
immediately on the upturned edges of the older Palæozoic rocks.
he Coalbrookdale coalfield the whole of the Carboniferous rocks
w the Coal Measures are represented by about 40 to 80 feet of
stone and 10 feet of calcareous sandstones and shales, and sand-
es. At Steeraways, five miles further west, on the S.E. flank
he Wrekin, the whole deposit is only 50 feet, showing the rapid
ning out of these beds.

There is little or no doubt that a ridge of land extended from
to west across England and Wales, and probably as far as the
klow Mountains, throughout Carboniferous times, too narrow
steep to supply any amount of detrital sediment, and probably
deepest part of the Carboniferous Limestone sea in the British area
a little north of this ridge. This pre-Carboniferous ridge, therefore,
ıs the south boundary of the great Pennine Carboniferous basin.

Of Carboniferous deposits eastward we know little or nothing,
south-east in Belgium the limestone is of very considerable
kness and is undivided by shales and sandstones, and the fauna
ractically that found in the limestones of Derbyshire, Clitheroe,
:oe, and Settle. At Clavier and Visé the limestone is succeeded
a series of shales, black limestones, sheets, and gannisters
aining the typical Pendleside fauna. On the western side of

M

the Pennine system, in North Wales,
consists of 1,700 feet of limestone, with 2
Holywell shale. 100 feet, the latter coi
Pendleside series. No Yoredale phase, (
area, and it was altogether outside the
sandbearing waters of the north.

Similarly the sequence in the Isle of
intercalated beds of detrital sediment.

Poolvash, Posidonomya shales :)
Poolvash Limestone. massive, she
Well-bedded Limestones of Scarle
And, as has been mentioned above (page 4
the Limestone is almost undivided by shale
west of the inlier of the older rocks cons
between Egremont and Penrith, masses o:
into well-marked beds, showing an appro
which characterises the whole of the
Ingleborough to the centre of Scotland.

Probably, therefore, in point of time,
in the Midlands corresponds to the Calcif
Carboniferous Limestone series of Scotla
aceous, and Calcareous divisions of Nortl
Scar Limestone. plus Yoredales, of Wensle
in the stratification is only what might he
to obtain in a marine area liable to be af
and the two types of rocks are essentiall;
story, and require no hypothetical barrie1
earth movements to explain them.

THE PENDLESIDE SI

In the north Midlands the thick and
stone is overlaid by a series of dark shales
limestones. These limestones gradually
in the series. and fine quartzose or gan:
plant remains, and an occasional marine

This series varies from a few to 1,:
formed the main subject of the paper

ioned above. Unfortunately this series was named Yoredale Series,
was supposed to be the equivalent of the Yoredale Series of
leydale, but lithologically, stratigraphically, and paleontologi-
the two series are quite distinct and are on different horizons.

.he Pendleside Series occupies a very limited area.   Its northern
dary being about a line from Greenhow Hill to Linton Mill, then
ig west to a point a little south of Giggleswick, thence across
ower end of the Furness district to Poolvash, Isle of Man, and
to Co. Meath, and across Ireland to Foynes Island, Co. Limerick.
To the south we know the series is represented in Leicestershire
few feet of shales, and is absent along the northern margin of
Coalbrookdale coalfield.   The series appears to be represented
orth Wales by the Holywell Shales.   The deposit is thickest at
le Hill, and here the greater thickness seems due to a greater
.nt of shales below the Pendleside Limestone, and to the greater
opment of the limestones.   This is a purely local thickening, and
n a few miles north, south, and west of Pendle the deposit is
ı thinner, and the limestones so much reduced that the officers
ə Survey did not think it worth while to map them, notwith-
ing the constancy of the bed, even though it was attenuated,
he strong paleontological evidence contained in it.

.he characters of the limestones of the Pendleside Series are
different from those of the real Yoredale Limestones, both in
re, chemical composition, and fossil contents.   In our paper
ıpra cit., pp. 394-401) one of our purposes was to show the
logical and chemical differences between the limestones of the
eside Series and those occurring in the dome-shaped hills of
e-in-Craven, because for some, to us, unaccountable reason
.tter had been correlated with the Pendleside Series, and there-
ctual details of the real Yoredale Limestones were not given at
ime.   It may be stated here that the Yoredale Limestones agree
ıracters with the various beds of the thick Massif Limestone.

ıt the base of the series, a series of passage beds containing
al shell material and fragments of limestone are found in places.
hers shales seem to come on regularly on the top of the Lime-
Massif, and elsewhere evidence seems to point to masses of

limestone being somewhat irregularly surrounded by sha
upper limit of the series is not however so clear, for towar
grit beds become stronger and better developed, and th
fauna occurs again and again even as high as the Lower Coal
or Gannister series.

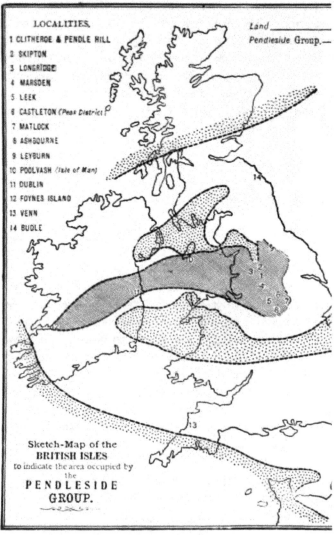

Fig. 1.

PALEONTOLOGY.

The important evidence furnished by a study of the paleon-
logy of the Carboniferous rocks for the correlation of the series
different districts has been very largely neglected. The extensive
emoirs on the fauna and flora of different districts published by
e Geological Surveys of other countries only bring home to us
ore forcibly the utter dearth of any such information in this country.

What is wanted now is a paleontological survey, and this could
e largely carried out by local geological societies. The accurate
entification of fossils is also an important desideratum. Many
ecies published in lists have been erroneously identified, from
rious causes, hence the necessity for local sub-committees to super-
tend the correct identification of specimens.

A study of the tables of fossils published in the appendices A and
to our paper (Q. J. Geol. Soc., Vol. LVII.) shows at once that the
una of the Carboniferous Limestone is identical with that of the
reat Scar Limestone plus the Yoredale Series, and further details
ave been published since in my report of the Committee on Life
ones in British Carboniferous Rocks (Brit. Ass. Rep., 1901).

In the same two papers it is noted that there does exist in
certain definite area, on the top of the Massif of Limestone, a
ries of black limestones, shales, and quartzose sandstones (the Pendle-
de Series), which contain a fauna peculiar and distinct from that
und in the Carboniferous Limestone of the Yoredale Series. It
hinted (pp. 379-401) that the fauna of the Pendleside Series bears
very striking similarity to that of the Culm Beds of Devonshire
nd Europe ; a view which I am persuaded will grow clearer as fresh
aleontological evidence turns up.

The percentage of fossils common to the Yoredale and the
endleside Series is low and practically the same as that common
o the Carboniferous Limestone and the Pendleside Series.

By far the greater number of the Brachiopoda of the Yoredales
nd the Limestone are absent in the Pendleside rocks. only some
2 species remaining.

The Actinozoa are only represented by a single species.

The characteristic fossils of the Pendleside Series are the cephalo-poda and lamellibranchs, several of which appear to be confined to the series.   I regard the following species as typical of the horizon :—

Dimorphoceras Gilbertsoni.

D. Looneyi.

Gastrioceras carbonarium.

G. Listeri.

Glyphioceras bilingue.

Gl. Davisi.

Gl. diadema.

Gl. reticulatum.

Gl. spirale.

Nomismoceras spirorbis.

Orthoceras Steinhaueri.

And the following lamellibranchs :- -

Chœnocardiola Footii.

Posidoniella lævis.

P. Kirkmani.

P. minor.

Posidonomya Becheri.

P. membranacea.

Leiopteria longirostris.

Pterinopecten papyraceus.

Aviculopecten prætenuis.

Not only is the Molluscan fauna of the Pendleside Series different from that of the Yoredales and Carboniferous Limestone, but the evidence afforded by the Vertebrate fauna is equally well marked. Dr. Traquair has shown that two totally different fish fauna, an upper and a lower, existed in Carboniferous times.   The great break between these two faunas in England comes on at a line which represents the very topmost limestones of the Yoredale Series in North Yorkshire, and the topmost beds of the limestone massif in South Yorkshire and Derbyshire.   The Red-beds Limestone at Leyburn has yielded a very rich fish fauna to Mr. J. Horne, and the majority of the forms found there occur also in the upper part of the massif limestone, not only in England, but in Ireland.   This

ie extended to Scotland comes at the top of the Upper Limestone
ries, and it is here that the great paleontological break occurs.

I pointed out in my paper (op. supra cit., page 380) that certain
nera and species are found at much lower horizons in the Carbon-
rous series in the north than in the south, and cited as good ex-
nples the various species of the family Nuculidæ. *Nuculana
tenuata* and *Nucula gibbosa* occur in the Calciferous Sandstone
ries of Fife, far below the Hurlet Limestone series, or the base of
e Lower Limestone series, and are also to be found recurring in
lcareous shales as high as the Upper Limestone series. *Nuculana
tenuata* appears to have come into the area some time before *Nucula
bbosa*. In the west of Scotland these species have not been found
:low the Beith Limestone series, the equivalent of the Hurlet. They
)pear to be absent in the Calciferous Sandstone series of Eskdale
id in the Tuedian series of Northumberland, but *N. attenuata*
)pears in the Carbonaceous division, and both shells are plentiful
, various horizons in the calcareous division.

Still further south, the lowest horizon in the Eden valley at
hich these two species have been found, is the shale over the Under-
t Limestone. Still further south the lowest horizon for *N. attenuata*
in the shales of the Pendleside series at Whitewell, while in South
orkshire *Nucula gibbosa* has not been found below the shales
:low the Third Grit at Eccup and Congleton Edge (Cheshire), and
North Staffordshire it occurs at one or two horizons in the Coal
:easures. These two species seem to occur at higher and higher
orizons as the beds pass south. We have proposed the term ISODIETIC
ir the line drawn across the strata representing the migration of
lese shells, which denotes a life zone due to conditions rather than
) time (Fig. 2.).

We have pointed out that other species and many families appear
) have migrated slowly south, and for further details would refer to
iges 380-385 of that paper.

The majority of the lamellibranchs which occur in the Calciferous
andstone series of Fife obey the same law. As I pointed out, this
,w obtains for byssiferous as well as for free lamellibranchs.

Since writing my paper a very interesting fact has been discovered, which shows that plants obey the same law. Mr. R. Kidston has determined some half-dozen specimens of plants, obtained by Mr. D. Tait, from the Pendleside Series of Pendle Hill as

> *Asterocalamites scrobiculatus,*
>
> *Lepidodendron Veltheimianum,*

both of which species occur in the Calciferous Sandstone Series of Fife, and apparently in that locality at that horizon only, and also a number of ferns from the *Posidonomya* beds of the Isle of Man, which he had never before met with above the Calciferous Sandstone Series.

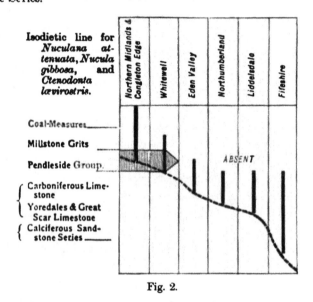

Fig. 2.

It is a remarkable fact that plants which occur very low down in the Carboniferous sequence of Fife should also occur at a much higher horizon further south. It will be interesting to ascertain if they occur, and at what horizon in the intervening country.

The fauna found in the Carboniferous Limestone massif is practically identical with that which occurs in the Great Scar and Yoredale Limestones. Some few species occur in the shales separating the limestones which have not been found in the limestone. A very natural state of things for a muddy environment would be extremely distasteful, if not fatal, to organisms which lived in a clear sea. I have never yet obtained any of the *Nuculidæ* in pure limestones, but the species of these genera always are found in muddy deposits. Of brachiopods, the *Lingulæ, Discina nitida, Orthis Michelini, Chonetes Laguessiana, Rhynchonella trilatera, Athyris ambigua* were those most able to live in muddy waters, and are not commonly found in pure limestones.

Productus giganteus, P. latissimus, Chonetes papilionacea, and *Amplexus coralloides,* and the great majority of the corals and Polyzoa are characteristic of the whole Carboniferous Limestone series, in which mass of rocks I have no evidence of life zones at present. For, although fossils are more numerous at the top, I cannot find that any species are confined to any definite horizon. It is probable that the rarity of fossils in the middle and lower beds of the thick massif limestone of the North Midlands is due to the fact that metasomatic changes have obliterated the fossils, but it is probable that a microscopic survey of the different beds of the massif might reveal some microscopic forms which had a limited vertical distribution. When one considers the repeated changes in the conditions of the sedimentary deposits which form the Yoredale phase of the Carboniferous rocks, it is a remarkable fact that the various limestones and marine shales are not characterised by definite fossils, but that the faunas of similar deposits at several horizons are identical. This condition of things is doubtless due to the very local character of the deposit of shales, and that the limestone fauna had not far to migrate southward when its present habitat was rendered unsuitable for it by the incursion of muddy water, and consequently had to travel back a very short distance to that area when muddy conditions ceased. In other words the Carboniferous Limestone fauna flourished, as a whole, right through the deposit of the massive beds, and advanced north or retreated south whenever clear conditions obtained in the area occupied by the Yoredale phase of rocks. Consequently the various limestones of the Yoredale Series contain the same faunas, and similarly the intercalated marine muds also contain the same faunas.

TABLE OF LIFE-ZONES SUGGESTED FOR THE BRITISH CARBONIFEROUS ROCKS.

	ZONES.	ENGLAND.	SCOTLAND.	IRELAND.	ISLE OF MAN.
UPPER COAL-MEASURES.	Zone of *Anthracomya calcifera*.	*Spirorbis*-limestones, Upper Coal-Measures.	Upper Coal-Measures of Ayrshire.	? Wanting.	Wanting.
	Zone of *Anthracomya Phillipsii*.	Upper Coal-Measures of Lancashire, Yorkshire, Staffordshire, and Bristol.	The Red Measures of Fifeshire.	? Wanting.	Wanting.
MIDDLE COAL-MEASURES.	Zone of *Naiadites modiolaris* and *Anthracomya modiolaris*; containing sub-zones of *Anthracomya Wardi, A. Adamsi, and A. Williamsoni*.	Middle Coal-Measures, universally.	Coal-Measures of Fife-shire.	Coal-Measures, Castle-comer and Leinster.	Wanting.
GANNISTER GROUP. LOWER COAL-MEASURES. MILLSTONE GRITS. PENDLESIDE GRIT.	Zone of *Pterinopecten papyraceus, Gastrioceras Listeri, G. carbonarium, Glyphioceras reticulatum, G. bilingue, Posidoniella laevis, P. minor*, with a sub-zone near the base of *Posidonomya Beskeri*, and *Posidonomya membranacea*.	Gannister Group of Lower Coal-Measures. Millstone Grit. Pendleside Group. ? The Pulm Measures of Venn and Swimbridge.	? Wanting. *Pterinopecten papyraceus*, said to be found above the Ell Coal, Wishaw and in the Lower Limestone Series of Kilbride.	Coal-Measures of Foynes Island (County Limerick). Upper Limestone Shales, County Dublin and County Meath.	The *Posidonomya*-schists of Poolvash.
CARBONIFEROUS LIMESTONE SERIES.	Zone of *Productus giganteus, Pr. Cora, Chonetes papilionacea, and Amplexus coralloides*.	The Carboniferous Limestone of Derbyshire and Staffordshire. Measures from the Scar Limestone to the Main Limestone, N.W. Yorkshire. Carboniferous and Calciferous divisions of Northumberland. Carboniferous Limestones of North and South Wales.	Carboniferous Limestone Series (Upper, Middle, Lower) of both the East and West of Scotland and Roxburghshire.	The Upper Limestone. The Calp. The Lower Limestone.	The limestones of Poolvash, Scarlett, and Ballasalla.

TABLE OF LIFE-ZONES SUGGESTED FOR THE BRITISH CARBONIFEROUS ROCKS.

APPENDIX C.

	Redesdale Ironstone and Limestone.	Four Laws Limestone at the Coomb.			Redesdale Ironstone and Limestone.	Four Laws Limestone at the Coomb.
BRACHIOPODA.				**LAMELLIBRANCHS**—*continued.*		
Athyris ambigua	*			*Edmondia unioniformis* ...	*	
,, *Royssii*	*			*Modiola Jenkinson* ...		*
Camarophoria crumena	*			*Parallelodon reticulatum*		*
Chonetes Buchiana ...	*			*Nucula gibbosa* ...	*	
,, *Laguessiana* ...	*			*Nuculana attenuata* ...	*	
Discina nitida	*			,, *breviostris* ...		
Dielasma sacculus	*	*		*Myalina pernoides* ...	*	
Lingula mytiloides ...	*			,, *Redesdalensis* ...	*	
,, *Scotica*	*			,, *Verneuillii* ...	*	
,, *squamiformis* ...				*Protoschizodus axiniformis*	*	
Orthis resupinata ...	*			,, *fragilis*	*	
Productus aculeatus ...	*			*Posidoniella elongata* ...	*	
,, *cora*	*			*Schizodus axiniformis* ...	*	
,, *giganteus* ...	*			*Solenomya costellata* ...		*
,, *longispinus* ...	*			*Sanguinalites clavatus* ...	*	
,, *latissimus* ...	*			,, *plicatus* ...	*	
,, *mesolobus* ...				,, *striatogranulosus*	*	
,, *punctatus* ...	*			,, *tricostatus* ...	*	*
,, *scabriculus* ...				,, *variabilis* ...	*	
,, *sinnatus* ...	*			,, v. *scriptus* ...	*	
,, *semireticulatus* ...	*			,, *rinetensis* ...		
,, *spinulosus* ...				*Sedgwickia ovata* ...	*	
,, *undatus* ...	*			*Pinna flabelliformis* ...	*	
Spirifer glaber	*			,, *mutica* ..	*	
,, *lineatus* ...	*			*Actinopteria persulcata* ...		
,, *ovalis* ...	*			*Aviculopecten*	Sp. 4	
,, *striatus* ...						
,, *trigonalis* ...	*			**GASTEROPODA.**		
Spiriferina octoplicata ...	*			*Euima Phillipsiana* ...		
,, var. *cristata*	*			*Euomphalus pentangulatus*		*
Orthotetes crenistria ...	*	*		,, *circus* ...		*
Stropstomena analoga ...	*			*Loxonema le Febvrei* ...		*
				,, *rugifera* ..		*
LAMELLIBRANCHS.				*Loxeilia Puzo*		*
Allorisma sulcata ...	*			*Macrocheilina acuta* ...		
Cardiomorpha parva ...	*			,, *imbricata* ...		*
Clinopistha abbreviata ...	*			,, *rectilira* ...		*
,, *parvula* ...	*			*Murchisonia telescopium*		*
Conocardium aliforme ...				*Naticopsis ampliata* ...		*
Ctenodonta undulata ...	*			,, *plicistria* ...		
,, *levirostris* ...	*			*Pleurotomaria altavittata*		*
Edmondia arcuata ...	*			,, *interstrialis* ...	*	
,, *Lowickensis*	*			,, *decipiens* M'Coy		
,, *Maccoyi* ...	*			*Platyschisma zonites* ...		
,, *oblonga* ...	*			*Dentalium ingens* ...		*
,, *Pentonensis* ...	*			*Bellerophon decussata* ...	*	
,, *rudis* ...	*	*		,, *Urei* ...	*	
,, *sulcata* ...	*	*		*Conularia quadrisulcata* ...	*	

APPENDIX C—*continued.*

	Redesdale Ironstone and Limestone.	Four Laws Limestone at the Coomb.			Redesdale Ironstone and Limestone.	Four Laws Limestone at the Coomb.
CEPHALOPODA.				**CRUSTACEA—***continued.*		
Orthoceras Gesneri	*	*		*Griffithides longispinus*... ...		*
„ *annulatum*				*Phillipsia gemninlifera* ...		
„ *attenuatum*... ...				„ *Derbiensis*		
„ *cylindricum* ...						
„ *sulcatum*			**PISCES.**		
Glyphioceras truncatum				*Chomatodus* sp.... ...		
G. diadema	*			*Cladodus mirabilis*		
Stroboceras bisulcatus ...	*			*Gyracanthus tuberculatus* ...		*
Solenacheilus	*			*Petalodus Hastingsii* ...	*	
c.f. *S. crassirenter*		*		*Psammodus porosus* ..	*	*
c.f. *Vestinautilus crateriformis*		*		*Rhizodus* scales	*	
Acanthonautilus bispinosus		*				
Thringoceras Hibernicum ..		*		**CRINOIDS.**		
CRUSTACEA.				*Ulocrinus nuciformis*	*	
Dithyrocaris glaber	*			*Forbesiocrinus* sp.		
„ *Dunni*	*			*Scytalecrinus* sp.		
„ *tricornis*	*			*Archeocidaris Urei*		

My thanks are due to the Geological Society for permission to reprint Fig. 1, p. 452, and Fig. 2, p. 456.

ON THE FISH FAUNA OF THE PENDLESIDE LIMESTONES.

BY EDGAR D. WELLBURN, L.R.C.P.E., F.R.I.P.H., F.G.S., ETC.

INTRODUCTION.

Since the appearance of the valuable and interesting memoirs of Dr. Wheelton Hind, F.G.S., a great interest in these limestones has been aroused in Yorkshire and elsewhere, and the author having worked in and collected a good number of fish remains from these rocks, considers that the Fish Fauna of the beds may prove of interest, especially as some of the fish are new to science, whilst the fauna as a whole is of importance as bearing on the question of the stratigraphical position of the rocks.

As a whole the fauna is very similar to that of the Millstone Grits* above, whereas it is very dissimilar to that of the Yoredales (Phillips) of North-west Yorkshire, but this is only mentioned here, as it is the intention of the writer to discuss this question at length later, when he hopes to adduce facts strongly supporting Dr. Hind's theory of the age and position of these limestones.

REMARKS ON THE FISH REMAINS.

FAMILY CLADODONTIDÆ.

GENUS CLADODUS

CLADODUS MIRABILIS Agassiz, 1843.

Several teeth† of this species have been found in the black limestones at Crimsworth Dean, near Hebden Bridge, Yorkshire.

CLADODUS sp.

The writer has found teeth of *Cladodus* in the Pendleside Limestones at Astbury, near Congleton, Cheshire, and others have occurred

---

* Wellburn, Geol. Mag., Dec. IV., Vol. VIII., No. 443, p. 216, 1901.

† Wellburn, Proc. Yorks. Geol. and Polytec. Soc., 1901, p. 175.

APPENDIX C—*continued*.

	Redesdale Ironstone and Limestone.	Four Laws Limestone at the Coomb.		Redesdale Ironstone and Limestone.	Four Laws Limestone at the Coomb.
CEPHALOPODA.			**CRUSTACEA**—*continued.*		
Orthoceras Gesneri	*	*	*Griffithides longispinus*...	...	
„ *annulatum* ...	...		*Phillipsia gemniulifera*	...	
„ *attenuatum*...	...		„ *Derbiensis* ...	...	
„ *cylindricum*	...		**PISCES.**		
„ *sulcatum* ...	...		*Chomatodus* sp....	...	
Glyphioceras truncatum	...		*Cladodus mirabilis*	...	
G. diadema	...		*Gyracanthus tuberculatus*	...	•
Stroboceras bisulcatus ...	... *		*Petalodus Hastingsii* ...	...	
Solenucheilus	...		*Psammodus porosus* ..	...,	
c.f. *S. crassiventer*	...		*Rhizodus* scales	...	
c.f. *Vestinautilus crateriformis*			**CRINOIDS.**		
Acanthonautilus bispinosus ...	...		*Ulocrinus nuciformis* ...	...	
Thringoceras Hibernicum ..			*Forbesiocrinus* sp. ...	...	
CRUSTACEA.			*Scytalecrinus* sp. ...	...	•
Dithyrocaris glaber	*		*Archeocidaris Urei* ...	...	•
„ *Dunni*	*				
„ *tricornis*	*				

My thanks are due to the Geological Society for permission to reprint Fig. 1, p. 452, and Fig. 2, p. 456.

ON THE FISH FAUNA OF THE PENDLESIDE LIMESTONES.

BY EDGAR D. WELLBURN, L.R.C.P.E., F.R.I.P.H., F.G.S., ETC.

INTRODUCTION.

Since the appearance of the valuable and interesting memoirs of Dr. Wheelton Hind, F.G.S., a great interest in these limestones has been aroused in Yorkshire and elsewhere, and the author having worked in and collected a good number of fish remains from these rocks, considers that the Fish Fauna of the beds may prove of interest, especially as some of the fish are new to science, whilst the fauna as a whole is of importance as bearing on the question of the stratigraphical position of the rocks.

As a whole the fauna is very similar to that of the Millstone Grits* above, whereas it is very dissimilar to that of the Yoredales (Phillips) of North-west Yorkshire, but this is only mentioned here, as it is the intention of the writer to discuss this question at length later, when he hopes to adduce facts strongly supporting Dr. Hind's theory of the age and position of these limestones.

REMARKS ON THE FISH REMAINS.

FAMILY CLADODONTIDÆ.

GENUS CLADODUS

CLADODUS MIRABILIS Agassiz, 1843.

Several teeth† of this species have been found in the black limestones at Crimsworth Dean, near Hebden Bridge, Yorkshire.

CLADODUS sp.

The writer has found teeth of *Cladodus* in the Pendleside Limestones at Astbury, near Congleton, Cheshire, and others have occurred

---

* Wellburn, Geol. Mag., Dec. IV., Vol. VIII., No. 443, p. 216, 1901.

† Wellburn, Proc. Yorks. Geol. and Polytec. Soc., 1901, p. 175.

in a band of crushed limestone at Burnside and Thorpe Fell, Yorkshire, but unfortunately not in such a condition as to render the determination of species at all certain.

FAMILY CESTRACIONTIDÆ.
GENUS ORODUS.

ORODUS ELONGATUS Davis (ex Agassiz M.S.), 1883.

Several teeth* have occurred in the limestones at Crimsworth Dean, near Hebden Bridge, Yorkshire, and the writer has also found them in the limestones at Astbury, near Congleton, Cheshire.

FAMILY ACANTHODIDÆ.
GENUS ACANTHODES.

ACANTHODES sp.

The writer has found specimens of *Acanthodes* in the limestones at Astbury, near Congleton, Cheshire, and also at Pule Hill, near Marsden, Yorkshire, but the state of preservation of the specimens was not such as to render the determination of species at all certain.

FAMILY DIPLACANTHIDÆ.
GENUS MARSDENIUS gen. nov.

Generic characters :—Body fusiform, laterally compressed. Fins, especially the caudal, well developed. Teeth in the form of broad based blunt cones, confluent at the base. Clavicular bones strongly developed. Fin spines robust, the pectoral one being much elongated, remaining ones broad, robust, and ornamented with well-marked longitudinal ridges ; two dorsal fin spines, the posterior being longer than the anterior. Several scutes, or free spines, on the ventral aspect. Scales minute.

NOTE.—The writer has long had specimens of this genus in his cabinets, but a detailed description has only now been rendered possible by recent finds. The name is taken from the locality where the type was found, viz., Marsden, Yorkshire.

Type : Author's col.

Locality : Pule Hill, near Marsden, Yorkshire.

---

* Wellburn, op. cit.

RSDENIUS SUMMITI, sp. nov.

Type : Imperfect fish, author's col.

The best specimen (the type) is in a limestone nodule. It shows sh of about 10 cms. in length : the caudal and ventral regions are l shown, but the head being crushed back on to the dorsal region s portion of the fish is not so well shown. The body is fusiform l laterally compressed. Of the fins, the caudal is strongly developed, lobes being prominent and well marked ; the pectoral, ventral, l anal fins are fairly well seen and appear to have been well eloped. The dorsal fins are not seen here, but another specimen ich shows their fin spines points to the fact that the posterior was the larger of the two. All the fins are covered with scales ilar to those on the body, but smaller, and, with the exception of caudal, all are provided with spines. In position the anal fin :lose to the caudal, the ventral being about one-third nearer the toral than the anal. The pectoral fin spine is the largest and st elongated ; the ventral spine is about half the length of the toral ; the anal rather smaller than the ventral ; the posterior sal about the size of the ventral ; whilst the anterior dorsal spine only about half the size of the posterior. On the ventral surface the fish are several very small recurved free spines. The pectoral ne appears to have been smooth, with the exception of a single ove which runs parallel with its anterior border. The spine slightly curved ; the remainder of the spines are straight, robust, d ornamented with well-marked longitudinal ridges, which have following arrangement :—The anterior one or two are more ongly marked than the others, and run parallel to the anterior der, whilst the more posterior ridges run longitudinally in a lewhat irregular manner, gradually converging towards the re strongly marked anterior ones. The body is covered with es, which are very minute, and the superficial ornamental layer ing been most removed, the scales mostly appear smooth, but and there the sculpture can be seen by the aid of a powerful . It is as follows :—On the principal flank the scales are sculptured l deep grooves which traverse the scale in an anterior-posterior ction, parallel with the superior and inferior borders, the scale

surface being divided into a number of wide rounded ridges, and
the grooves cutting deeply into the posterior margin gives it a scalloped
appearance. On the posterior flank the scales appear to have been
smooth. The head bones are not well seen in any specimen, but
what evidence there is shows that they were of the ordinary Acan-
thodian type. The bones of the shoulder girdle were strongly
developed, and are very similar in form to those of Parixius.

Form and Loc.: Pendleside Limestones, Pule Hill, near Marsden,
Yorkshire.

MARSDENIUS ACUTA sp. nov.

Type : Portion of fish, author's col.

There are several specimens of this fish in the author's cabinets.
They show the same general characters as the last of the type species,
but a much smaller fish is indicated ; the type specimen would point
to a fish of about 7 cms. in length. The fish differs, however,
from the last species in the character and ornamentation of its fin
spines, and also the sculpture of its scales. The pectoral fin spines
are more elongated and slender ; they gradually taper to a fine point,
and have a groove and ridge running parallel to the anterior border.
The other spines are similar to those of the last species, with the exception that the ridges which sculpture the spines are relatively
much finer and more numerous and run in a more irregular manner.
The clavicular bones are very strongly developed, and one specimen
in the author's collection is of great interest, as it shows the form
and characters of the mandible, with its dentition. The mandible
is very similar in its general characters to that of the fish *Acanthodopsis Wardi* Egerton of the Coal Measures. The dentition consists
of well-marked, blunt, broad, low cones, confluent at their bases.
The scales are very minute, and here and there a scale sculpture
of very fine transverse striæ may be made out by the aid of a powerful
lens.

The name "acuta" is given to the fish on account of the pointed
character of its pectoral fin spines.

Form and Loc.: Pendleside Limestones, Pule Hill, near Marsden,
Yorkshire.

MARSDENIUS sp. ?

The author has collected and seen specimens of this genus, but unfortunately not in such a condition as to render the determination of species at all certain, from the Pendleside Limestones of the following localities, viz.:—Pule Hill, Marsden, Yorkshire ; Dane Valley, Staffs, (half a mile above the bridge, and immediately below the Kinderscout bed of the Millstone Grits).

Remarks :—The above specimens, in certain characters, appear to indicate that the species is new, but the writer deems it better to wait for more perfect specimens.

Remarks on the genus *Marsdenius* :—From the foregoing remarks it will be seen that the conformation of the mandible and the character and arrangement of the teeth, and also that of the pectoral fin spines of *Marsdenius* is very similar to that of *Acanthodopsis Wardi* Egerton of the Coal Measures, whereas, on the other hand, the general characters and arrangement of the fin spines as a whole, and the great and characteristic development of the clavicular bones, also the seutes or ventral free spines, strongly ally the fish to many of the Diplacanthidæ, and this being so the author has ventured to place the genus in that family.

FAMILY RHIZODONTIDÆ.

GENUS RHIZODOPSIS Young, 1866.

RHIZODOPSIS SAUROIDES Williamson, 1837.

There are several fragmentary specimens of this fish in the collections of Mr. Barns, F.G.S., Higher Broughton, Manchester, and also in that of the author.

Loc. : Pule Hill, Marsden, Yorkshire.

GENUS STREPSODUS Young, 1866.

STREPSODUS SAUROIDES Binney sp., 1841.

Fragmentary remains in the collections of Mr. Barns and the author.

Loc.: Pule Hill, Marsden, Yorkshire.

N

FAMILY COELACANTHIDÆ.

GENUS COELACANTHUS Agassiz, 1844.

CŒLACANTHUS HINDI sp. nov.

Type : Portion of fish, author's col.

The specimen shows several of the bones of the head and the anterior portion of the body. Of the head bones most are seen from the inner surface, but they appear to have been ornamented with ridges, which run more or less parallel to the borders of the bones. On the operculum—which is half as high as broad—there is an indication of faint, more irregular ridges and granulation between the principal ridges, whilst on the jugular plates the ridges appear to have run in a fairly regular manner. The latter element (jugular) gradually tapers anteriorily to a fine point. Behind the operculum are a number of long slender rays, which are probably the remains of the pectoral fin. Further back are a number of neural spines, which show the usual characters of Cœlacanthus. Well-marked and characteristic branchio-stegal rays (detached) are also shown. The body is covered with large, much rounded scales, the posterior or exposed portion of which is ornamented in a highly characteristic manner, with strongly marked ganoine coated ridges, which run in a very regular manner in half circles parallel to the rounded posterior border.

Remarks :—The form and very striking ornamentation of the scales are so dissimilar to that of any known cœlacanth that I treat the fish as new, and propose the specific name *Hindi*, after Dr. Wheelton Hind, F.G.S., of Stoke-on-Trent, on account of the very valuable work which he has accomplished in respect to the Pendleside Limestones. Mr. J. Ward, F.G.S., who has seen the specimen, also regards it as new.

Loc.: Bank of River Hamps. near Waterhouses, Staffs.

FAMILY PALÆONISCIDÆ.

GENUS ELONICHTHYS Giebel, 1848.

ELONICHTHYS OBLIQUUS Wellburn

There are many more or less imperfect specimens of this fish in the author's collection, but as the type was found in the Mill-

ne Grits, he purposes to describe it later, along with some other
w fishes from these rocks.

Loc.: Pule Hill, near Marsden, and Todmorden, both in York-
ire.

LONICHTHYS AITKENI Traquair.

Fragmentary remains of this fish have been found in the following
calities :—Pule Hill and Crimsworth Dean, both Yorkshire.

LONICHTHYS sp. ?

One nodule in the author's collection shows a mass of scales
a small *Palæoniscid* fish which, from the general characters of
e scales, should be placed in this genus, but on account of the small
e of scales it is very difficult to make out their sculpture, and so
say anything definite about the species. By the aid of a powerful
is the following ornamentation may be made out on some of the
ger flank scales, viz.: There are many closely-arranged fine ridges
striæ which, commencing at the superior border, run down
e scale parallel to the anterior border, then turning above the
terior inferior angle, they sweep across the scale towards the posterior
rder, which is entire. Some of the smaller, presumably posterior
nk scales, are smooth.

Remarks.—The scale sculpture appears to differ from that
any known species, but it seems to be better to defer the deter-
nation of species until later, in the hope that better specimens
iy be found.

Loc. : Pule Hill, Marsden, Yorkshire.

GENUS RHADINICHTHYS.

Some small, more or less smooth, scales I place in this genus
m their general resemblance to the smoother scales of the fish
RADINICHTHYS MONENSES Egerton.

Loc.: Dane Valley, Staffs.

GENUS ACROLEPIS.

ROLEPIS HOPKINSI McCoy.

Fragmentary remains.

Loc.: Crimsworth Dean and Pule Hill, Yorkshire ; Dane Valley,
affs.

FAMILY PLATYSOMIDÆ.

GENUS PLATYSOMUS.

The writer has in his collection a scale which he found in a limestone nodule at Whitewell, near Clitheroe, which certainly belongs to the above genus. The scale is high, narrow, with a well-developed articular peg ; in fact, in general form and characters it very closely resembles many of the upper and lower flank scales of *Platysomus fosteri* Handk. and Atthey, of the Coal Measures. The sculpture consists of fine, regular, closely-arranged, oblique ridges.

I cannot conclude without expressing my thanks to Dr. Wheelton Hind, F.G.S., and Mr. Barns, F.G.S., for having granted me the privilege of examining the specimens in their collections.

TABLE SHOWING THE DISTRIBUTION OF THE FISH REMAINS IN THE PENDLESIDE LIMESTONES.

	Pule Hill, Marsden, Yorkshire.	'Himsworth Dean, near Hebden Bridge, Yorkshire.	Whitewell, near Clitheroe, Yorkshire.	Todmorden, Yorkshire.	Burnsall and Thorpe Fell, Yorkshire.	Bank of River Hodder, near Stoneyhurst, Lancashire.	Dove-holes Tunnel, near Barton.	Bank of River Hamps, near Waterhouses, Staffordshire.	Dane Valley, Staffordshire.	Astbury, near Congleton, Cheshire.
Cladodus mirabilis Agassiz.		*			*					
Cladodus sp.?	*	*				*				*
Orodus elongatus Davis	*						?			*
Acanthodes sp.	*									*
Marsdenius summiti gen. et sp. nov.	*									
Marsdenius acuta sp. nov.	*								*	
Marsdenius sp.?	*									
Strepsodus sauroides Binney	*									
Rhizodopsis sauroides Williamson	*	*		*						
Coelacanthus Hindi sp. nov.										
Elonichthys Aitkeni Traquair	*			*				*		
Elonichthys ?sp Wel Mrn	*									
Elonichthys sp.?	*								*	
Rhadinichthys sp.?		*								
Acrolepis Hopkinsi McCoy	*	*	*						*	
Platysomus sp.?										

ON THE GENUS CŒLACANTHUS AS FOUND IN THE YORKSHIRE COAL
MEASURES, WITH A RESTORATION OF THE FISH.

BY EDGAR D. WELLBURN, L.R.C.P.E., F.R.I.P.H., F.G.S., ETC.

INTRODUCTION.

The author having collected a very large number of specimens
of these fishes from the Yorkshire Coal Measures, many being nearly
perfect, whilst several of the species are new to science, he thought
that it might be of interest to place on record the result of his finds,
especially as they have enabled him to complete a restoration of
the fish, showing nearly the whole of its anatomy ; and it is further
worthy of note that one of his specimens shows the internal skeleton
of the lobe of the pectoral fin,* and enables him to confirm Dr. A.
Smith Woodward's† description of the anatomy of this fin, and this
is of particular interest, as, besides the author's, there is only one
other specimen from the Talbragar Beds (Jurrassic ?), New South
Wales, which throws any light on this interesting and important
point.

GENERAL ANATOMY OF THE FISH.

FORM AND PROPORTIONS.

In form the fish is deeply and irregularly fusiform. Of the
total length the head occupies one-fourth, the body two-fourths,
and the tail the remaining fourth. The greatest depth of the body
is at a point immediately posterior to the first dorsal fin, the depth
here being about one-fourth the total length of the fish.

HEAD.—EXTERNAL ANATOMY.

The head is peculiarly characteristic in form, one of the chief
peculiarities being the slope of the cranial roof bones. The cranial

---

* Wellburn, Geol. Mag., Dec. IX., Vol. VIII., No. 440, p. 71.

† Woodward, A. S., Mem. Geol. Survey of New South Wales, Palæontology
No. 9, 1895.

roof is divided into a posterior or parieto-occipital (Pa.) and an anterior or frontal (Fr.) portion. These meet at an obtuse angle. The parieto-occipital moietie runs parallel to the base of the skull, whilst the frontal runs downwards to the snout, which is blunt and rounded, but none of the many specimens examined by the writer show any of the sutures by which this ethmoidal (E.) region was probably subdivided.

The parieto-occipital region comprises a pair of large bones (Pa.), the parietals. These meet in the middle line, and are flanked postero-externally by a pair of triangular bones (Sq.), which appear to represent the squamosals fused with the post temporals. The frontal region comprises a pair of long narrow bones, which are divided down the middle line by a suture, and are flanked on each outer margin by a series of quadrate membrane bones—the para frontals (Pa. Fr.)—whilst on each side, immediately posterior to the transverse suture which divides the bones of the cranial roof, are two small bones—the posterior frontals (P.F.).   The orbit was large, prominent, and surrounded by a ring of delicate sclerotic plates, and was situated at a point about the junction of the anterior and middle thirds of the length of the cranial roof. Above are the para frontals ; in front there is a triangle-shaped bone which probably represents the fused anterior frontal and anterior orbital (A.O.) ; behind there is an irregularly shaped bone—the posterior orbital (P.O.). On the cheek, behind and below the latter bone, there are two triangular shaped elements, which appear to be the equivalent of the cheek plates of *Megalichthys* and *Rhizodopsis*. The uppermost of these two plates (X and X[1]) is deeply triangular, and about twice the size of the lower one. From the anterior border of the larger cheek plate (X), commencing at a point a little above its anterior inferior angle, a long narrow sub-orbital element (S.O.) runs forward below, then circling upwards in front of the orbit, joins the cranial roof bones above, whilst below this latter a long narrow bone, ornamented on its external surface, runs straight forward to the snout, from a point at the anterior inferior angle of the upper cheek plate. This latter bone (Mx.) is considered by Zittel and Reis to be the palatine, but, as pointed out by Huxley, " it has more the

appearance of an external element." Again, in the author's collection
there are specimens which to him prove that the bone was external,
and that it is the maxilla. Of the premaxilla (P.Mx.), its characters
are somewhat doubtful, as it is not well shown in any specimen,
the head here being generally seen in a crushed condition. The
mandible is well seen in many specimens, both *in situ* and detached.
Its structure is somewhat complex, and it consists of the following
parts, viz., the greater portion of each mandibular ramus is formed
by a long, narrow, articulo-angular element (Art. An.), ornamented
on its external surface ; its inferior margin is nearly straight, its
superior arched in advance of the articular facette, behind which
there is a short extension. A small toothless dentary element (D.)
meets this in front, reaching to the symphysis, and bounded below
by a thin infradentary (I.D.). A long, deep, lamina sphenial bone,
tapering in front, but with a straight dentigerous border in the
greater part of its length, is opposed to the dentary and articular-
angular on their inner face, this forming the wall of a vacuity existing
between the upper portion of the two outer elements.

The branchio-stegal apparatus is represented by a pair of large
opercular (Op.) bones, which fill in a more or less triangular space
between the parietals, the cheek plates, and the shoulder girdle,
and by two elongated jugular (J.) plates which fill in the space in
the gular region between the mandibular rami.

INTERNAL CRANIAL ANATOMY.

The chondrocranium is extensively ossified, but there is no
interorbital septum, and the base is formed by a long slender para-
sphenoid bone, which exhibits a spatulate expansion anteriorily.
The hyomandibular and pterygo-quadrate arcade are fused into
a continuous triangular, lamelliform bone on each side, articulating
with the hinder portion of the cranium above, and is below provided
with a ginglymoid condyle for the articulation of the mandible.
The bone terminates in an attenuated angle in front, and its superior
portion inclines inwards to form the roof of the mouth. This surface
is finely granular, whilst the outer surface is smooth. In front of
the pterygo-quadrates are a pair of small palatine bones, and in

lvance of these is a large robust azygous element, which probably presents the coalesced vomers. Its surface is covered with numerous lall tubercules, which form a dense rasp-like surface. (This is ell seen in one specimen in the author's collection.)

DENTITION.—One detached sphenial bone shows on its upper rder a few irregularly arranged, detached, low, blunt, ill-defined vations, which may possibly have had a dentary function, but th the exception of the rasps of granules mentioned above, the iter has seen no evidence of teeth in these fishes, and he considers to be highly probable that these rasp-like surfaces constituted eir dentition, at any rate in the Coal Measure species.

CERATOHYALS.—These were robust and connected on each le to the hyomandibular by elongated elements, which probably present the stylohyals.

BRANCHIAL ARCHES.—There are about five on each side, which e delicately and deeply channelled on the hinder aspect, as in *lypterus*. Each arch consists of a pair of much arcuated elements, hilst a single large copula, with a spatulate hinder extremity, unites l the lower extremities of the arches in the median line.

AXIAL SKELETON.—The axial skeleton extends beyond the caudal l to form a supplemental caudal fin. The notochord must have en persistent, as its situation is always—so far as the writer has en—represented by a blank space. Reis,[*] however, mentions rpocentra as occurring in *C. hassiæ*. The neural arches are slender, ld the two halves of each arch are firmly joined to the neural spines, hich are long and slender. The hæmal arches are similar to the ural, but in the abdominal region their appended spines are short ld rudimentary, whilst posteriorily, in the caudal region, they rrespond in development to the neural spines.

BODY.

The body is covered with deeply overlapping cycloidal scales, e exposed posterior portion of which being ornamented with noine, whilst the anterior covered portion is smooth, and shows ne concentric lines of groth.

[*] Die Cœlacanthinen—Palæontographica, Vol. XXXV., 1888.

THE LATERAL LINE is very rarely seen. It probably runs (as shown by Reis[*]) from a point near the junction of the clavicle with the supra-clavicular bone to a point well back on the body prolongation of the caudal fin.

PECTORAL ARCH.—The membrane bones of the shoulder girdle are, although slender, always conspicuous, and seem to have been covered by the skin. There are a pair of long, slender, gently curved clavicles (CL), which exhibit a robust post-clavicular process (P.Cl.Pr.). They articulate above with a small supra-clavicle (S.Cl.), while a long, slender infra-clavicle (I.CL) overlaps its lower spatulate extremity. The latter element curves sharply forwards and inwards, terminating in a triangular expansion where it meets its fellow of the opposite side in a median suture.

FINS.

The paired fins are well developed and obtusely lobate.

PECTORAL FIN.—A specimen in the writer's collection shows that the internal skeleton of the lobe of this fin consists of several superficially ossified basal supports (Ba.S.), which are jointed at their proximal extremities to the post-clavicular process (P.Cl.Pr.), whilst distally each is opposed to the proximal ends of one or more of the dermal rays (D.R.) of the supports. The anterior four are elongated and more or less uniform in thickness. The fifth is more hourglass-shaped, while the sixth is more robust and widely expanded distally. The dermal rays of the fin increase in length from the anterior border to the middle of the lobe, whence they decrease backwards, and finally become extremely fine. All the rays are slender and closely articulated distally.

PELVIC FINS.—These fins are supported by a pair of basipterygia (Ax. 4), having the following characters, viz., distally, where they would join the basal supports, they are broad and expanded the proximal half form a thin, more or less triangular expansion the expansion being strengthened by three thickenings, one being central and two lateral, while springing from the centre of the bone are two inwardly directed precesses which are loosely apposed in

---

[*] Op. cit.

ıe middle line. The lobe of this fin is always (as far as known) ¡presented by a blank space, the baseosts having been destitute , or very slightly ossified. The dermal rays are similar in character ɩ those of the pectoral fin.

UNPAIRED FINS.

With the single exception of the first dorsal, the fins are lobate, ιe lobe being, however, more acute than that of the paired fins.

FIRST DORSAL FIN.—This fin shows no lobation as its dermal rays 'e opposed (as is well shown in a specimen in the author's collection) by ɩeir proximal ends, to the upper surface of a stout, well-ossified :onost (Ax. 1), whose upper end is somewhat expanded, whilst its wer is forked. The dermal rays are more robust and fewer in ımber than those of the other fins, but otherwise they show similar ɩaracters.

SECOND DORSAL FIN.—This fin, as well as the anal, has a forked :onost (Ax. 2), which is, however, less robust than that of the first ɩrsal, and each fin is (as mentioned above) somewhat acutely lobate, ɩe lobe showing as a blank space in the fossil state, the baseosts ɩving been unossified, or very slightly so. The dermal rays of both ɩs are similar in characters to those of the paired fins. In position ɩe first dorsal fin is opposite to the space between the pectoral and ɩntral fin, the second dorsal opposite to the space between the ventral ɩd anal fins ; the latter fin arises close to the caudal.

CAUDAL FIN.—This fin is always a conspicuous feature of these shes ; it is composed of a strongly-developed, symmetrical, principal ɩ (C F), and a small, feebly-developed, supplemental one (S C F). he principal caudal is supported above and below by a series of ng, slender interspinous bones (I.S.), which equal, and are directly ɩposed to, the blunt distal extremities of the neural and hæmal ɩines of the axial skeleton. A single stout dermal ray is connected ith the distal end of each of these elements by an overlapping ·ticulation. The sparse dermal rays of the supplemental caudal fin ɔpear to be in direct contact with the spines of the axial skeleton.

SWIM-BLADDER.—In the abdominal region of these fishes the ɩsified air or swim-bladder (S.W.) is always a conspicuous feature ;

it was of large size, extending from a point immediately posterior to the pectoral fins to a point a short distance anterior to the anal fin. It sometimes shows a single anterior aperture, by which its internal cavity communicated with the œsophagus. Its walls are formed of a longitudinal series of large, imbricating, bony laminæ, composed of superposed lamellæ. The inner face exhibits (as pointed out by Von Zittel), and as shown by specimens in the author's collection, a large reticulating rugæ, very similar to the network (made known by Owen) in the lung-like air-bladder of the recent fish *Polypterus*, and it appears to the writer to be highly probable that the air-bladder of *Cœlacanthus* played a similar, if not identical, part in the economy of the fossil fish to that played by the lung-like air-bladder of the recent fish, and this seems to be rendered more certain when we study the life-history of the fishes, as both seem to have thrived best, and to have been most plentiful under the same muddy surroundings : again it appears likely that the air-bladder of *Cœlacanthus*, as well as *Polypterus*, was able, under certain dried-up conditions, to perform for the time being the functions of a lung.

DISTRIBUTION AND RANGE.

Cœlacanthus has a very wide distribution in the Yorkshire Coal Measures, being found in nearly all the fish-bearing localities : in fact, it is by far the most characteristic fish in these Measures. It has a range from the Halifax Soft Bed Coal at the base of the Lower Coal Measures to the Stanley Scale Coal, one of the uppermost beds in the Middle Measures, i.e., to one of the uppermost coal seams found in Yorkshire.

REMARKS.

The vast number of specimens which have enabled the writer to complete the restoration have been mainly found in the cannel coal, at Tingley, near Leeds, where the fish was, until quite recently, found in great abundance and perfection, but now, unfortunately, owing to the non-working of the cannel coal in this locality, specimens are not to be found.

Another point may be mentioned here, viz., that the Yorkshire Measures have yielded eleven different species (see table of distribution).

TABLE SHOWING THE DISTRIBUTION OF CŒLACANTHUS IN THE YORKSHIRE COAL MEASURES.

	LOWER MEASURES					MIDDLE MEASURES					
	Halifax Soft Bed, Halifax.	Halifax Hard Bed, Halifax.	Better Bed Coal, Low Moor and Wyke.	Black Bed Coal, Low Moor.	Crow Coal, Leeds.	Cannel above Middleton Main, Tong, near Bradford.	Cannel Coal, Tingley.	Thick Coal, Barnsley.	Cannel above Thick Bed, Barnsley.	Kent Thick Coal, Barnsley.	Stanley Scale Coal, Wakefield.
Cœlacanthus Tingleyensis Davis			*		*		*				
,, *elegans* Newberry	*	*	*	*	*	*	*	*	*	*	*
,, *Phillipsi* Agassiz		*									
,, *robustus* Newberry			?								
,, *granulostriatus* Wellburn			*								
,, *dielans* Wellburn			*								
,, *tuberculatus* Wellburn			*								
,, *spinatus* Wellburn			*								
,, *corrugatus* Wellburn			*					*			
,, *Woodwardi* Wellburn			*								
,, *elongatus* Huxley							*				

six of which are new to science. It had been the intention of the writer
to figure and describe these latter, but, unfortunately, owing to lack
of time it has been impossible to do so. He hopes to describe them
later in a separate paper.

EXPLANATION OF FIGURE.

Pa.	Parieto-occipitals.	Cl.	Clavicle.
Fr.	Frontals.	I.CL	Infra-clavicle.
E.	Ethmoidals.	N.C.I.	Notochordal inter-
Or.	Orbit.		space.
Sc.Pl.	Sclerotic ring.	N.Sp.	Neural spines.
A.O.	Anterior Orbital and	H.Sp.	Haemal spines.
	anterior frontals	I.S.	Interspinous bones.
	(fused).	D.R.	Dermal rays.
P.O.	Posterior Orbital.	P.F.	Pectoral fin.
S.O.	Sub-orbital.	V.F.	Ventral fin.
P.F.	Post frontal.	A.F.	Anal fin.
Sq.	Squamosal and post-	D.F. 1st.	First dorsal fin.
	temporals (fused).	D.F. 2nd.	Second dorsal fin.
Mx.	Maxilla.	C.F.	Caudal fin.
P.Mx.	Premaxilla.	S.C.F.	Supplemental caudal
X. and X²	Cheek-plates.		fin.
Op.	Operculum.	Ba.	Basal supports.
Art. An.	Articulo-angular.	Ax. 1 2 3 and 4	Axonosts.
D.	Dentary.	S.W.	Swim-bladder.
I.D.	Infradentary.	N.W.	Neural arches.
J.	Jugular.	H.A.	Haemal arches.
S.Cl.	Supra-clavicle.		

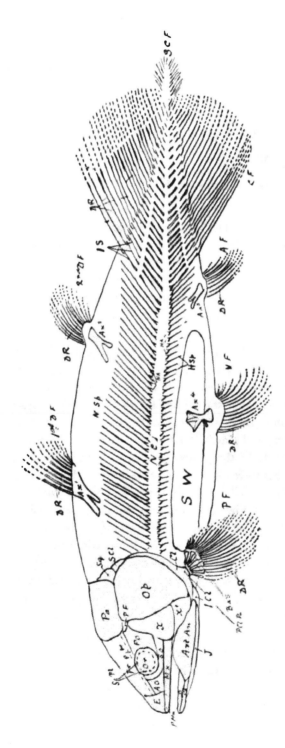

A RESTORATION OF CŒLACANTHUS FROM SPECIMENS COLLECTED IN THE YORKSHIRE COAL MEASURES.

Edgar D. Wellburn del.

A STRIATED SURFACE AT SANDSEND.

BY JOHN W. STATHER, F.G.S.

(Read November 14th, 1901.)

Though glacial beds cover the solid rocks of the Yorkshire coast from Redcar to Bridlington, glaciated surfaces are rare. This arises from the texture and comparatively soft nature of the Jurassic and Cretaceous strata, of which East Yorkshire is built. Hard bands do, however, occur here and there in the Oolitic beds, and in two localities, Filey and Robin Hood's Bay, striated surfaces have been observed upon them. The object of the following note is to record an additional locality.

Sandsend is a small fishing village, picturesquely situated on the Yorkshire coast, two and a half miles north of Whitby, at the point where the Sandsend beck enters the sea. South of the beck, the high cliffs overlooking the sea are for the most part composed of boulder clay and gravel, and there is evidence that in the neighbourhood of Upgang the drifts fill in a pre-glacial valley, believed by Mr. G. Barrow to be the pre-glacial channel of the Esk. North of the beck the high ground consists of Jurassic rocks covered by drift. These beds are well exposed on the coast, and can also be seen in the series of old quarries connected with the abandoned Sandsend alum works. Throughout these old workings the "Dogger" (Oolite, a very variable bed) overlies the alum shale, and is represented by a conspicuous band of evenly-bedded, hard, ferruginous sandstone, thirty feet thick, flaggy towards the bottom, but massive and exceedingly hard at the top. The boulder clay which overlies the "Dogger" is eight feet thick, and is of the usual East Coast type.

In October, 1899, the drift sections exposed along the edge of these old quarries were examined by the writer, and at a point overlooking the "Deep Grove" quarry, 250 feet above ordnance datum, indications of a glaciated surface were seen. A few square

⸋ards of the upper surface of the "Dogger" sandstone had, for
quarrying purposes, been cleared of its covering of drift, and on
the hard sandstone surface thus exposed striæ and other indications
of glacial action were clearly visible. One slab of rock, about a yard
square, washed clean by the rain, showed the striæ particularly
well, and the whole of the adjacent surface, so far as it could be seen,
had the rounded and smoothed appearance peculiar to glaciated
rock. The direction of the striæ was 35 degrees west of north.

OTHER LOCALITIES.

Perhaps it may be of interest, in connection with the above
note, to recall a few particulars regarding the striated surfaces
previously recorded on the Yorkshire coast.

In 1891 Mr. G. W. Lamplugh recorded striæ on the *south* side
of Filey Car Naze. Direction of striæ, 20 degrees east of north.

In 1896 Messrs. Sheppard and Muff described a striated surface
at Robin Hood's Bay. Direction of striæ, exactly north.

In 1896 the writer recorded a large striated surface at Filey
on the *north* side of the Car Naze. Direction of striæ, 24 degrees
east of north.

STRIATED SURFACES ON THE YORKSHIRE COAST.

Locality.	Height above O.D.	Direction of Striæ.
SANDSEND (*8 miles N.W. of Robin Hood's Bay*) ...	250 feet.	35° west of north.
ROBIN HOOD'S BAY (*20 miles N.N.W. of Filey*)... ...	400 feet.	Exactly north.
FILEY	40 feet.	24° east of north.

INFERENCES.

From the above it will be seen that the direction of the striæ
varies with the locality in a very significant way. And the variation
is the more remarkable when it is remembered, that at each of the
three localities the direction of the ice-flow, as shown by the markings,
is from a quarter that is now the open sea ; and that, although the

o

adjacent land seems to have had some influence in deflecting the movement, the course of the main ice-stream has been almost parallel to the coast, and therefore transverse to the slope of the land. This direction would have been impossible unless the North Sea basin had been choked with ice to a level somewhat higher than that of the adjacent shores. It is clear from the striæ that the general flow has been southward along the sea-floor, but with a constant tendency to creep in westward upon the coast. It is also clear that the upland buttress of the Oolitic rocks, which has its eastern corner on the coast at Whitby, has caused the ice-flow to swerve through an angle of 60 degrees in rounding it.

These conclusions are strongly confirmatory of the views as to the "East British" ice-sheet which Mr. Kendall has deduced, from the study of the glacial marginal phenomena of the interior of the Oolitic uplands, views which are too well known to this Society to need recapitulation here.

REFERENCES.

G. BARROW. Geological Survey Memoir. The Geology of North Cleveland, p. 69.

G. W. LAMPLUGH. "The Larger Boulders of Flamborough Head." Proc. Yorks. Geol. and Polyt. Soc., Vol. XI., p. 401.

T. SHEPPARD and H. B. MUFF. "Notes on the Glacial Geology of Robin Hood's Bay." Glacialist's Mag., Vol. IV., p. 52.

J. W. STATHER. "A Glaciated Surface at Filey." Proc. Yorks. Geol. and Polyt. Soc., Vol. XIII., p. 346.

NOTES ON THE IGNEOUS ROCKS OF THE ENGLISH LAKE DISTRICT.

BY ALFRED HARKER, M.A., F.R.S., F.G.S.

I.—THE ORDOVICIAN VOLCANIC SERIES.

This series is divisible stratigraphically and petrographically to several distinct groups, the distribution of which in the district s been indicated by Mr. Marr.[*]

(a) *Falcon Crag Andesite Group.*—These rocks are found chiefly in the country south of Keswick and eastward to the St. John's Vale. They are pyroxene-andesites, and thin slices show under the microscope sometimes hypersthene (converted to bastite), sometimes augite, sometimes both. To the eye the rocks are pale greenish to dark grey, with a compact ground-mass, usually containing scattered minute porphyritic felspars, or more rarely abundant crystals up to ¼ inch in length. Many of the flows are amygdaloidal. There are rare flows of more basic lavas, besides some tuffs.

(b) *Eycott and Ullswater Basalt Group.*—The most widely distributed group of all. The basalts, like the andesites, often contain altered hypersthene; they are sometimes rather rich in magnetite, but olivine does not occur. These rocks are usually darker and denser than the andesites, but a discrimination is not always possible in the field. Part of the rocks are porphyritic, and in some there are conspicuous crystals of felspar (bytownite or labradorite) which on Eycott Hill reach a diameter of an inch or more. Amygdaloidal varieties are found. Basic tuffs are sometimes intercalated among the lavas, and in some places there are andesitic flows included in this group.

(c) *Scawfell Tuff and Breccia Group, with Kentmere-Coniston Slate-Band.*—This group presents itself under two different phases. (1) In the central mountains it consists mainly of

_____

[*] Proc. Geol. Assoc., 1900, Vol. XVI., pp. 453-459, with map (Plate XIII.)

well-banded fine tuffs and v
are finely laminated and som
they have been converted into
several contain fragments of
are then of coarse texture.
vary. About Sty Head and s
able thickness of breccia is chara
of pink compact rhyolite: in o
prevail. Both tuffs and bree
garnets easily visible to the
metamorphic origin: but som
the district also contain garne
a primary constituent. Garn
are especially frequent in a pu
immediately below the Scaw
by Mr. Marr to be intrusive:
in this place. (2) The fragment
reappears further south in a b
is extensively quarried for slates
as regards their general appear
Most of the rocks of this group
both basic and acid varieties a
the preponderance of one or ot
the constituent fragments.

Shap Andesite Group.—This high
a strip running from Torver.
granite. The rocks are an
hypersthene being apparently

Shap Rhyolite and Yewdale B
forms a strip immediately sou
rhyolites are of very compact a
any conspicuous porphyritic
quartz). They are sometimes g
sometimes duller with a pink or
scopic structure is variable,
secondary changes; some var

are microspherulitic ; some show partial replacement by secondary quartz. In places, and especially at Great Yarlside, nodular varieties occur, the nodules often being one or two inches in diameter and sometimes six inches. They represent large spherulites (and probably in some cases lithophyses), which have been greatly altered, the original structures being obscured or obliterated, and the spherulites partly or wholly replaced by quartz and other substances. In places there is a considerable thickness of tuffs and breccias at the base of the rhyolite group. These fragmental accumulations are not all of acid composition, and they differ from the tuffs, &c., of the Scawfell group in containing sometimes a certain amount of detrital material, chiefly sand-grains ; while some beds again are calcareous tuffs.

II.—INTRUSIVE ROCKS, OLDER SUITE.

The rocks included here are closely associated with the Volcanic ries, and probably belong to the same great period of igneous ivity, though their intrusion succeeded the extrusion of the volcanic ks. The age of the intrusions is, however, a matter of inference her than demonstration, and much more information is desirable. is at least certain that a considerable number of the rocks are er than the epoch of the principal crust-movements.

a) *Granophyres.* The Ennerdale and Buttermere granophyre, occupying a considerable area on the west side of the district, is a pink fine-textured rock with indistinct quartz-grains and crystals of felspar. Thin slices show it to consist mainly of micropegmatite. There are chloritic pseudomorphs after augite and biotite. A number of smaller intrusions of granophyre (most of them, if not all, augitic) occur in the district. Some, such as that of Blea Crag, Langstrathdale, contain small garnets. The well-known Armboth and Helvellyn dykes, also garnetiferous in places, are microspherulitic rocks, with porphyritic crystals of quartz and felspar which serve as nuclei for the spherulitic growths.

b. *Microgranites.*—A well-known rock of this type forms two conspicuous masses at the foot of St. John's Vale, and is quarried at Threlkeld. This is a fine-textured grey rock with small porphyritic felspars scattered through it. These are microcline. Flakes of biotite are also present, and the general mass of the rock is a microcrystalline aggregate of quartz and felspars. Smaller intrusions of microgranite, with dykes and sills, are found in several parts of the district.

c. *Quartz-porphyries and Quartz-porphyrites.*—These are included together because it is scarcely possible to draw any line between the truly acid and the sub-acid rocks. Dykes and sills are found at numerous localities, and occasionally more irregular intrusions (e.g. Wansfell). The rocks are in nowise remarkable. Small crystals of quartz and felspar are visible in a compact grey or pink ground-mass, which under the microscope may be either microcrystalline or cryptocrystalline. The ferro-magnesian element, which is never abundant, may be biotite or sometimes augite, and both are often chloritised.

d. *Dolerites or Diabases.*—The distribution of these rocks is sufficiently indicated on the Geological Survey Map. They form dykes and sills, and exceptionally a small boss (e.g. Castle Head, Keswick). They are mostly ordinary dolerites without olivine, often considerably altered with production of chlorite, &c. Usually they appear in the field as dull, dark-coloured medium to fine-grained rocks, without porphyritic elements. The Castle Head rock, however, has porphyritic augites, and is in places micaceous.

e. *Basic Picrites or Olivine-Picrites.*—These rocks are of limited distribution, forming irregular, sheet-like or laccolitic masses at Little Knott and Great Cockup in the Skiddaw Slate area. They have been described as hornblende-picrites and are black, rather coarsely crystalline rocks, composed mainly of hornblende. Thin slices reveal spots which probably represent destroyed olivine.

III.—INTRUSIVE ROCKS, YOUNGER SUITE.

The granites of the Lake District, with their related dykes, are
e referred to the Old Red Sandstone period or to the interval
wing the Silurian, and their intrusion seems to have been con-
ed in some way with the crust movements which have imparted
uliar character to the district.

) *Granites.*—There are three considerable masses of granite, and
these present very different petrographical characters. The
Skiddaw granite is seen in three distinct areas, which are
inliers probably of a large concealed mass. In the southern
area (in Sinen Gill), and the middle and largest one (in the
Caldew Valley), the rock is a biotite-granite, with only occasionally a little white mica ; in the northern area (near the
junction of Grainsgill with the Caldew) both micas are essential
constituents. The normal rock is a medium-grained granite
of light grey colour. In Grainsgill, however, it gives place to
a greisen, composed essentially of quartz and white mica.
The Eskdale and Wastdale granite is a moderately coarse
rock, either reddish or grey, with dark mica only. In thin
slices it is found that microline and microperthite often play
an important part, and in some varieties of the rock micropegmatite. The third granite, that of Shap, forms a plug-
like mass, not, like the other two, an irregular sheet. It is
a biotite-granite with large red crystals of orthoclase in a
medium-grained ground. Thin slices show that sphene is
rather abundant. The rock is never micropegmatitic. A
characteristic feature of the Shap granite is the frequent
occurrence in some places of ovoid patches, an inch or two in
diameter or larger, of a black fine-grained modification, much
richer in biotite and sphene than the normal rock.

) *Quartz-porphyries.*—A number of dykes and some sills, especially
in the area about the Shap granite, are to be referred to this
suite of intrusions, and are doubtless related to the granite.
They have biotite as their ferro-magnesian element, and, in
addition to porphyritic quartz, some of them contain relatively
large crystals of felspar.

Mica-Lamprophyres.—These dyke-rocks, sometimes termed "mica-traps," are too well known to need description, their richness in biotite being a very striking feature. It may be mentioned that, while the typical lamprophyres are strongly contrasted with the often associated acid dykes just mentioned, we find in certain places varieties intermediate between the two (e.g., Stakeley Folds, near Shap Wells, and Long Sleddale).

The remaining intrusions probably belong to a still later period. There is no direct evidence of their age, except the fact that they are younger than the great crust-movements ; but their very close petrographical resemblance to Tertiary Rocks in other parts of Britian raises a certain presumption that they belong to the same period.

The Carrock Fell intrusions.—The principal rocks here are two, gabbro and granophyre. The gabbro is a highly variable rock. Usually it is of medium or moderately coarse grain, only exceptionally very coarse, with large lustrous crystals of augite. The central part of the mass is a quartz-gabbro, the quartz sometimes visible in interstitial grains, but commonly in micropegmatite detected only in thin slices. The augite has a strong striation parallel to the basal plane. This relatively acid rock passes gradually through normal gabbro (i.e., without quartz), into an extremely basic rock at the edge of the mass. Here felspar is reduced to a minimum, and about one-fourth of the rock is made up of titaniferous iron-ore. The granophyre has the dull confused aspect characteristic of a micropegmatitic rock, and contains scattered crystals of oligoclase and abundant little dark specks, which are augite-crystals.

Icterite and Andesite Dykes.—A few dykes have been observed, and possibly more remain to be detected, which resemble known Tertiary dykes, and are perhaps to be referred to that period. They are apparently less basic on the whole than the older dolerites, and they are in a fresher condition.

The foregoing include all the more important igneous rocks of the district. Since one object of these brief notes is to assist in the identification of Lake District boulders in other areas, it is desirable to point out which of the rocks have sufficiently distinctive characters to be recognised with tolerable certainty. The following include the most important for this purpose : the porphyritic hypersthene-basalts (Eycott type), the amygdaloidal andesites, volcanic breccias with pink rhyolite-fragments, garnetiferous breccias and tuffs, the banded tuffs and the hornstones, some rhyolites and especially the nodular rhyolites, the Threlkeld microgranite, the Armboth type of grano-phyre, the olivine-diorite (Little Knott type), the three different kinds of granite, and the different varieties of the Carrock Fell gabbro. The quartz-gabbro, it may be mentioned, is often indistinguishable from the coarse type of the Whin Sill in Teesdale, and it is possible that some records of " Whin Sill " boulders require reconsideration. Some of the metamorphosed rocks bordering the granitic intrusions are at least as distinctive as many of the igneous rocks themselves. The chiastolite-slate of Sinen Gill and the more highly metamorphosed rocks, with andalusite, cordierite, &c., in Grainsgill and the Caldew Valley should be noticed ; and again the metamorphosed basaltic lavas near the Eskdale granite. The metamorphosed volcanic rocks are even more characteristic near the Shap granite, the basalts on the north side becoming black splintery rocks with hornblende, pyrites, and other new-formed minerals, and the andesites on the west being converted into dark glossy rocks rich in minute flakes of biotite.

LIST OF THE PRINCIPAL PUBLICATIONS DEALING WITH THE PETROLOGY OF THE ENGLISH LAKE DISTRICT.

BY ALFRED HARKER, M.A., F.R.S., F.G.S.

[For list of chemical analyses and partial analyses and a number of specific gravity determinations see Naturalist, 1899, pp. 53–58. 143–154. 156.]

I.—THE ORDOVICIAN VOLCANIC SERIES.

a *Falcon Crag Andesite Group.*—J. Clifton Ward, Q.J.G.S. (1875), XXXI. 407. and Geol. N. Part. Lake Distr. (1876), 13–19: Teall. British Petrography (1888), 282 : Hutchings, Geol. Mag. (1891), 462. 463 (tuff).

b *Eycott and Ullswater Basalt Group* (including some andesites and tuffs).—Ward. Q.J.G.S. (1875), XXXI., 406. 407. Geol. N. Part. Lake Distr., 20–23. and Monthly Microsc. Journ. (1877), XVII. 239–246. Pl. CLXXXVII ; Bonney. Geol. Mag. (1885), 76–80 ; Teall. Brit. Petr., 225–227 ; Hutchings. Geol. Mag. (1891), 538–543 (basalts and andesites : Harker and Marr. Q.J.G.S. (1893), XLIX., 360–365 and Pl. XVII., figs. 1–5 (metamorphism of basalts and tuffs near Shap granite) : Harker. Petrology for Students (1895 and 1897), figs. 43. 60. 67 : and Q.J.G.S. (1894). L. 331–334 (metamorphism by gabbro). Also Hutchings, Proc. Liverp. Geol. Soc. (1901), IX., 106–110 (cleaved tuffs of Buttermere and Honister).

c *Kentmere-Coniston Slate-Band.* — Sorby. Q.J.G.S. (1880). XXXVI., Proc. 74–76 : Hutchings. Geol. Mag. (1892). 154–161. 218–223 : Proc. Liverp. Geol. Soc. (1901). IX., 108. 111. 112 (cleaved tuffs of Elterwater and Tilberthwaite).

d *Shap Andesite Group* (with metamorphism near Shap granite).— Harker and Marr. Q.J.G.S. (1891), XLVII., 293–301. Pl. XI., figs. 4–6 : Harker. Petr. for Stud., fig. 40.

(e) *Shap Rhyolite Group.*—Rutley, Q.J.G.S. (1884), XL., 345,
Pl. XVIII., fig. 6, and Felsitic Lavas Engl. and Wales (Mem.
Geol. Sur., 1885), 12–15, Pl. II., figs. 1, 2 ; Teall, Brit.
Petr., Pl. XXXVIII ; Harker and Marr, Q.J.G.S. (1891),
XLVII., 301–309 (with metam. by Shap granite) ; Rutley,
Q.J.G.S. (1894), L., 10–13, Pl. I., figs. 1, 2 ; Harker, Petr. for
Stud., fig. 33. See also Hutchings, Geol. Mag. (1895), 314–
317, on basic tuff in this group metamorphosed by Shap
granite.

II.—INTRUSIVE ROCKS, OLDER SUITE.

(a) *Granophyres* (Ennerdale and Buttermere mass and Armboth
dykes).—Ward, Geol. N. Part. Lake Distr., 31–35 ; Teall,
Brit. Petr., 323–342 ; Harker, Naturalist (1889), 209, 210,
and (1890), 239, 240.

(b) *Microgranites.*—Ward, l.c., 33, 34 ;   Harker, Naturalist
(1890), 240, 241.

(c) *Quartz-porphyrite.*—Hutchings, Geol. Mag. (1891), 537, 538
(also intrusive porphyrite at Shap Wells, 544).

(d) *Diabases, &c.*—Ward, l.c., 37, 38 ; Teall, l.c., 224, 225 ;
Harker, Naturalist (1890), 242 ; Hutchings, Geol. Mag.
(1891), 538 ; Postlethwaite, Q.J.G.S. (1893), XLIX., 531–
535.

(e) *Diorites, &c.*—Ward, l.c., 36. Bonney, Q.J.G.S. (1885), XLI.,
511–515, Pl. XVI., fig. 2 ; Postlethwaite, Q.J.G.S. (1892),
XLVIII., 508–513 ; Harker, Petr. for Stud. (1895), 56,
57, and (1897) 64.

III.—INTRUSIVE ROCKS, YOUNGER SUITE.

(a) *Granites.*—Ward, Q.J.G.S. (1875), XXXI., 568–602, Pl.
XXX., XXXI., and (1876) XXXII., 1–11, Pl. I. (especially
metamorphism), and Geol. N. Part. Lake Distr., 6–12
(metam.). and 30, 31 ; Phillips, Q.J.G.S. (1880), XXXVI.,
9–10, Pl. I., figs. 3–5, and (1882) XXXVIII., 216–217 (dark
patches in Shap granite) ; Teall, Brit. Petr., 322, 323 (Eskdale) ; Pl. XXXV., fig. 1 (Shap) ; Harker, Naturalist, 1890,

... ... Harker and Marr. Q.J.G.S. (1891.
XLVII. 277–265. Pl. XL. figs. 1–3 Shap : Harker. Q.J.G.S.
1895. LI. 125–148 Skiddaw granite and Grainsgill greisen :
See also in metamorphosed Skiddaw slates. Rosenbusch, Die
Steiger Schiefer 1877. 211–213. translated in Naturalist
1892. 114. 180 : Yeal. Le. Pl. XXXIII. fig. 2 : Harker. Geol.
Mag. 1894. 168. 171. and Petr. for Stud. figs. 64. 65 ;
also in metamorphosed Coniston Flags near Shap granite.
Emmanus, Geol. Mag. 1891. 458–462 and (1892). 40–45.
i.—7.

' *quartz-porphyries* with transitional varieties between these
and the amphibolites.—Harker and Marr. Q.J.G.S. (1891).
XLVII. 265–291.

· **Mica Lamprophyres.**—Bonney and Houghton. Q.J.G.S. (1879).
XXXV. 165–171. Ward Geol. N. Part. Lake Distr..
22 : Harker and Marr. l.c. : Harker. Geol. Mag. 1892.
195–206 . Naturalist 1889. 210. 211 : and Petr. for Stud.
fig. 35.

‹ **Tertiary Soi sections.**—Ward Q.J.G.S. (1876). XXXII.
16. 17. 18–26 : Teschmann Geol. Mag. 1882. 210–212
(picrite): Yeal. Geol. Mag. 1883. 109 (granophyre): Brit.
Petr. 178–180 (gabbro). Pl. XLVII. fig. 5 (granophyre :
Harker. Q.J.G.S. 1894. L. 316–324. Pl. XVII. (gabbro.
and 1895 LI. 125–148 granophyre : also Petr. for Stud.
figs. 40. 41. On small dykes and veins cutting the larger
masses see Green. Q.J.G.S. 1889. XLV. 298–304. Pl.
XII (spherulitic trachyte. and Harker. Geol. Mag. (1894.
321–334. with Petr. for Stud. fig. 41 varioltic andesite.

· **Tertiary.**—None of the supposed Tertiary dolerite dykes seem
to have been described from the Lake District proper:
but see Q.J.G.S. 1891. XLVII. 525 from Eden valley.

The Meetings and Field Excursions during the year have fully maintained their scientific value and interest, and the condition of the Society is thoroughly satisfactory.

A Special Meeting was held under the presidency of Dr. Forsyth at the Church Institute, Leeds, on the evening of April 25th, for the purpose of hearing a paper by Mr. Wm. Ackroyd, F.I.C., on " Salt Circulation and its Bearing on Geological Problems." Mr. Ackroyd pointed out that salt was very widely distributed over the land by means of the rainfall. The Widdop Reservoir, for instance, which held some 640 million gallons of water, contained practically 55 tons of salt, which was probably renewed more than once in the course of the year. the salt coming down in the rainfall at the rate of about 172 lb. per acre per year. This salt was taken into the atmosphere from the sea, was carried over the land by the wind, and, falling with the rain, was borne back to the sea by the rivers. A distinction, he said, was to be drawn between cyclic sea salt and salt derived freshly from the earth, the amount of the latter being exceedingly small, a fact to be considered in criticising Professor Joly's calculation as to the age of the earth, which, briefly stated, was that the total of the salts in the sea, divided by the total borne down by the rivers, equalled the age of the earth. According to Professor Joly's calculation, that represented a period of ninety million years, but the rate of denudation was so slow that fully eight thousand million years would be required for the process, which, said the lecturer, was too long even for the most exacting biologist. Incidentally, Mr. Ackroyd referred to the saltiness of the Dead and Caspian Seas, which he attributed chiefly to cyclic sea salt, and only in a small degree to denuded earth salt ; and, in conclusion, pointed out that in the solvent denudation method of arriving at the age of the earth too little allowance had been made for the mass of salt carried by the rivers which had been derived from the ocean by atmospheric transportation, and consequently too high a figure had been adopted for the variable proportion of salt derived from the earth.

The paper was followed by an interesting discussion, in which several of the members took part. The Chairman (Dr. Forsyth) mentioned the occurrence of a higher percentage of salt on the small islets and promontories on the west of Scotland, due to the nearness of the sea and prevalence of certain winds, which would give to the streams that drained them a higher analytical value of NaCl.

Mr. F. W. Branson, F.I.C., spoke on the American geological examinations of the passage of chlorine from the sea to inland areas.

Mr. P. F. Kendall, F.G.S., defended Professor Ramsey's theory of the formation of the Caspian Sea, and gave a very interesting account of personal observations on the occurrence inland of salt spray, and other cases he had noticed on the Yorkshire coast during a high wind.

A hearty vote of thanks to the Reader of the paper concluded a very interesting meeting.

The first General Meeting and Field Excursion for 1901 was held at Keswick for the examination in situ of the rocks in that neighbourhood which are found as boulders in the drift deposits of Yorkshire.

A representative gathering of members assembled at the Keswick Hotel on Thursday evening, June 27th, with every prospect of favourable weather and an instructive meeting. On Friday morning, June 28th, an early start was made by train for Threlkeld, under the guidance of Mr. J. Postlethwaite, F.G.S., and some time was spent in the granite quarry, and noting the interesting junctions of the intruded rock with the Skiddaw Slates, which show very few signs of metamorphism. A visit was also paid to the works where flagstones are manufactured.

A traverse was then made, under the leadership of Mr. P. F. Kendall, F.G.S., to the Glenderaterra Valley, to see the Skiddaw Granite and note the extensive metamorphic changes in the Skiddaw Slates. Excellent exposures of the chiastolite slate and spotted schist were seen, and the granite was found well exposed in Synen Gill. Junctions with the schist were examined, and a tongue of more finely grained granite noted. The party walked back round the shoulder of Skiddaw to Keswick. After dinner the General Meeting was held at the Keswick Hotel, under the presidency of Mr. Percy F. Kendall. F.G.S. The following new members were elected :—Rev. W. Johnson (York), Rev. Henry Canham, F.G.S. (Leathley), Dr. Tempest Anderson, F.G.S. (York), Messrs. Edmund Spence (Clapham). Norman McLeod R. Wilson, A.M.I.C.E. (Northallerton), Simeon Walton (Elland), Joe Sagar (Halifax), W. Denison Roebuck, F.L.S. (Leeds). The Chairman delivered an interesting address on the work to be done during the meetings. He pointed out that the primary object and justification of a meeting so far from Yorkshire was the fact that in the Keswick district there was a series of rocks which were found as erratics in the Yorkshire Drifts. He also pointed out the great interest of the glacial geology of the district. and said that there was another subject of great interest, viz., the origin of the lake-basins.

Mr. C. S. Middlemiss, F.G.S., of the Indian Geological Survey. gave an address on the Geology of the Himalayas. He commented on the practice of some geologists, who had done good work in England, of expressing opinions on the geology of districts of which they could know nothing, and pointed out that several eminent geologists had expounded the theory that the Himalayas were of very modern origin, one—Sir Henry Howarth—going so far as to say that the range was upheaved in post-glacial times. Mr. Middlemiss gave an account of certain characteristic districts, and contended that the system was not so simple as was suggested, but was like an ancient house, built in different times and with different materials, and there was no evidence to sustain the theory of a sudden and wholly modern upheaval.

An interesting discussion followed, in which Messrs. B. Hobson (Owens College), W. Lower Carter, C. W. Fennell, F. F. Walton, J. W. Stather, and the Chairman took part.

On Saturday morning the party, about 30 in number, took the noted Buttermere round, via Borrowdale and Honister Pass. Castle-head, an old volcanic neck rising 300 feet above the lake, was noted. and a halt made at Falcon Crag to examine the lava beds exposed on its slopes. A detour was made at Grange to examine the junction between the Skiddaw Slates and the Borrowdale Series at Hollows

Farm. During the drive down Borrowdale excellent examples o
roches moutonnées were noted. At Seatoller a very interesting rocl
containing epidote was examined. Thence the party strolled u
the glen, the grand view from the summit of Honister Pass was dul
enjoyed, and Buttermere was reached rather late owing to th
numerous stops by the way. Here the delta separating the lakes (
Buttermere and Crummock Water was examined, and a visit pai
to Sour Milk Ghyll, where a series of Granophyre specimens wei
obtained showing an extraordinary variation in structure. Tl
party returned to Keswick by way of the Vale of Newlands.

The General Meeting was resumed, under the presidency of M
Kendall, on Saturday evening. A paper was read by Mr. Joh
Postlethwaite on "The Geology of Keswick and District." In tl
course of the discussion which followed, the Chairman referred t
some interesting points in the geology and physical geography of tl
neighbourhood. Alluding to the controversy as to the origin of tl
lake-basins, he did not think that either the advocates of ice-excavatt
lake-basins or those who claimed that they were ponded back t
banks of moraine had satisfactorily proved their views. Probab
each theory was right in certain cases. An interesting piece of e\
dence with regard to the rapidity of denudation was the prese1
comparative levels of the twin lakes, Buttermere and Crummac!
Derwentwater and Bassenthwaite. The level of Bassenthwai
was lower than that of Derwentwater by no less than 21 feet, showir
that the westward outlet of Bassenthwaite had been lowered by th
amount since the separation of the two lakes by the delta of the Gret
There was a similar difference to be noticed in the levels of Butte
mere and Crummack Lake, which gave an interesting suggestic
as to the remote period at which these lakes were separated. Tl
proceedings closed by a hearty vote of thanks to the Chairman. tl
Readers of papers, and the Leaders of the party in the field.

On Monday morning, July 1st, the party travelled to Troutbe
in order to visit Eycott Hill. From the station the way lay for
mile or two across meadows, but as the lower slopes of the hill we
ascended a beautiful view of the Lake Hills was obtained. Crossir
a stretch of moorland, Eycott Hill was climbed and the fine vie

red. As the hill was ascended the party had the opportunity
amining a fine series of lavas, probably of the age of the Borrow-
Series, some pink, some greenish, and containing large crystals
spar. From the summit a fine view was obtained of the base-
: Carboniferous beds of Mell Fell resting on the Borrowdale Series.
Caldew Valley was ascended to view the contorted and meta-
hosed Skiddaw Slates, and to examine exposures of the granite
greissen. Returning over Carrock Fell, the diabase, grano-
e, and gabbro were examined, and the party descended at Stone
The gabbro of Carrock Fell contains many fragments of Eycott
rock, into which it was intruded. The composition of this rock
s considerably, at the centre there being about 60 per cent. of
, with about 5 per cent. of the oxides of iron and titanium. But,
ie proceeds outwards, the percentage of these bases rises steadily
the silica falls,. until the proportions become about equal at the
ers of the mass.
On Tuesday the party drove from Keswick through the Vale
addle. Junctions of the Skiddaw Slates and Volcanic Series
examined at the foot of Nest Brow, near Causeway Foot. Well-
rved *roches moutonnées* were seen in St. John's Vale and a quarry
ι old volcanic neck at Bridge End, close to Thirlmere Dam. The
ι then drove along the eastern side of Thirlmere to the top of Dun-
Raise, where several examples of drift mounds were examined.
fine view over Grasmere was duly admired, and the origin of the
:alled Dunmail Raise was eagerly discussed. The return journey
by the western side of Thirlmere to Armboth, and specimens
ie Armboth Dyke were secured. Shoulthwaite Gill was then
ided, and some bedded ash beds with faults were well seen near
Crag. From Shoulthwaite the conveyances took the party back
e Keswick Hotel, and so finished a most inspiring and instructive
excursion, and the earnest thanks of the members were voted
essrs. J. Postlethwaite, F.G.S., and P. F. Kendall, F.G.S., for
excellent arrangements and admirable leadership.
The Second General Meeting and Field Excursion was held at
urn, from July 26th to 29th, for the investigation of Wensley-
with the special view of examining the typical Yoredale Beds of

Professor Phillips.   The party met at Leyburn on Friday, July 26
and after luncheon were detained by a heavy thunderstorm, but
inverting the order of the programme the time was well utilised
a visit to the interesting private museum of Mr. Wm. Horne, F.G
who was the chosen leader of the excursion.   Mr. Horne, who has b
a diligent collector for many years of antiquities and natural hist
specimens, has a great many objects of interest on exhibition, includ
many interesting local fossils and antiquarian finds, and a v
pleasant and profitable hour was spent in his sanctum.   Ab
three o'clock a start was made to see the geology of the neighbourh(
Under the guidance of Mr. Horne, the party paid a visit to the Bl
Stone Quarries, near Leyburn.   These quarries have been wor
in a band of cherty shale, which forms excellent road metal. '
extensive operations of past years are shown by the large spoil he
but only lately has the work been resumed.   A clear section
exposed, showing well-jointed, regularly bedded rock, with line
cavities containing rotten stone at intervals.   Few fossils were fou
and those were of a fragmentary nature.   The extensive quarr]
the Main Limestone was then visited, and the evenly bedded, v
jointed rock examined.   The party then crossed the golf links to
"Shawl" which is a natural limestone terrace, extending abov
mile, and laid out as a picturesque promenade.   A haze prever
the magnificent views along the dale from being enjoyed, and
the saying is, "We viewed the mist, but missed the view."

On returning to Leyburn a visit was paid to a chert qu
worked by Mr. Horne.   Here a magnificent bed of chert about '
feet in thickness is exposed, and a considerable time was spen
examining the bed and discussing its formation.   The layers v
much contorted and included wedges of encrinital limestone,
mostly converted into silica, but the general aspect of the bed le
the belief that it must have been largely formed by siliceous organi
in situ, and not as a whole have been due to chemical alteration.
a careful microscopical examination alone can the secrets of its
stitution be explained.

After dinner at the Golden Lion Hotel, the General Meeting
held under the presidency of Mr. Wm. Horne, F.G.S.   The Chair

n his address gave some interesting information about the beds
rhich had been examined, noting among other things that they
iad yielded the oldest known Labyrinthodont. A paper was read
iy the Rev. W. Lower Carter, M.A., F.G.S., on "The Yoredales and
heir Southern Equivalents in Yorkshire." The paper was a historical
ccount of the Lower Carboniferous rocks and their distribution
iorth of the Craven Fault, showing a thinning out eastwards and
outhwards. The extension southwards was, however, cut off
.bruptly by the faults, and the massive limestones were represented
in the south side of the faults by a great series of shales and thin
iedded limestones. Mr. Tiddeman's views were quoted to explain.
his huge discrepancy. He believed that the Craven Fault was a line
if movement in Carboniferous times, and that the sea bed was sinking
inequally during the deposition of the Lower Carboniferous beds,
he south side of the fault sinking much more rapidly than the
iorthern side. The physical conditions appeared to be a barrier
eef on the north side of the fault, and a deep sea studded with coral
slands to the south. An interesting discussion followed, and a vote
if thanks was passed to the Chairman for his leadership and conduct
if the meeting.

On Saturday, July 27th, the party left Leyburn by an early
rain for Redmire station. As the fields were being crossed to
3olton Castle a heavy thunderstorm broke, necessitating a rapid
etreat to cover. When the weather had cleared, Bolton Castle,
he stronghold of the Scropes and the prison of Mary Queen
if Scots, was visited. Mr. Horne conducted the party over this
nteresting ruin and explained its chief architectural and historical
ioints. Fine weather enabled the party to cross the fields and visit
he lower falls of the Yore at Aysgarth in comfort. Here the lowest
ieds of limestone in the dale are seen. The river has cut back a
iicturesque gorge through the level bedded limestone strata, and fine
iranching corals were seen in the rock. Thin bands of shale between
he limestone beds contained quantities of shells. Luncheon was
irovided at the Palmer Flat Hotel, and a visit paid to the church,
ioted for its fine old roodloft from Jervaulx Abbey, which is now
onverted into choir stalls. The party then walked to the upper fall,

below which the river is crossed by a fine stone bridge of 70 feet sp
built in the time of Queen Elizabeth, and subsequently widened
carry the carriage road. In the shales above the fall numere
corals and other fossils were obtained in good preservation. Crossi
the river, Bear Park was entered, and the charming Alpine gare
belonging to Mr. Thomas Bradley was exhibited. Here a beauti
collection of Alpine plants is successfully cultivated. An ancie
inscribed stone, built into the wall of the house, was examined. I
supposed to have formed part of an altar at Jervaulx Abbey. ?
best thanks of the members were voted to Mr. Bradley for his courte
kindness.

Monday, July 29th, was bright and clear, after a series of he
thunderstorms on Saturday night and Sunday, and the party w
enabled to see the waterfalls at their best. The train was taken
Askrigg, where the ruins of Fors Abbey, now farm buildings, w
visited. The monks migrated from this spot when Jervaulx ʼ
built, but a small window and some of the ancient masonry still
shown as part of the cowhouse. A walk down the field brought
party to the junction of the Yore and the Bain, which is a str
stream flowing from Semmer Water. Thence Mr. Horne took
members to see an old Norman bridge, built by the monks, and
ancient stone footbridge with a curiously narrowed outlet to prev
the cattle going over. Near this was the pretty Bow Bridge i
Mill Gill falls were next visited, and were much admired, there be
a flush of water from the moors. Mill Gill runs up the hillside ri
up to the Main Limestone, and exposes the whole thickness of
Yoredale beds in its course. This section was taken by Profes
Phillips as the typical section for his Yoredale Series, and therei
is of great historical interest. Time did not permit of the Gill be
ascended, but several of the beds were seen during the day. ʼ
train was then taken to Hawes, and a pleasant walk brought
party to Hardraw Force, which is a fine fall which has cut bac
picturesque gorge. The shales being weathered back under the li
stone ledge over which the water falls, it was easy to pass under
fall and return over the fallen rocks on the other side. A st
back, under a warm sun, concluded a very pleasant day's excurs

The membership at the close of the current year is 180, being two less than at the same period of last year. We have to record, with deep regret, the loss of four members by death. Mr. Charles Wheatley, of Mirfield, who had been a member since 1840, and passed away at the advanced age of 88 years ; Mr. Arthur Briggs, J.P., of Rawdon, who had been a member since 1875 ; Alderman William Gaukroger, J.P., of Halifax, who was an active man in public and political circles, and a prominent freemason. He was a life member of our Society, and was elected in 1883. Mr. Richard Taylor Manson, F.G.S., L.R.C.P., of Darlington, who was a widely respected medical and scientific man, and joined our Society in 1896. Eight members have severed their connection with the society by resignation, but ten new members have been elected.

On April 8th our honoured President (the Marquis of Ripon) and the Marchioness of Ripon celebrated their golden wedding. It was thought fitting by our Council that they should present an address of congratulation to His Lordship, who has for the long period of 43 years presided over the affairs of our Society. It was therefore resolved that an illuminated address should be prepared, and that Messrs. G. Bingley, F. W. Branson, and W. L. Carter should be a sub-committee to carry out the necessary arrangements. The following were the terms of the address drawn up by the sub-committee :—

1851—1901.

To

THE MOST HONOURABLE THE MARQUIS OF RIPON,

K.G., G.C.S.I., LL.D., D.C.L., F.R.S.

May it please your Lordship,

We, the members of the Council of the Yorkshire Geological and Polytechnic Society, desire to present to your Lordship our sincere and respectful congratulations on your Golden Wedding.

We have much satisfaction in recalling that your Lordship's connection with this Society, as a member, has continued over the long period of 45 years. During the 64

years of the Society's existence it has only had two Presidents.
and has been honoured by your occupancy of the chair
for no less than 43 years, to the manifest advantage of the
Society.

Your Lordship's deep interest in all scientific questions
is well known, and especially have you done much to further
the cause of science and scientific education in Yorkshire.
Our Society has repeatedly profited by your suggestive
addresses from the Chair, and is greatly indebted to your
Lordship for your unvarying sympathy and interest in
geological work.

We earnestly hope that your Lordship and the noble
Marchioness will be yet spared for many years in health
and happiness, and that you will long continue to occupy
the office of President.

We are,

Your Lordship's obedient and grateful servants.

WALTER ROWLEY, *Vice-President.*

WM. CASH, *Treasurer.*

W. LOWER CARTER, *Hon. Secretary.*

This address was engrossed with the signatures of all the members of Council and the Local Secretaries in facsimile and illuminated
by Messrs. Goodall & Suddick, and was bound in blue morocco. The
address was embellished by a number of drawings in neutral tint
illustrative of Yorkshire geology. At the head was placed a device of
crossed hammers, with the initials Y.G.P.S. in the four quadrants.
Five views of notable Yorkshire geological scenery, from photographs
by Mr. Godfrey Bingley and Mr. James Bedford, were appended:
Norber perched boulder; Malham Cove; Draughton Quarry;
Idol Rock, Brimham; and Thorniwick Bay, Flamborough. Drawings of three typical Yorkshire fossils, *Phillipsia seminifera, Amal-
theus spinatus,* and *Woodocrinus macrodactylus,* were added.

Lord Ripon, being in London at the time, expressed his willing-
ness to receive the address at his residence, Chelsea Embankment,
where it was presented by the Hon. Secretary in person.

An invitation was conveyed to our Society from the Leeds Scientific Societies to co-operate with them in arranging for a lecture on " The Australian Alps " by Professor Stirling, President of the Victorian Geological Survey. The Council thereupon resolved to unite with the Leeds Societies in their arrangements on equal terms. The lecture was given in the Leeds Philosophical Hall, on Tuesday, October 22nd, and was very interesting and successful.

The Underground Waters' Committee has held several meetings during the year, and carried on its investigations as vigorously as was possible. The following report was presented by Mr. A. R. Dwerryhouse, B.Sc., F.G.S., the Secretary of the B.A. Committee, at the Glasgow Meeting of the British Association :—

" The Committee are carrying out the investigation in conjunction with a committee of the Yorkshire Geological and Polytechnic Society.

" The work of investigating the flow of underground water in Ingleborough, described in the report presented to the Association at the Bradford Meeting, was resumed by the Committee on November 10th, 1900, when it was determined to study the underground course of a small stream known as Hard Gill.

" This stream rises, on the south side of Ingleborough, in a spring at 1,600 feet above the sea, and flows for a distance of about half a mile over boulder clay.

" It then reaches the bare limestone and commences to sink near the eastern corner of the croft at Crina Bottom.

" In wet weather the stream is not entirely absorbed at this point, but flows on past the house at Crina Bottom, and enters the rock at Rowan Tree Hole (Rantree Hole on 6 inch map).

" At the time of the experiments the water of Hard Gill was entirely absorbed between the point where the 1,200 feet contour crosses the stream and the eastern corner of the croft, and consequently the investigation of Rowan Tree Hole, the primary object of the excursion, had to be abandoned.

" It was found, however, that the bulk of the water was absorbed at the point where the 1,200 feet line crosses the stream, and consequently it was determined to introduce one pound of fluorescein into the open joint down which the water was flowing.

" This was done at 2 p.m. on November 11th, and before 7 a.m. on the 12th the water of the large spring at the reservoir in the Greta Valley was strongly coloured.

" After introducing the fluorescein a general survey was made of the direction of the joints in the limestone in the neighbourhood of the sink and on the clints above Crina Bottom. with the following results :—

Joint at ' sink '	N. 55° W.
On ' clints ' near sink	N. 55° W.
On ' clints ' above and to the west of ..	(main) N. 50° W.
Crina Bottom	(secondary) S. 25° W.

" The spring at the reservoir is thrown out close to the line of junction of the Carboniferous Limestone with the underlying Silurian rocks, and the line from the sink where the fluorescein was introduced to the spring runs N. 55° W.—that is, in the direction of the master joints in the limestone.

" Thus, again, it has been demonstrated that the direction of underground flow is determined by that of the master joints in the limestone.

" After a considerable though unavoidable delay, the work was resumed on June 21st, 1901, when Alum Pot, on the Ribblesdale side of Ingleborough, was the scene of operations.

" The joints in the neighbourhood of Alum Pot are more complicated than in the parts of the district previously investigated, there being three sets of joints, all more or less irregular in places.

" Close to Alum Pot there are two sets running S. 5° W. and N. 80° E. respectively.

" Thirty yards higher up Alum Pot Beck they run due N. and S. and N. 80° E., the north and south joints being the stronger and more continuous.

" On the ' clints ' 100 yards above the Pot there are three sets of joints, as follows, viz.—

Master	N. 10° E.
Secondary	{ N. 35° E. { N. 85° E.

" One pound of fluorescein was put into the stream flowing into Alum Pot on Friday, June 21st, at 7 p.m.

" There was not much water flowing at the time, and a few days afterwards several important springs in the neighbourhood run dry, including that at Turn Dub, on the opposite bank of the Ribble, which is the reputed outlet of the Alum Pot stream.

" The springs commenced to flow again a few days later ; but although they were carefully watched, as was also the river itself, no trace of colour was seen.

" It was therefore concluded that either the fluorescein had passed into one of the other river basins or had become so diluted as to be invisible.

" This experiment having proved inconclusive, a further one was commenced on Thursday, September 5th, the results of which are not yet known.

" Owing to the long delay caused by the drought and other circumstances beyond their control, the Committee have been unable to complete the work during the present year, and therefore ask to be reappointed and to be allowed to retain the unexpended balance of the grant made at the Bradford Meeting."

Since the Glasgow meeting the fluorescein put in at Long Churn, the water of which falls into Alum Pot, on September 5th, has emerged on the opposite side of the River Ribble, in a pool called Turn Dub, on September 17th. Between these dates Turn Dub had run dry, but on September 17th rain fell and the pool filled with greenish water, but did not run over. The following day the water of the Ribble was coloured by the stream flowing out of Turn Dub. The Sub-Committee is arranging to continue the investigations, and to have the complicated underground water courses communicating with Long Churn surveyed as soon as possible.

The Rev. W. Lower Carter, M.A., F.G.S., was appointed the representative Governor of the Yorkshire College, and Mr. A. R. Dwerryhouse, F.G.S., delegate to the B.A. Corresponding Societies Committee at the Glasgow meeting.

The Hon. Secretary was invited to be one of the Secretaries of the Geological Section at the Glasgow meeting, but was unable to be present at the meeting and so could not accept the appointment.

The Proceedings, Vol. XIV., Part I., was delayed in its iss
owing to unforeseen causes, but was issued to the members
December 30th, 1900. Part II., illustrated by 28 plates, includi
a splendid series of half-tone blocks of Carboniferous fossil plan
was published in October.

The Council suggest the following arrangements for the Gene
Meetings and Field Excursions in 1902 :—

(1) South East Scotland—for the examination of the igne
rocks, with a view to the identification of erratics, and t
glacial deposits.

(2) Whitby—for examination of the glacial features of the m
lands between Whitby and Pickering.

(3) The Annual Meeting to be held at Hull.

The arrangements of dates and leaders were deferred until t
Spring Council Meeting.

Our Proceedings as usual have been forwarded to leadi
Scientific Societies in various parts of the world, and publicati
in exchange have been received from the following Societies:

British Association.
Royal Dublin Society.
Royal Geographical Society.
Royal Society of Edinburgh.
Royal Physical Society of Edinburgh.
Royal Society of New South Wales.
Department of Mines, Sydney, N.S.W.
Department of Mines, Adelaide, S. Australia.
Nova Scotian Institute of Science.
Royal Institution of Cornwall, Truro.
Bristol Naturalists' Society.
Cambridge Philosophical Society.
Essex Naturalists' Field Club.
Edinburgh Geological Society.
Geological Association, London.
Geological Society of London.
Leeds Philosophical and Literary Society.
Liverpool Geological Society.
Liverpool Geological Association.
Hampshire Field Club.
Hull Geological Society.

Herefordshire Natural History Society.

Manchester Geological Society.

Manchester Geographical Society.

Manchester Literary and Philosophical Society.

Nottingham Naturalists' Society.

Rochdale Literary and Scientific Society.

University Library, Cambridge.

Yorkshire Naturalists' Union.

Yorkshire Philosophical Society, York.

American Philosophical Society, Philadelphia, U.S.A.

American Museum of Natural History, New York, U.S.A.

Academy of Natural Sciences, Philadelphia, U.S.A.

Brooklyn Institute of Arts and Sciences, Brooklyn, U.S.A.

Boston Society of Natural History, Boston, U.S.A.

Kansas University, Lawrence, Kansas.

Wisconsin Geological and Natural History Survey, Madison, Wis., U.S.A.

Geological Survey of Minnesota, Minneapolis, Minn., U.S.A.

Chicago Academy of Sciences.

Museum of Comparative Zoology at Harvard College, Cambridge, Mass.

New York Academy of Sciences, New York.

United States Geological Survey, Washington, D.C.

Elisha Mitchell Scientific Society, University of N. Carolina, Chapel Hill, U.S.A.

New York State Library, Albany, U.S.A.

Wisconsin Academy of Sciences, Arts, and Letters.

Smithsonian Institution, Washington, D.C.

L'Academie Royale Suedoise des Sciences, Stockholm.

Societé Imperiale Mineralogique de St. Petersburg.

Societé Imperiale des Naturalistes, Moscow.

Comité Geologique de la Russie, St. Petersburg.

Instituto Geologico de Mexico.

Sociedad Cientifica "Antonio Alzate," Mexico City.

Australian Museum, Sydney.

Australian Association for the Advancement of Science, Sydney.

Natural History Society of New Brunswick.

L'Academie Royale des Sciences et des Lettres de Danemark, Copenhague.

Kaiserliche Leopold-Carol. Deutsche Akademie der Naturforscher, Halle-a-Saale.

Geological Institution, Royal University Library, Upsala.

Imperial University of Tokyo, Japan.

W. LOWER CARTER.

THE YORKSHIRE GEOLOGICAL AND POLYTECHNIC SOCIETY.

Statement of Receipts and Expenditure, 1st November, 1900, to 1st November, 1901.

REVENUE ACCOUNT.

Receipts.	£	s.	d.	Expenditure.	£	s.	d.
1900.				**1901.**			
Nov. 1. To Balance in Treasurer's hands	5	5	11	Nov. 1. By F. Carter (Circulars and Stationery)	9	16	1
1901.				,, Chorley & Pickersgill (Printing Proceedings), £32 11 0			
Nov. 1. To Annual Subscriptions	64	7	0	Do. do. £25 5 4 }	57	16	4
,, Halifax Corporation Interest	9	18	8	,, Chorley & Pickersgill (Postages of Proceedings)	3	5	4
,, Transfer from Capital A/c	6	6	0	,, Postages and Petty Cash	10	6	1
,, Sale of Proceedings	6	5	3	,, Expenses of Meetings	9	7	2
,, Balance of Upper Wharfedale Exploration Fund A/c	0	18	4	,, Goodall & Suddick (Address to Lord Ripon)	6	6	0
,, Balance due to Treasurer	3	15	10				
	£96	**17**	**0**		**£96**	**17**	**0**

CAPITAL ACCOUNT.

	£	s.	d.		£	s.	d.
To Life Members' Subscriptions	6	6	0	By Transfer to General Account	6	6	0
To Halifax Corporation Bond	350	0	0				

Examined and found correct, 8th November, 1901,
J. H. HOWARTH, Halifax, *Auditor.*

RECORDS OF MEETINGS.

Council Meeting, Philosophical Hall, Leeds, 21st March, 1901.
Chairman :—Mr. P. F. Kendall.
Present :—Messrs. W. Cash, G. Bingley, A. R. Dwerryhouse,
. D. Wellburn, W. Ackroyd, J. J. Wilkinson, C. W. Fennell, J. E.
edford, and W. L. Carter (Hon. Secretary).

The minutes of the previous Council Meeting were read and
infirmed.

Letters of regret for non-attendance were read from Messrs.
. Law, W. Gregson, W. Simpson, J. H. Howarth, J. W. Stather,
. H. Parke, and F. W. Branson.

A letter was read from Lord Masham accepting the office of
ice-President.

The Hon. Secretary reported that he had received an invitation
om the Council of the British Association to be one of the Secre-
iries of Section C at the Glasgow meeting, but that he had not been
ole to accept the honour.

Mr. Bingley announced that a lecture on " The Australian Alps "
y Mr. Stirling, the President of the Victorian Survey, would be
iven on a Tuesday evening in October or November, and invited
ir Society to unite in the arrangements.

Resolved, that this Society unite with the Leeds Scientific
ocieties on equal terms in the arrangements for the lecture.

Short reports of the present position of the Underground Waters'
nvestigation were made by the Secretary and Mr. Dwerryhouse,
nd it was resolved that the Committee have the right to add the
ames of landowners or others interested in the investigations to
he Committee.

The following arrangements for the General Meetings and Field
Excursions were made :—

(1) That the first meeting be held in the Lake District, with
Keswick as the centre, from June 28th to July 2nd,
with Mr. John Postlethwaite, F.G.S., as leader.

(2) That the second meeting be held at Leyburn on July 26th
and 27th, with Mr. William Horne, F.G.S., as leader.

(3) That the Annual Meeting be held at Bradford.

The following accounts were passed for payment :—

	£	s.	d.
F. Carter (stationery and circulars) ..	2	13	3
Chorley & Pickersgill (printing Proceedings)	32	11	0
	£35	4	3

The report of Mr. W. Gregson, F.G.S., the British Association
delegate, was read by the Secretary

Resolved, that the report be accepted and that Mr. Gregson
be thanked for his services.

The Rev. W. Lower Carter, M.A., was elected Representative
Governor of the Yorkshire College.

Mr. A. R. Dwerryhouse was unanimously appointed delegate
to the B.A. Corresponding Societies Committee at the Glasgow
meeting.

It was resolved that an illuminated address be presented to
the Marquis of Ripon on the occasion of his Golden Wedding on
April 8th : that the address be in book form, and that Messrs. G.
Bingley, F. W. Branson, and the Secretary form a sub-committee
to carry out the arrangements.

Resolved, that a Special Meeting be held at Leeds on Thursday,
April 25th, to hear a paper by Mr. W. Ackroyd, F.I.C.

---

*Meeting of the Underground Waters' Committee (Ingleborough
Sub-Committee).*

Chairman :—Mr. P. F Kendall.

Present :—Messrs. A. R. Dwerryhouse, W. Ackroyd, and W.
L. Carter (Hon. Sec.).

Letters of regret for absence were received from Messrs. J. H.
Howarth, F. W. Branson, and C. W. Fennell.

The minutes of the previous Committee Meeting were read and
confirmed.

Mr. Dwerryhouse reported the results of the experiment made with fluorescein at Crina Bottom, which was discharged the next day in the springs supplying the Ingleton reservoir. The line of underground water movement appeared to be that of the master joints.

The following accounts were passed for payment.—

				£	s.	d.
Reynolds & Branson	..	..	..	0	17	2
Chorley & Pickersgill	..	..	..	3	11	9
New Inn, Clapham	..	..	..	1	1	5
F. Carter (stationery)..	..	..	..	0	13	6
				£6	3	10

Resolved, that flourescein be put into Alum Pot on Saturday, March 30th, and that Messrs. P. F. Kendall, A. R. Dwerryhouse, J. H. Howarth, F. W. Branson, W. Ackroyd, and W. L. Carter form a Sub-Committee to carry out the arrangements.

The question of adding names to the Committee was considered and postponed.

General Meeting, Chemist's Room, Church Institute, Leeds, April 25th, 1901.

Chairman :—D. Forsyth, Esq., M.A., D.Sc.

A paper was read by Mr. William Ackroyd, F.I.C., on "Salt Circulation and its Bearing on Geological Problems."

A discussion followed, in which Messrs. P. F. Kendall, F. W. Branson, the Chairman, and other members took part.

The proceedings closed with a hearty vote of thanks to the Reader of the paper.

General Meeting and Field Excursion, Keswick, June 28th to July 2nd, 1901.

June 28th.—The party visited Threlkeld granite quarry under the leadership of Mr. John Postlethwaite, F.G.S., and inspected the granite and its junctions with the Skiddaw Slate. The flag works was also visited

A traverse was then made to the Glenderaterra Valley, and the alternation of the snales into chiastolite slate and spotted schist was traced from schist passing in as the granite was approached. In Sinen wall the Skiddaw Granite and its junctions with the schist were examined. The party returned to Keswick by way of Latrigg.

The second Meeting was held at the Keswick Hotel, under the presidency of Mr. Perry F. Kendal, F.G.S.

The following new members were elected :—

     Rev. W. Johnson, York.

     Dr. Tempest Anderson, F.G.S., York.

     Mr. Edmund Spence, Clapham.

     Rev. Henry Graham, F.G.S., Leathley.

     Mr. Norman McLeod Ramsay Wilson, A.M.I.C.E., North-
        allerton.

     Mr. Spence Walker, Eland.

     Mr. Joe Sugar, Halifax.

     Mr. W. Denison Roebuck, F.L.S., Leeds.

The Chairman delivered an address on the work to be done during the meeting.

An address was delivered by Mr. C. S. Middlemiss, B.A., F.G.S., of the Indian Geological Survey, on "The Geology of the Himalayas."

After a discussion a vote of thanks was awarded to the Lecturer.

The remaining business was adjourned until Saturday evening.

June 9th.—The party left for Borrowdale and Buttermere by waggonette. The lower lava at Falcon Crag was examined, and the junction between the Skiddaw Slates and the Borrowdale Series was seen at Hollows Farm, near Grange. The route over Honister Pass was then taken to Buttermere, where an exposure of the granophyre was examined. The return journey to Keswick was made by way of Newlands.

At the adjourned meeting an address was given by Mr. John Postlethwaite, F.G.S., on "The Geology of the Keswick District." This was followed by a discussion, and hearty votes of thanks were passed to the Lecturer and Leader of the Excursions, and to the Chairman for his valued assistance.

July 1st.—The party travelled by train to Troutbeck, and went
' waggonette to Eycott Hill, where the lavas and ashes were ex-
nined. Thence the route was taken to Carrock Fell, where ex-
sures of ilmenite, diorite, trachyte, and greissen were examined.

July 2nd.—The party drove from Keswick through the Vale
Naddle, and examined the junction of the Skiddaw Slates and
e Volcanic Series at the foot of West Brow, near Causeway Foot.
n old volcanic neck at Bridge End, close to Thirlmere dam, was
xt visited, and the drive continued to Dunmail Raise, evidences
glacial action being noted on the way. The return journey was
ken by the western side of Thirlmere to examine the Armboth
yke. This concluded a most profitable and interesting excursion.

General Meeting and Field Excursion, Leyburn, July 26th
, 29th, 1901.

July 26th.—The party met at Leyburn, and visited Mr. Horne's
useum. Mr. William Horne, F.G.S., then led them to the Black
tone quarries, and the Main Limestone below was examined. The
turn route was along Leyburn Shawl, and a chert quarry in Ley-
arn was visited.

The General Meeting was held at the Golden Lion Hotel, Leyburn,
nder the presidency of Mr. W. Horne, F.G.S.

An address was given by the Chairman on the local geological
nd antiquarian object of interest.

A paper was read t the Rev. W. Lower Carter, M.A., F.G.S,
n " The Yoredales and their southern equivalents in Yorkshire."

A discussion followed in which Messrs. R. H. Tiddeman, J. J.
Wilkinson, C. T. Whitmell, and others took part.

The proceedings concluded with a vote of thanks to the Chair-
an and the Reader of the paper.

July 27th.—Train was taken to Redmire station, and a visit
aid to Bolton Castle. Thence the fields were crossed to Aysgarth
alls, and luncheon was provided at the Palmer Flat Hotel. After
isiting Aysgarth Church, and inspecting the old rood screen, the

Q

A traverse was then made to the Glenderaterra Valley, and the alterations of the slates into chiastolite slate and spotted schist was traced, mica schist coming in as the granite was approached. In Synen Gill the Skiddaw Granite and its junctions with the schist were examined. The party returned to Keswick by way of Latrigg.

The General Meeting was held at the Keswick Hotel, under the presidency of Mr. Percy F. Kendall, F.G.S.

The following new members were elected :—

    Rev. W. Johnson, York.
    Dr. Tempest Anderson, F.G.S., York.
    Mr. Edmund Spence, Clapham.
    Rev. Henry Canham, F.G.S., Leathley.
    Mr. Norman McLeod Ramsay Wilson, A.M.I.C.E., North-allerton.
    Mr. Simeon Walton, Elland.
    Mr. Joe Sagar, Halifax.
    Mr. W. Denison Roebuck, F.L.S., Leeds.

The Chairman delivered an address on the work to be done during the meeting.

An address was delivered by Mr. C. S. Middlemiss, B.A., F.G.S., of the Indian Geological Survey, on "The Geology of the Himalayas."

After a discussion a vote of thanks was awarded to the Lecturer.

The remaining business was adjourned until Saturday evening.

June 29th.—The party left for Borrowdale and Buttermere by waggonette. The lower lava at Falcon Crag was examined, and the junction between the Skiddaw Slates and the Borrowdale Series was seen at Hollows Farm, near Grange. The route over Honister Pass was then taken to Buttermere, where an exposure of the granophyre was examined. The return journey to Keswick was taken by way of Newlands.

At the adjourned meeting an address was given by Mr. John Postlethwaite, F.G.S., on "The Geology of the Keswick District." This was followed by a discussion, and hearty votes of thanks were passed to the Lecturer and Leader of the Excursions, and to the Chairman for his valued assistance.

July 1st.—The party travelled by train to Troutbeck, and went by waggonette to Eycott Hill, where the lavas and ashes were examined. Thence the route was taken to Carrock Fell, where exposures of ilmenite, diorite, trachyte, and greissen were examined.

July 2nd.—The party drove from Keswick through the Vale of Naddle, and examined the junction of the Skiddaw Slates and the Volcanic Series at the foot of West Brow, near Causeway Foot. An old volcanic neck at Bridge End, close to Thirlmere dam, was next visited, and the drive continued to Dunmail Raise, evidences of glacial action being noted on the way. The return journey was taken by the western side of Thirlmere to examine the Armboth Dyke. This concluded a most profitable and interesting excursion.

General Meeting and Field Excursion, Leyburn, July 26th to 29th, 1901.

July 26th.—The party met at Leyburn, and visited Mr. Horne's museum. Mr. William Horne. F.G.S., then led them to the Black Stone quarries, and the Main Limestone below was examined. The return route was along Leyburn Shawl, and a chert quarry in Leyburn was visited.

The General Meeting was held at the Golden Lion Hotel, Leyburn, under the presidency of Mr. W. Horne, F.G.S.

An address was given by the Chairman on the local geological and antiquarian object of interest.

A paper was read t the Rev. W. Lower Carter, M.A., F.G.S, on " The Yoredales and their southern equivalents in Yorkshire."

A discussion followed in which Messrs. R. H. Tiddeman, J. J. Wilkinson, C. T. Whitmell, and others took part.

The proceedings concluded with a vote of thanks to the Chairman and the Reader of the paper.

July 27th.—Train was taken to Redmire station, and a visit paid to Bolton Castle. Thence the fields were crossed to Aysgarth Falls, and luncheon was provided at the Palmer Flat Hotel. After visiting Aysgarth Church, and inspecting the old rood screen, the

Q

fossiliferous limestones above the upper fali were worked, yieldin
many corals. Bear Park was then visited, by the kind permission c
Mr. Thomas Bradley, and his interesting Alpine garden examined.

July 29th.—Train was taken to Askrigg, and Mill Gill and Bo'
Bridge Falls were visited. The party went on to Hawes and walke
to Hardraw Foss. Owing to Sunday's heavy rains the waterfal
were seen at their best. The members separated at Hawes Statior
with heartiest thanks to Mr. William Horne for his genial and in
structive leadership.

———

Council Meeting. Philosophical Hall, Leeds, 31st October, 1901
Chairman :—Mr. Godfrey Bingley.

Present :—Messrs. P. F. Kendall, J. E. Wilson, J. J. Wilkinsor
W. Cash, J. H. Howarth, W. Ackroyd, W. Simpson, C. W. Feunel
E. D. Wellburn, J. E. Bedford. A. R. Dwerryhouse, F. W. Bransor
and W. L. Carter (Hon. Sec.).

Letters of regret for non-attendance were received from Messn
W. Rowley, J. W. Stather, J. J. Wilkinson, and G. H. Parke.

The minutes of the preceding Council Meeting were read an
confirmed.

Annual Meeting.—The Secretary read a letter from Lord Ripot
agreeing to the suggestion that the Annual Meeting should be hel
at Bradford on November 14th, but regretting that, owing to hi
physician's orders, he would not be able to preside.

The Secretary reported that, in consultation with the Bradfor
Local Secretary, Mr. J. E. Wilson, he had invited the Rev. W. H
Keeling, M.A., Headmaster of the Bradford Grammar School, t
preside, and that he had accepted the chairmanship of the Meeting
The Headmaster had also kindly granted the use of a room at th
Grammar School for the Meeting.

It was arranged that Messrs. J E. Wilson and Dr. J. Monckmai
should lead an Excursion on the morning of November 14th, to se
the lake-deposits at Leventhorpe, walk to Chellow Dean, and acros
to Shipley. Luncheon to be at the Royal Hotel. Messrs. Wilson an
Monckman to arrange for opening a face in the gravels.

These arrangements were approved.

It was then decided that the Annual Meeting should commence at 3.15 p.m. Papers were accepted from Dr. Wheelton Hind and Mr. J. W. Stather; report of the U.W.C. by Mr. Dwerryhouse, illustrated by a large model; lantern slides, illustrating the Keswick Excursion. by Mr. Godfrey Bingley; and papers by Professor Hughes and Mr. E. D. Wellburn. The dinner to be at the Royal Hotel at 6 o'clock.

The Secretary read letters from Messrs. J. E. Marr, F.R.S., E. G. Garwood, F.G.S., and R. H. Tiddeman, F.G.S., regretting their inability to attend the Annual Meeting.

Dr. Monckman's offer to exhibit a series of rock specimens was accepted.

The following accounts were passed for payment :—

	£	s.	d.
Chorley & Pickersgill (blocks for Proceedings)	25	5	4
Goodall & Suddick (illuminated address) ..	6	6	0
F. Carter (circulars, stationery. and stamps)	7	2	10
	£38	14	2

The list of Officers and Council for the forthcoming year was considered, and it was decided to renominate the members who had served for the present year, with the following exceptions :—

Mr. William Cash intimated that he could not continue to fill the office of Treasurer, and desired the Council to nominate someone in his place. Mr. Cash was warmly and unanimously urged to continue in his office, but preferred to resign.

Mr. James H. Howarth, F.G.S., was proposed and seconded for nomination in Mr. Cash's place.

A resolution of sincere regret at Mr. Cash's resignation, and of hearty recognition of his long and valued services, was unanimously adopted.

Mr. William Simpson, F.G.S., was nominated as Auditor instead of Mr. J. H. Howarth.

After some consideration it was decided that the following arrangements should be made for 1902 :—

(1) South-east Scotland—for examination of igneous rocks and glacial deposits.

2 Whitby—for the examination of the glacial features of the moraines between Whitby and Pickering.

3 Annual Meeting at Hull.

The Secretary read the Annual Report, which, with a few emendations, was accepted for presentation to the Annual Meeting.

Meeting of the Underground Waters' Committee (Ingleborough Sub-committee), Leeds, 31st October, 1891.

Chairman:—Mr. P. F. Kendall.

Present:—Messrs. J. H. Howarth, F. W. Branson, W. Ackroyd, A. R. Dwerryhouse, C. W. Fennell and W. L. Carter (Hon. Sec.).

The minutes of the preceding Committee Meeting were read and confirmed.

Mr. Dwerryhouse read the report which was presented to the B.A. Meeting at Glasgow.

Mr. Dwerryhouse reported that the fluorescein put into Long Churn had appeared at Turn Dub, on the opposite side of the Ribble.

Mr. Kendall reported that, on a recent visit to the lower part of Crina Bottom Valley, Jenkin Beck appeared as a small trickly stream, and was joined by a copious spring coming out of the eastern side of the valley. A little lower down, the stream entered a small hole in the western side of the valley. By fouling the stream it was found to emerge in 2¼ minutes about 20 yards further down the valley, and by levelling it was ascertained that this point of emergence was at the exact altitude of the visible outcrop of the Ordovician rock lower down the valley.

It was decided to put down bore holes near Turn Dub to ascertain the depth of the drift.

Mr. Dwerryhouse suggested that Long Churn and its passages should be surveyed and a plan made of them, and proposed that Mr. Theodore Ashley, of Leeds, should be invited to undertake this survey, his out-of-pocket expenses being refunded by the Committee. This was approved.

It was decided to retest the Shooting Box stream, and to find out the relation of Footnaws Hole to the Alum Pot drainage.

On account of the unusually dry season it was thought that inquiries should be made about the state of the springs at Malham, and the Secretary was instructed to write to Mr. Walter Morrison for information. If Gordale were dry it was thought very advisable to test it again with both fluorescein and sulphate of ammonia, in order to ascertain whether both of the springs at the base of Gordale Scar were affected. Also it was thought advisable to test the spring that sinks at Gordale Scar, and to see whether the water emerges at Janet's Cove as reported.

——— -

Annual General Meeting, the Grammar School, Bradford, November 14th, 1901.

Chairman :—Rev. W. H. Keeling, M.A., Headmaster of the Bradford Grammar School.

The Hon. Secretary read letters of regret for non-attendance from the Marquis of Ripon, Dr. A. Goyder (President of the Bradford Literary and Philosophical Society), the Rev. E. Maule Cole, M.A., and Messrs. J. Norton Dickons, W. Gregson, D. Forsyth, and C. W. Fennell.

The Annual Report was read by the Hon. Secretary.

The Financial Statement was presented by the Treasurer, Mr. W. Cash.

Resolved :—" That the Report and Financial Statement be adopted and printed in the Proceedings." Proposed by Mr. Walter Rowley, seconded by Mr. Edwin Hawkesworth, and carried.

The following new members were elected :—

Mr. Joseph Lomas, F.G.S., Birkenhead.

Mr. John Naughton, Harrogate.

Mr. Arthur N. Briggs, Bradford.

Resolved :—" That the thanks of the Society be given to the Officers, Members of Council, and Local Secretaries, for their conduct of the affairs of the Society during the past year, and that the names nominated be and hereby are elected to serve for the year 1901-2." Proposed by Mr. J. W. Sutcliffe, seconded by Mr. J. Lomas, and carried.

President :

The Marquis of Ripon, K.G.

Vice-Presidents :

Earl Fitzwilliam, K.G.

Earl of Wharncliffe.

Earl of Crewe.

Viscount Halifax.

H. Clifton Sorby, LL.D., F.R.S.

Walter Morrison, J.P.

W. T. W. S. Stanhope, C.B.

James Booth, J.P., F.G.S.

F. H. Bowman, D.Sc., F.R.S.E.

W. H. Hudleston, F.R.S., F.G.S.

J. Ray Eddy, F.G.S.

David Forsyth, D.Sc., M.A.

Walter Rowley, F.G.S., F.S.A.

Lord Masham.

Sir Christopher Furness, M.P., D.L.

Treasurer :

J. H. Howarth, F.G.S.

Hon. Secretary :

William Lower Carter, M.A., F.G.S.

Hon. Librarian :

Henry Crowther, F.R.M.S.

Auditor :

W. Simpson, F.G.S.

Council :

W. Ackroyd, F.I.C.	W. Cash, F.G.S.
J. T. Atkinson, F.G.S.	P. F. Kendall, F.G.S.
J. E. Bedford, F.G.S.	R. Law, F.G.S.
Godfrey Bingley.	G. H. Parke, F.L.S., F.G.S.
F. W. Branson, F.I.C.	F. F. Walton, F.G.S.
A. R. Dwerryhouse, F.G.S.	E. D. Wellburn, F.G.S.

CPSIA information can be obtained
at www.ICGtesting.com
Printed in the USA
BVHW08*1241021018
529052BV00009B/903/P